Taking Sides:
Clashing Views on
Environmental Issues, 16/e

by Thomas A. Easton

http://create.mcgraw-hill.com

ISBN-10: 1259161137   ISBN-13: 9781259161131

Contents

Preface

Most fields of academic study evolve over time. Some evolve in turmoil, for they deal in issues of political, social, and economic concern. That is, they evolve in controversy.

It is the mission of the *Taking Sides* series to capture current, ongoing controversies and make the opposing sides available to students. This book focuses on environmental issues, from the philosophical to the practical. It does not pretend to cover all such issues, for not all provoke controversy or provoke it in a suitable fashion. But there is never any shortage of issues that can be expressed as pairs of opposing essays that make their positions clear and understandable.

The basic technique—presenting an issue as a pair of opposing essays—has risks. Students often display a tendency to remember best those essays that agree with the attitudes they bring to the discussion. They also want to know what the "right" answers are, and it can be difficult for teachers to refrain from taking a side, or from revealing their own attitudes. Should teachers so refrain? Some do not, but of course they must still cover the spectrum of opinion if they wish to do justice to the scientific method and the complexity of an issue. Some do, though rarely so successfully that students cannot see through the attempt.

For any *Taking Sides* volume, the issues are always phrased as yes/no questions. Which answer—yes or no—is the correct answer? Perhaps neither. Perhaps both. Perhaps we will not be able to tell for another hundred years. Students should read, think about, and discuss the readings and then come to their own conclusions.

An *Introduction* at the beginning of each issue provides historical background and context, recent developments, and a brief description of the debate. An *Exploring the Issue* section at the end of each issue offers *Critical Thinking* questions, *Is There Common Ground*, *Additional Resources*, and *Internet References* for further exploration of the issue.

An *Instructor's Resource Guide with Test Questions* (multiple choice and essay) is available through the publisher for the instructor using *Taking Sides* in the classroom. For more information on other McGraw-Hill Create™ titles and collections, visit www.mcgrawhillcreate.com.

Editor of This Volume

THOMAS A. EASTON is a professor of science at Thomas College in Waterville, Maine, where he has been teaching environmental science, science, technology, and society, emerging technologies, and computer science since 1983. He received a BA in biology from Colby College in 1966 and a PhD in theoretical biology from the University of Chicago in 1971. He writes and speaks frequently on scientific and futuristic issues. His books include *Focus on Human Biology*, 2nd ed., coauthored with Carl E. Rischer (HarperCollins, 1995), *Careers in Science*, 4th ed. (VGM Career Horizons, 2004), *Taking Sides: Clashing Views in Science, Technology and Society* (McGraw-Hill, 11th ed., 2014), *Taking Sides: Clashing Views in Energy and Society* (McGraw-Hill, 2nd ed., 2011), and *Classic Editions Sources: Environmental Studies* (McGraw-Hill, 5th ed., 2014). Dr. Easton is also a well-known writer and critic of science fiction.

Academic Advisory Board Members

Members of the Academic Advisory Board are instrumental in the final selection of articles for the *Taking Sides* series. Their review of the articles for content, level, and appropriateness provides critical direction to the editor(s) and staff. We think that you will find their careful consideration reflected in this book.

Correlation Guide

The *Taking Sides* series presents current issues in a debate-style format designed to stimulate student interest and develop critical thinking skills. Each issue is thoughtfully framed with an issue summary, an issue introduction, and a postscript. The pro and con essays—selected for their liveliness and substance—represent the arguments of leading scholars and commentators in their fields.

Taking Sides: Clashing Views on Environmental Issues, 16/e is an easy-to-use reader that presents issues on important topics such as *energy, environmental philosophy, and sustainability*. For more information on *Taking Sides* and other McGraw-Hill Create™ titles and collections, visit www.mcgrawhillcreate.com.

This convenient guide matches the issues in **Taking Sides: Environmental Issues** with the corresponding chapters in three of our best-selling McGraw-Hill Environmental Science textbooks by Enger/Smith and Cunningham/Cunningham.

Taking Sides: Clashing Views on Environmental Issues, 16/e	Environmental Science: A Study of Interrelationships, 13/e by Enger/Smith	Principles of Environmental Science: Inquiry & Applications 7/e by Cunningham/Cunningham	Environmental Science: A Global Concern, 13/e by Cunningham/Cunningham
Do We Need the Precautionary Principle?	**Chapter 3:** Environmental Risk: Economics, Assessment, and Management	**Chapter 1:** Understanding Our Environment **Chapter 15:** Environmental Policy and Sustainability	**Chapter 8:** Environmental Health and Toxicology **Chapter 24:** Environmental Policy, Law, and Planning
Are There Limits to Growth?	**Chapter 7:** Populations: Characteristics and Issues **Chapter 15:** Water Management	**Chapter 4:** Human Populations **Chapter 7:** Food and Agriculture **Chapter 10:** Water: Resources and Pollution	**Chapter 6:** Population Biology **Chapter 7:** Human Populations **Chapter 9:** Food and Hunger **Chapter 10:** Farming: Conventional and Sustainable Practices **Chapter 17:** Water Use and Management
Should We Be Pricing Ecosystem Services?	**Chapter 6:** Kinds of Ecosystems and Communities	**Chapter 6:** Environmental Conservation: Forests, Grasslands, Parks, and Nature Preserves **Chapter 7:** Food and Agriculture **Chapter 11:** Environmental Geology and Earth Resources **Chapter 14:** Economics and Urbanization	**Chapter 23:** Ecological Economics
Does Designating "Wild Lands" Harm Rural Economies?	**Chapter 19:** Environmental Policy and Decision Making	**Chapter 16:** Environmental Policy and Sustainability	**Chapter 24:** Environmental Policy, Law, and Planning
Does Excessive Endangered Species Act Litigation Threaten Species Recovery, Job Creation, and Economic Growth?	**Chapter 3:** Environmental Risk: Economics, Assessment, and Management	**Chapter 5:** Biomes and Biodiversity	**Chapter 11:** Biodiversity: Preserving Species
Can "Green" Marketing Claims Be Believed?	**Chapter 19:** Environmental Policy and Decision Making	**Chapter 16:** Environmental Policy and Sustainability	**Chapter 24:** Environmental Policy, Law, and Planning
Do We Need Research Guidelines for Geoengineering?			
Should We Continue to Rely on Fossil Fuels?	**Chapter 8:** Energy and Civilization **Chapter 9:** Energy Sources	**Chapter 13:** Energy	**Chapter 20:** Sustainable Energy
Is Shale Gas the Solution to Our Energy Woes?	**Chapter 8:** Energy and Civilization **Chapter 9:** Energy Sources	**Chapter 13:** Energy	**Chapter 20:** Sustainable Energy
Is Renewable Energy Green?	**Chapter 8:** Energy and Civilization	**Chapter 13:** Energy	**Chapter 20:** Sustainable Energy
Are Biofuels a Reasonable Substitute for Fossil Fuels?	**Chapter 8:** Energy and Civilization **Chapter 9:** Energy Sources	**Chapter 13:** Energy	**Chapter 20:** Sustainable Energy

Taking Sides: Clashing Views on Environmental Issues, 16/e	Environmental Science: A Study of Interrelationships, 13/e by Enger/Smith	Principles of Environmental Science: Inquiry & Applications 7/e by Cunningham/Cunningham	Environmental Science: A Global Concern, 13/e by Cunningham/Cunningham
Is Hydropower a Sound Choice for Renewable Energy?	**Chapter 8:** Energy and Civilization: Patterns of Consumption **Chapter 10:** Renewable Energy Sources	**Chapter 9:** Climate **Chapter 11:** Water: Resources and Pollution **Chapter 13:** Energy	**Chapter 17:** Water Use and Management **Chapter 20:** Sustainable Energy
Do We Have a Population Problem?	**Chapter 7:** Populations: Characteristics and Issues **Chapter 15:** Water Management	**Chapter 4:** Human Populations **Chapter 7:** Food and Agriculture **Chapter 11:** Water: Resources and Pollution	**Chapter 6:** Population Biology **Chapter 7:** Human Populations **Chapter 9:** Food and Hunger **Chapter 10:** Farming: Conventional and Sustainable Practices **Chapter 17:** Water Use and Management
Does Commercial Fishing Have a Future?		**Chapter 7:** Food and Agriculture	**Chapter 9:** Food and Hunger
Does the World Need High-Tech Agriculture?	**Chapter 14:** Agricultural Methods and Pest Management	**Chapter 4:** Human Populations **Chapter 7:** Food and Agriculture **Chapter 15:** Economics and Urbanization	**Chapter 9:** Food and Hunger **Chapter 10:** Farming: Conventional and Sustainable Practices
Should Society Impose a Moratorium on the Use and Release of "Synthetic Biology" Organisms?	**Chapter 2:** Environmental Ethics	**Chapter 16:** Environmental Policy and Sustainability	**Chapter 24:** Environmental Policy, Law, and Planning
Is Bisphenol A a Potentially Serious Health Threat?		**Chapter 8:** Environmental Health and Toxicology	**Chapter 8:** Environmental Health and Toxicology
Should Agricultural Animal Wastes Be Exempt from the Requirements of Superfund Legislation?	**Chapter 18:** Environmental Regulations: Hazardous Substances and Wastes **Chapter 19:** Environmental Policy and Decision Making	**Chapter 8:** Environmental Health and Toxicology **Chapter 14:** Solid and Hazardous Waste **Chapter 16:** Environmental Policy and Sustainability	**Chapter 8:** Environmental Health and Toxicology **Chapter 21:** Solid, Toxic, and Hazardous Waste **Chapter 24:** Environmental Policy, Law, and Planning
Should the United States Reprocess Spent Nuclear Fuel?	**Chapter 10:** Nuclear Energy	**Chapter 13:** Energy **Chapter 16:** Environmental Policy and Sustainability	**Chapter 19:** Conventional Energy

Topic Guide

This topic guide suggests how the selections in this book relate to the subjects covered in your course.

All the articles that relate to each topic are listed below the bold-faced term.

Agriculture
Does the World Need High-Tech Agriculture?

Biodiversity
Does Excessive Endangered Species Act Litigation Threaten Species Recovery, Job Creation, and Economic Growth?

Biofuels
Are Biofuels a Reasonable Substitute for Fossil Fuels?

Business
Can "Green" Marketing Claims Be Believed?
Does Designating "Wild Lands" Harm Rural Economies?
Does Excessive Endangered Species Act Litigation Threaten Species Recovery, Job Creation, and Economic Growth?
Should Agricultural Animal Wastes Be Exempt from the Requirements of Superfund Legislation?

Conservation
Does Designating "Wild Lands" Harm Rural Economies?

Economics
Can "Green" Marketing Claims Be Believed?
Does Designating "Wild Lands" Harm Rural Economies?
Does Excessive Endangered Species Act Litigation Threaten Species Recovery, Job Creation, and Economic Growth?
Should We Be Pricing Ecosystem Services?

Ecosystems
Does Commercial Fishing Have a Future?
Does Designating "Wild Lands" Harm Rural Economies?
Should We Be Pricing Ecosystem Services?

Ecosystem Services
Should We Be Pricing Ecosystem Services?

Endangered Species
Does Excessive Endangered Species Act Litigation Threaten Species Recovery, Job Creation, and Economic Growth?

Energy
Are Biofuels a Reasonable Substitute for Fossil Fuels?
Do We Need Research Guidelines for Geoengineering?
Is Hydropower a Sound Choice for Renewable Energy?
Is Renewable Energy Green?
Is Shale Gas the Solution to Our Energy Woes?
Should the United States Reprocess Spent Nuclear Fuel?
Should We Continue to Rely on Fossil Fuels?

Environmental Law
Does Designating "Wild Lands" Harm Rural Economies?

Does Excessive Endangered Species Act Litigation Threaten Species Recovery, Job Creation, and Economic Growth?
Should Agricultural Animal Wastes Be Exempt from the Requirements of Superfund Legislation?

Environmental Philosophy
Are There Limits to Growth?
Do We Have a Population Problem?
Do We Need the Precautionary Principle?
Is Renewable Energy Green?
Should We Be Pricing Ecosystem Services?

Environmental Policy
Do We Need Research Guidelines for Geoengineering?
Does Designating "Wild Lands" Harm Rural Economies?
Should Agricultural Animal Wastes Be Exempt from the Requirements of Superfund Legislation?
Should Society Impose a Moratorium on the Use and Release of "Synthetic Biology" Organisms?
Should We Be Pricing Ecosystem Services?

Fisheries
Does Commercial Fishing Have a Future?

Food
Does Commercial Fishing Have a Future?
Does the World Need High-Tech Agriculture?

Fossil Fuels
Is Shale Gas the Solution to Our Energy Woes?
Should We Continue to Rely on Fossil Fuels?

Fracking
Is Shale Gas the Solution to Our Energy Woes?

Genetically Modified Crops
Does the World Need High-Tech Agriculture?

Geoengineering
Do We Need Research Guidelines for Geoengineering?

Global Collapse
Are There Limits to Growth?

Global Warming
Do We Need Research Guidelines for Geoengineering?

Human Health and the Environment
Is Bisphenol A a Potentially Serious Health Threat?

Introduction

Environmental Issues: The Never-Ending Debate

When I teach Environmental Science, I often begin by explaining the roots of the word "ecology," from the Greek *oikos* (house or household), and assigning the students to write a brief paper about their own household. How much, I ask them, do you need to know about the place where you live? And why?

The answers vary. Some of the resulting papers focus on people—roommates if the "household" is a dorm room, spouses and children if the students are older, parents and siblings if they live at home—and the needs to cooperate and get along, and perhaps the need not to overcrowd. Some pay attention to houseplants and pets, and occasionally even bugs and mice. Some focus on economics—possessions, services, and their costs, where the checkbook is kept, where the bills accumulate, the importance of paying those bills, and of course the importance of earning money to pay those bills. Some focus on maintenance—cleaning, cleaning supplies, repairs, whom to call if something major breaks. For some the emphasis is operation—garbage disposal, grocery shopping, how to work the lights, stove, fridge, and so on. A very few recognize the presence of toxic chemicals under the sink and in the medicine cabinet and the need for precautions in their handling. Sadly, a few seem to be oblivious to anything that does not have something to do with entertainment.

Not surprisingly, some students object initially that the exercise seems trivial. "What does this have to do with the environment?" they ask. Yet the course is rarely very old before most are saying, "Ah! I get it!" That nice, homey microcosm has a great many of the features of the macrocosmic environment, and the multiple ways people can look at the microcosm mirror the ways people look at the macrocosm. It's all there, as is the question of priorities: What is important? People or fellow creatures or economics or maintenance or operation or waste disposal or food supply or toxics control or entertainment? Or all of the above?

And how do you decide? I try to illuminate this question by describing a parent trying to teach a teenager not to sit on a woodstove. In July, the kid answers, "Why?" and continues to perch. In August, likewise. And still in September. But in October or November, the kid yells "Ouch!" and jumps off in a hurry.

That is, people seem to learn best when they get burned.

This is surely true in our homely *oikos*, where we may not realize our fellow creatures deserve attention until houseplants die of neglect or cockroaches invade the cupboards. Economics comes to the fore when the phone gets cut off, repairs when a pipe ruptures, air quality when the air conditioner breaks or strange fumes rise from the basement, waste disposal when the bags pile up and begin to stink or the toilet backs up. Toxics control suddenly matters when a child or pet gets into the rat poison or household cleaning supplies.

In the larger *oikos* of environmental science, such events are paralleled by the loss of a species, or an infestation by another, by floods and droughts, by lakes turned into cesspits by raw sewage, by air turned foul by industrial smokestacks, by groundwater contaminated by toxic chemicals, by the death of industries and the loss of jobs, by famine and plague and even war.

If nothing is going wrong, we are not very likely to realize there is something we should be paying attention to. And this too has its parallel in the larger world. Indeed, the history of environmental science and environmentalism is in important part a history of people carrying on with business as usual until something goes obviously awry. Then, if they can agree on the nature of the problem (Did the floor cave in because the joists were rotten or because there were too many people at the party?), they may learn something about how to prevent recurrences.

The Question of Priorities

There is of course a crucial "if" in that last sentence: *If people can agree. . . .* It is a truism to say that agreement is difficult. In environmental matters, people argue endlessly over whether anything is actually wrong, what its eventual impact will be, what if anything can or should be done to repair the damage, and how to prevent recurrence. Not to mention who's to blame and who should take responsibility for fixing the problem! Part of the reason is simple: Different things matter most to different people. Individual citizens may want clean air or water or cheap food or a convenient commute. Politicians may favor sovereignty over international cooperation. Economists and industrialists may think a few coughs (or cases of lung cancer, or shortened lifespans) a cheap price to pay for wealth or jobs.

No one now seems to think that protecting the environment is not important. But different groups—even different environmentalists —have different ideas of what "environmental responsibility" means. To a paper company cutting trees for pulp, it may mean leaving a screen of trees (a "beauty strip") beside the road and minimizing erosion. To hikers following trails through or within view of the same tract of land, that is not enough; they want the trees left alone. The hikers may also object to seeing the users of trail bikes and all-terrain-vehicles on the trails. They may even object to hunters and anglers, whose activities they see as diminishing the wilderness experience. They may therefore push for protecting the land as limited-access wilderness. The hunters and anglers object to that, of course, for they want to be able to use their vehicles to bring their game home, or to bring their boats to their favorite rivers and lakes. They also argue, with some justification, that their license fees support a great deal of environmental protection work.

To a corporation, dumping industrial waste into a river may make perfect sense, for alternative ways of disposing of waste are likely to cost more and diminish profits. Of course, the waste renders the water less useful to wildlife or downstream humans, who may well object. Yet telling the corporation it cannot dump may be seen as depriving it of property. A similar problem arises when regulations prevent people and corporations from using land—and making money—as they had planned. Conservatives have claimed that environmental regulations thus violate the Fifth Amendment to the U.S. Constitution, which says "No person shall . . . be deprived of . . . property, without due process of law; nor shall private property be taken for public use, without just compensation."

One might think the dangers of such things as dumping industrial waste in rivers are obvious. But scientists can and do disagree, even given the same evidence. For instance, a chemical in waste may clearly cause cancer in laboratory animals. Is it therefore a danger to humans? A scientist working for the company dumping that chemical in a river may insist that no such danger has been proven. Yet a scientist working for an environmental group such as Greenpeace may insist that the danger is obvious since carcinogens do generally affect more than one species.

Scientists are human. They have not only employers but also values, often rooted in political ideology and religion. They may feel that the individual matters more than corporations or society, or vice versa. They may favor short-term benefits over long-term benefits, or vice versa.

And scientists, citizens, corporations, and government all reflect prevailing social attitudes. When America was expanding westward, the focus was on building industries, farms, and towns. If problems arose, there was vacant land waiting to be moved to. But when the expansion was done, problems became more visible and less avoidable. People could see that there were "trade-offs" involved in human activity: more industry meant more jobs and more wealth, but there was a price in air and water pollution and human health (among other things).

Nowhere, perhaps, are these trade-offs more obvious than in Eastern Europe. The former Soviet Union was infamous for refusing to admit that industrial activity was anything but desirable. Anyone who spoke up about environmental problems risked jail. The result, which became visible to Western nations after the fall of the Iron Curtain in 1990, was industrial zones where rivers had no fish, children were sickly, and life expectancies were reduced. The fate of the Aral Sea, a vast inland body of water once home to a thriving fishery and a major regional transportation route, is emblematic: Because the Soviet Union wanted to increase its cotton production, it diverted for irrigation the rivers which delivered most of the Aral Sea's fresh water supply. The sea then began to lose more water to evaporation than it gained, and it rapidly shrank, exposing sea-bottom so contaminated by industrial wastes and pesticides that wind-borne dust is now responsible for a great deal of human illness. The fisheries are dead, and freighters lie rusting on bare ground where once waves lapped.

The Environmental Movement

The twentieth century saw immense changes in the conditions of human life and in the environment that surrounds and supports human life. According to historian J. R. McNeill, in *Something New Under the Sun: An Environmental History of the Twentieth-Century World* (W. W. Norton, 2000), the environmental impacts that resulted from the interactions of burgeoning population, technological development, shifts in energy use, politics, and economics in that period are unprecedented in both degree and kind. Yet a worse impact may be that we have come to accept as "normal" a very temporary situation that "is an extreme deviation from any of the durable, more 'normal,' states of the world over the span of human history, indeed over the span of earth history." We are thus not prepared for the inevitable and perhaps drastic changes ahead.

Environmental factors cannot be denied their role in human affairs. Nor can human affairs be denied their place in any effort to understand environmental change. As McNeill says, "Both history and ecology are, as fields of knowledge go, supremely integrative. They merely need to integrate with each other."

The environmental movement, which grew during the twentieth century in response to increasing awareness of human impacts, is a step in that direction. Yet environmental awareness reaches back long before the modern environmental movement. When John James Audubon (1785–1851), famous for his bird paintings, was young, he was an enthusiastic slaughterer of birds (a few of which he used as models for the paintings). Later in life, he came to appreciate that birds were diminishing in numbers, as were the American bison, and he called for conservation measures. His was a minority voice, however. It was not till later in the century that George Perkins Marsh warned in *Man and Nature* (1864) that "We are, even now, breaking up the floor and wainscoting and doors and window frames of our dwelling, for fuel to warm our bodies and seethe our pottage, and the world cannot afford to wait till the slow and sure progress of exact science has taught it a better economy." The Earth, he said, was given to man for "usufruct" (to use the fruit of), not for consumption or waste. Resources should remain to benefit future generations. Stewardship was the point, and damage to soil and forest should be prevented and repaired. He was not concerned with wilderness as such; John Muir (1838–1914; founder of the Sierra Club) was the first to call for the preservation of natural wilderness, untouched by human activities. Marsh's ideas influenced others more strongly. In

1890, Gifford Pinchot (1865–1946) found "the nation . . . obsessed by a fury of development. The American Colossus was fiercely intent on appropriating and exploiting the riches of the richest of all continents." Under President Theodore Roosevelt, he became the first head of the U.S. Forest Service and a strong voice for conservation (not to be confused with preservation; Gifford's conservation meant using nature but in such a way that it was not destroyed; his aim was "the greatest good of the greatest number in the long run"). By the 1930s, Aldo Leopold (1887–1948), best known for his concept of the "land ethic" and his book, *A Sand County Almanac* (1949), could argue that we had a responsibility not only to maintain the environment but also to repair damage done in the past.

The modern environmental movement was kick-started by Rachel Carson's *Silent Spring* (Houghton Mifflin, 1962). In the 1950s, Carson realized that the use of pesticides was having unintended consequences—the death of non-pest insects, food-chain accumulation of poisons and the consequent loss of birds, and even human illness—and meticulously documented the case. When her book was published, she and it were immediately vilified by pesticide proponents in government, academia, and industry (most notably, the pesticides industry). There was no problem, the critics said; the negative effects if any were worth it, and she—a *woman* and a nonscientist—could not possibly know what she was talking about. But the facts won out. A decade later, DDT was banned and other pesticides were regulated in ways unheard of before Carson spoke out. Other issues have followed or are following a similar course.

The situation before Rachel Carson and *Silent Spring* is nicely captured by Judge Richard Cudahy, who in "Coming of Age in the Environment," *Environmental Law* (Winter 2000), writes, "It doesn't seem possible that before 1960 there was no 'environment'—or at least no environmentalism. I can even remember the Thirties, when we all heedlessly threw our trash out of car windows, burned coal in the home furnace (if we could afford to buy any), and used a lot of lead for everything from fishing sinkers and paint to no-knock gasoline. Those were the days when belching black smoke meant a welcome end to the Depression and little else."

Historically, humans have felt that their own well-being mattered more than anything else. The environment existed to be used. Unused, it was only wilderness or wasteland, awaiting the human hand to "improve" it and make it valuable. This is not surprising at all, for the natural tendency of the human mind is to appraise all things in relation to the self, the family, and the tribe. An important aspect of human progress has lain in enlarging our sense of "tribe" to encompass nations and groups of nations. Some now take it as far as the human species. Some include other animals. Some embrace plants as well, and bacteria, and even landscapes.

The more limited standard of value remains common. Add to that a sense that wealth is not just desirable but a sign of virtue (the Puritans brought an explicit version of this with them when they colonized North America; see Lynn White, Jr., "The Historical Roots of Our Ecological Crisis," *Science*, March 10, 1967), and it is hardly surprising that humans have used and still use the environment intensely. People also tend to resist any suggestion that they should restrain their use out of regard for other living things. Human needs, many insist, must come first.

The unfortunate consequences include the loss of other species. Lions vanished from Europe about 2000 years ago. The dodo of Mauritius was extinguished in the 1600s (see the American Museum of Natural History's account at www.amnh.org/exhibitions/expeditions/treasure_fossil/Treasures/Dodo/dodo.html?acts). The last of North America's passenger pigeons died in a Cincinnati zoo in 1914 (see www.amnh.org/exhibitions/expeditions/treasure_fossil/Treasures/Passenger_Pigeons/pigeons.html?acts). Concern for such species was at first limited to those of obvious value to humans. In 1871, the U.S. Commission on Fish and Fisheries was created and charged with finding solutions to the decline in food fishes and promoting aquaculture. The first federal legislation designed to protect game animals was the Lacey Act of 1900. It was not until 1973 that the U.S. Endangered Species Act was adopted to shield all species from the worst human impacts.

Other unfortunate consequences of human activities include dramatic erosion, air and water pollution, oil spills, accumulations of hazardous (including nuclear) waste, famine, and disease. Among the many "hot stove" incidents that have caught public attention are the following:

- The Dust Bowl—in 1934 wind blew soil from drought-stricken farms in Oklahoma all the way to Washington, DC;
- Cleveland's Cuyahoga River caught fire in the 1960s;
- The Donora, Pennsylvania, smog crisis—in one week of October 1948, 20 died and over 7,000 were sickened;
- The London smog crisis in December 1952—4000 dead,
- The Torrey Canyon, Exxon Valdez, and—most recently—the 2010 BP Macondo oil spills, which fouled shores and killed seabirds, seals, and fish;
- Love Canal, where industrial wastes seeped from their burial site into homes and contaminated ground water;
- Union Carbide's toxics release at Bhopal, India—3,800 dead and up to 100,000 ill, according to Union Carbide; others claim a higher toll;
- The Three Mile Island, Chornobyl, and Fukushima nuclear accidents;
- The decimation of elephants and rhinoceroses to satisfy a market for tusks and horns;
- The loss of forests—in 1997, fires set to clear Southeast Asian forest lands produced so much smoke that regional airports had to close;

- Ebola, a virus which kills nine tenths of those it infects, apparently first struck humans because growing populations reached into its native habitat;
- West Nile Fever, a mosquito-borne virus with a much less deadly record, was brought to North America by travelers or immigrants from Egypt;
- Acid rain, global climate change, and ozone depletion, all caused by substances released into the air by human activities.

The alarms have been raised by many people in addition to Rachel Carson. For instance, in 1968 (when world population was only a little over half of what it is today), Paul Ehrlich's *The Population Bomb* (Ballantine Books) described the ecological threats of a rapidly growing population and Garrett Hardin's influential essay, "The Tragedy of the Commons," *Science* (December 13, 1968) described the consequences of using self-interest alone to guide the exploitation of publicly-owned resources (such as air and water). (In 1974, Hardin introduced the unpleasant concept of "lifeboat ethics," which says that if there are not enough resources to go around, some people must do without). In 1972, a group of economists, scientists, and business leaders calling themselves "The Club of Rome" published *The Limits to Growth* (Universe Books), an analysis of population, resource use, and pollution trends that predicted difficult times within a century; the study was redone as *Beyond the Limits to Growth: Confronting Global Collapse, Envisioning a Sustainable Future* (Chelsea Green, 1992) and again as *Limits to Growth: The 30-Year Update* (Chelsea Green, 2004), using more powerful computer models, and came to very similar conclusions; Graham Turner, "A Comparison of *The Limits to Growth* with Thirty Years of Reality," *Global Environmental Change* (August 2008), notes that the *Limits to Growth* projections have been very much on track with actual events.

Among the most recent books is Jared Diamond's *Collapse: How Societies Choose to Fail or Succeed* (Viking, 2005), which uses historical cases to illuminate the roles of human biases and choices in dealing with environmental problems. Among Diamond's important points is the idea that in order to cope successfully with such problems, a society may have to surrender cherished traditions.

The following list of selected U.S. and U.N. laws, treaties, conferences, and reports illustrates the national and international responses to the various cries of alarm:

1967 The U.S. Air Quality Act set standards for air pollution.

1968 The U.N. Biosphere Conference discussed global environmental problems.

1969 The U.S. Congress passed the National Environmental Policy Act, which (among other things) required federal agencies to prepare environmental impact statements for their projects.

1970 The first Earth Day demonstrated so much public concern that the Environmental Protection Agency (EPA) was created; the Endangered Species Act, Clean Air Act, and Safe Drinking Water Act soon followed.

1971 The U.S. Environmental Pesticide Control Act gave the EPA authority to regulate pesticides.

1972 The U.N. Conference on the Human Environment, held in Stockholm, Sweden, recommended government action and led to the U.N. Environment Programme.

1973 The Convention on International Trade in Endangered Species of Wild Fauna and Flora (CITES) restricted trade in threatened species; because enforcement was weak, however, a black market flourished.

1976 The U.S. Resource Conservation and Recovery Act and the Toxic Substances Control Act established control over hazardous wastes and other toxic substances.

1979 The Convention on Long-Range Transboundary Air Pollution addressed problems such as acid rain (recognized as crossing national borders in 1972).

1982 The Law of the Sea addressed marine pollution and conservation.

1982 The second U.N. Conference on the Human Environment (the Stockholm +10 Conference) renewed concerns and set up a commission to prepare a "global agenda for change," leading to the 1987 Brundtland Report (*Our Common Future*).

1983 The U.S. Environmental Protection Agency and the U.S. National Academy of Science issued reports calling attention to the prospect of global warming as a consequence of the release of greenhouse gases such as carbon dioxide.

1987 The Montreal Protocol (strengthened in 1992) required nations to phase out use of chlorofluorocarbons (CFCs), the chemicals responsible for stratospheric ozone depletion (the "ozone hole").

1987 The Basel Convention controlled cross-border movement of hazardous wastes.

1988 The U.N. assembled the Intergovernmental Panel on Climate Change, which would report in 1995, 1998, and 2001 that the dangers of global warming were real, large, and increasingly ominous.

1992 The U.N. Convention on Biological Diversity required nations to act to protect species diversity.

1992 The U.N. Conference on Environment and Development (also known as the Earth Summit), held in Rio de Janeiro, Brazil, issued a broad call for environmental protections.

1992   The U.N. Convention on Climate Change urged restrictions on carbon dioxide release to avoid climate change

1994   The U.N. Conference on Population and Development, held in Cairo, Egypt, called for stabilization and reduction of global population growth, largely by improving women's access to education and health care.

1997   The Kyoto Protocol attempted to strengthen the 1992 Convention on Climate Change by requiring reductions in carbon dioxide emissions, but U.S. resistance limited success.

2001   The U.N. Stockholm Convention on Persistent Organic Pollutants required nations to phase out use of many pesticides and other chemicals. It took effect May 17, 2004, after ratification by over fifty nations (not including the United States and the European Union).

2002   The U.N. World Summit on Sustainable Development, held in Johannesburg, South Africa, brought together representatives of governments, nongovernmental organizations, businesses, and other groups to examine "difficult challenges, including improving people's lives and conserving our natural resources in a world that is growing in population, with ever-increasing demands for food, water, shelter, sanitation, energy, health services and economic security."

2003   The World Climate Change Conference held in Moscow, Russia, concluded that global climate is changing, very possibly because of human activities, and the overall issue must be viewed as one of intergenerational justice. "Mitigating global climate change will be possible only with the coordinated actions of all sectors of society."

2005   The U.N. Millennium Project Task Force on Environmental Sustainability released its report, *Environment and Human Well-Being: A Practical Strategy*.

2005   The U.N. Millennium Ecosystem Assessment released its report, *Ecosystems and Human Well-Being: Synthesis* (www.millenniumassessment.org/en/index.aspx) (Island Press).

2005   The U.N. Climate Change Conference held in Montreal, Canada, marked the taking effect of the Kyoto Protocol, ratified in 2004 by 141 nations (not including the U.S. and Australia, which finally ratified the Protocol in 2007).

2007   The Intergovernmental Panel on Climate Change (IPCC) released its Fourth Assessment Report, asserting that global warming is definitely due to human releases of carbon dioxide, the effects on both nature and humanity will be profound, and mitigation, though possible, will be expensive. The Fifth Assessment Report is due in 2013–2014.

2009   In December, the U.N. Climate Change Conference sought increased commitments to reducing carbon dioxide emissions. A follow-up meeting was held in 2010. Both failed to draw binding commitments from the developed world. The U.S. has not yet ratified the Kyoto Protocol. Another meeting in June 2011 sought even stronger commitments, but without success. Rather than agree to international cooperation, governments appear to be choosing to act alone.

2012   The United Nations Conference on Sustainable Development, called Rio + 20 because it commemorated the 1992 Rio de Janeiro conference (see above), saw global commitment to sustainable development and recognition that a sustainable future requires changes in the way societies consume and produce.

Rachel Carson would surely have been pleased by many of these responses, for they suggest both concerns over the problems identified and determination to solve those problems. But she would just as surely have been frustrated, for a simple listing of laws, treaties, and reports does nothing to reveal the endless wrangling and the way political and business forces try to block progress whenever it is seen as interfering with their interests. Agreement on banning chlorofluorocarbons was relatively easy to achieve because CFCs were not seen as essential to civilization and there were substitutes available. Restraining greenhouse gas emissions is harder because we see fossil fuels as essential and though substitutes may exist, they are so far more expensive.

The Globalization of the Environment

Years ago, it was possible to see environmental problems as local. A smokestack belched smoke and made the air foul. A city sulked beneath a layer of smog. Bison or passenger pigeons declined in numbers and even vanished. Rats flourished in a dump where burning garbage produced clouds of smoke and runoff contaminated streams and groundwater and made wells unusable. Sewage, chemical wastes, and oil killed the fish in streams, lakes, rivers, and harbors. Toxic chemicals such as lead and mercury entered the food chain and affected the health of both wildlife and people.

By the 1960s, it was becoming clear that environmental problems did not respect borders. Smoke blows with the wind, carrying one locality's contamination to others. Water flows to the sea, carrying sewage and other wastes with it. Birds migrate, carrying with them whatever toxins they have absorbed with their food. In 1972, researchers were able to report that most of the acid rain falling on Sweden came from other countries. Other researchers have shown that the rise and fall of the Roman Empire can be tracked in Greenland, where glaciers preserve lead-containing dust deposited over the millennia—the amount rises as Rome flourished, falls with the Dark Ages, and rises again with the Renaissance and Industrial Revolution. Today we know that pesticides and other chemicals can show up in places (such as the Arctic) where they have never been used, even years after their use has been discontinued. The 1979 Convention on Long-Range Transboundary Air Pollution has been strengthened several times with amendments to address persistent organic pollutants, heavy metals, and other pollutants.

We have become aware of new environmental problems that exist only in a global sense. Ozone depletion, first identified in the stratosphere over Antarctica, threatened to increase the amount of ultraviolet light reaching the ground, and thereby increase the incidence of skin cancer and cataracts, among other things. The cause was the use by the industrialized world of chlorofluorocarbons (CFCs) in refrigeration, air conditioning, aerosol cans, and electronics (for cleaning grease off circuit boards). The effect was global. Worse yet, the cause was rooted in northern lands such as the United States and Europe, but the worst effects seemed likely to strike lands where the sun shines brightest—in the tropics, which are dominated by developing nations. Fortunately, the world was able to ban CFCs and in time the damage to the ozone layer will heal.

A similar problem arises with global warming, which is also rooted in the industrialized world and its use of fossil fuels. The expected climate effects will hurt worst the poorer nations of the tropics, and perhaps worst of all those on low-lying South Pacific islands, which are expecting to be wholly inundated by rising seas. People who depend on the summertime melting of winter snows and mountain glaciers (including the citizens of California) will also suffer, for already the snows are less and the glaciers are vanishing. According to the Global Humanitarian Forum's "Human Impact Report: Climate Change—The Anatomy of a Silent Crisis" (May 29, 2009; see www.ghf-ge .org/human-impact-report.pdf) global warming is already affecting over 300 million people and is responsible for 300,000 deaths per year. A serious issue of justice or equity is therefore involved.

Both the developed and the developing world are aware of the difficulties posed by environmental issues. In Europe, "green" political parties play a major and growing part in government. In Japan, some environmental regulations are more demanding than those of the United States. Developing nations understandably place dealing with their growing populations high on their list of priorities, but they play an important role in UN conferences on environmental issues, often demanding more responsible behavior from developed nations such as the United States (which often resists these demands; it has refused to ratify international agreements such as the Kyoto Protocol, for example).

Western scholars have been known to suggest that developing nations should forgo industrial development because if their huge populations ever attain the same per-capita environmental impact as the populations of wealthier lands, the world will be laid waste. It is not hard to understand why the developing nations object to such suggestions; they too want a better standard of living. Nor do they think it fair that they suffer for the environmental sins of others.

Are global environmental problems so threatening that nations must surrender their sovereignty to international bodies? Should the U.S. or Europe have to change energy supplies to protect South Pacific nations from inundation by rising seas? Should developing nations be obliged to reduce birth rates or forgo development because their population growth and industrialization are seen as exacerbating pollution or threatening biodiversity?

Questions such as these play an important part in global debates today. They are not easy to answer, but their very existence says something important about the general field of environmental studies. This field is based in the science of ecology. Ecology focuses on living things and their interactions with each other and their surroundings. It deals with resources and limits and coexistence. It can see problems, their causes, and even potential solutions. And it can turn its attention to human beings as easily as it can to deer mice.

Yet human beings are not mice. We have economies and political systems, vested interests, and conflicting priorities and values. Ecology is only one part of environmental studies. Other sciences—chemistry, physics, climatology, epidemiology, geology, and more—are involved. So are economics, history, law, and politics. Even religion can play a part.

Unfortunately, no one field sees enough of the whole to predict problems (the chemists who developed CFCs could hardly have been expected to realize what would happen if these chemicals reached the stratosphere). Environmental studies is a field for teams. That is, it is a holistic, multidisciplinary field.

This gives us an important basic principle to use when evaluating arguments on either side of any environmental issue: Arguments that fail to recognize the complexity of the issue are necessarily suspect. On the other hand, arguments that endeavor to convey the full complexity of an issue may be impossible to understand. A middle ground is essential for clarity, but any reader or student must realize that something important may be being left out.

Current Environmental Issues

In 2001, the National Research Council's Committee on Grand Challenges in Environmental Sciences published *Grand Challenges in Environmental Sciences* (National Academy Press, 2001) in an effort to reach "a judgment regarding the most important environmental research challenges of the next generation—the areas most likely to yield results of major scientific and practical importance if pursued vigorously now." These areas include the following:

- Biogeochemical cycles (the cycling of plant nutrients, the ways human activities affect them, and the consequences for ecosystem functioning, atmospheric chemistry, and human activities)
- Biological diversity
- Climate variability
- Hydrologic forecasting (groundwater, droughts, floods, etc.)
- Infectious diseases
- Resource use
- Land use
- Reinventing the use of materials (e.g., recycling)

Similar themes appeared when *Issues in Science and Technology* celebrated its twentieth anniversary with its Summer 2003 issue. The editors noted that over the life of the magazine to date, some problems have hardly changed, nor has our sense of what must be done to solve them. Others have been affected, sometimes drastically, by changes in scientific knowledge, technological capability, and political trends. In the environmental area, the magazine paid special attention to:

- Biodiversity
- Overfishing
- Climate change
- The Superfund program
- The potential revival of nuclear power
- Sustainability

Many of the same basic themes were reiterated when *Science* magazine (published weekly by the American Association for the Advancement of Science) published in November and December 2003 a four-week series on the "State of the Planet," followed by a special issue on "The Tragedy of the Commons." In the introduction to the series, H. Jesse Smith began with these words: "Once in a while, in our headlong rush toward greater prosperity, it is wise to ask ourselves whether or not we can get there from here. As global population increases, and the demands we make on our natural resources grow even faster, it becomes ever more clear that the well-being we seek is imperiled by what we do."

Among the topics covered in the series were:

- Human population
- Biodiversity
- Tropical soils and food security
- The future of fisheries
- Freshwater resources
- Energy resources
- Air quality and pollution
- Climate change
- Sustainability
- The burden of chronic disease

Many of the topics on these lists are covered in this book. There are of course a great many other environmental issues—many more than can be covered in any one book such as this one. I have not tried to deal here with invasive species, the depletion of aquifers, floodplain development, urban planning, or many others. My sample of the variety available begins with the more philosophical issues. For instance, there is considerable debate over the "precautionary principle," which says in essence that even if we are not sure that our actions will have unfortunate consequences, we should take precautions just in case. This principle plays an important part in many environmental debates, from those over the wisdom of hydraulically fracturing (fracking) deep rock layers to increase supplies of natural gas and oil or the value of hydroelectric power or the hazards of "synthetic biology" to the folly (or wisdom) of reprocessing nuclear waste.

I said above that many people believed (and still believe) that nature has value only when turned to human benefit. One consequence of this belief is that it may be easier to convince people that nature is worth protecting if one can somehow calculate a cash value for nature "in the raw." Some environmentalists object to even trying to do this, on the grounds that economic value is not the only value, or even the value that should matter. Related to this question of value is that of whether certain human activities should take precedence over environmental protection. Should profits and jobs come before protecting wilderness or endangered species?

Should we be concerned about the environmental impacts of specific human actions or products? Here we can consider using fossil fuels, as well as "synthetic biology," so far of concern only in a theoretical way, the hormone-like effects of some pesticides and other chemicals on both wildlife and humans. World hunger is a major problem, with people arguing over whether organic farming can help. There is also great concern over the future of ocean fisheries. Waste disposal is a problem area all its own. It encompasses hazardous waste and nuclear waste.

What solutions are available? Some are specific to particular issues, as shale gas, renewable energy, biofuels, or hydroelectric power may be to the problems associated with fossil fuels. Some are more general, as we might expect as soon as we hear someone speak of population growth as a primary cause of environmental problems (there is some truth to this, for if the human population were small enough, its environmental impact—no matter how sloppy people were—would also be small).

Some analysts argue that whatever solutions we need, government need not impose them all. Private industry may be able to do the job if government can find a way to motivate industry, as with the idea of tradable pollution rights.

The overall aim, of course, is to avoid disaster and enable human life and civilization to continue prosperously into the future. The term for this is "sustainable development," and it was the chief concern of the U.N.

World Summit on Sustainable Development, held in Johannesburg, South Africa, in August 2002. Exactly how to avoid disaster and continue prosperously into the future are the themes of the U.N. Millennium Ecosystem Assessment report, *Ecosystems and Human Well-Being: Synthesis* (www.millenniumassessment.org/en/) (Island Press, 2005). The main findings of this report are that over the past half century meeting human needs for food, fresh water, fuel, and other resources has had major negative effects on the world's ecosystems; those effects are likely to grow worse over the next half century and will pose serious obstacles to reducing global hunger, poverty, and disease; and although "significant changes in policies, institutions, and practices can mitigate many of the negative consequences of growing pressures on ecosystems, . . . the changes required are large and not currently under way." Also essential will be improvements in knowledge about the environment, the ways humans affect it, and the ways humans depend upon it, as well as improvements in tech-

nology, both for assessing environmental damage and for repairing and preventing damage, as emphasized by Bruce Sterling in "Can Technology Save the Planet?" *Sierra* (July/ August 2005). Sterling concludes, perhaps optimistically, that "When we see our historical predicament in its full, majestic scope, we will stir ourselves to great and direly necessary actions. It's not beyond us to think and act in a better way. Yesterday's short-sighted habits are leaving us, the way gloom lifts with the dawn." George Musser, introducing the special September 2005 "Crossroads for Planet Earth" issue of *Scientific American* in "The Climax of Humanity," notes that the next few decades will determine our future. Sterling's optimism may be fulfilled, but if we do not make the right choices, the future may be very bleak indeed.

Thomas A. Easton
Thomas College

Unit 1

UNIT

Environmental Philosophy

*E*nvironmental debates are rooted in questions of values—what is right? What is just?—and are inevitably political in nature. It is worth stressing that people who consider themselves to be environmentalists can be found on both sides of most of the issues in this book. They differ in what they see as their own self-interest and even in what they see as humanity's long-term interest.

Understanding the general issues raised in this section is useful preparation for examining the more specific controversies that follow in later sections.

Selected, Edited, and with Issue Framing Material by:
Thomas Easton, *Thomas College*

ISSUE

Do We Need the Precautionary Principle?

YES: Alexandros Khoury, from "Is It Time for an EU Definition of the Precautionary Principle?" *King's Law Journal* (February 2010)

NO: Jonathan H. Adler, from "The Problems with Precaution: A Principle without Principle," *The American Enterprise* (May 25, 2011)

Learning Outcomes

After reading this issue, you will be able to:

- Define the Precautionary Principle.
- Explain why accepted definitions of the Precautionary Principle may not be adequate.
- Describe the values that affect the positions people take on the Precautionary Principle.
- Explain why some people see the Precautionary Principle as an obstacle to progress.

ISSUE SUMMARY

YES: Alexandros Khoury argues that the Precautionary Principle is valuable because it provides a framework for dealing with risk assessment and management. Used properly, it may highlight and address environmental and public health concerns without dictating courses of action. Precautionary *policies* should come only after citizen involvement in identifying concerns and desired levels of protection.

NO: Jonathan H. Adler argues that although the basic idea of taking precautions is good, the regulatory system already takes many precautions. The Precautionary Principle is so vague, ill-defined, and value-ridden that it provides little guidance and holds the potential to do more harm than good. In addition, by focusing on the risks of action, it ignores the risks of inaction, and it is worth noting that despite the ill effects of novel technologies, the world is far better off because of those technologies.

The traditional approach to environmental problems has been reactive. That is, first the problem becomes apparent—wildlife or people sicken and die, drinking water or air tastes foul. Then researchers seek the cause for the problem and regulators seek to eliminate or reduce that cause. The burden is on society to demonstrate that harm is being done and a particular cause is to blame.

An alternative approach is to presume that all human activities—construction projects, new chemicals, new technologies, etc.—have the potential to cause environmental harm. Therefore, those responsible for these activities should prove in advance that they will not do harm and should take suitable steps to prevent any harm from happening. A middle ground is occupied by the "Precautionary Principle," which has played an increasingly important part in environmental law ever since it first appeared in Germany in the mid-1960s. In the international scene, it has been applied to climate change, hazardous waste management, ozone depletion,

biodiversity, and fisheries management. In 1992, the Rio Declaration on Environment and Development, listing it as Principle 15, codified it thus:

> In order to protect the environment, the precautionary approach shall be widely applied by States according to their capabilities. When there are threats of serious or irreversible damage, lack of full scientific certainty shall not be used as a reason for postponing cost-effective measures to prevent environmental degradation.

Other versions of the principle also exist, but all agree that when there is reason to think—but not necessarily absolute proof—that some human activity is or might be harming the environment, precautions should be taken. Furthermore, the burden of proof should be on those responsible for the activity, not on those who may be harmed. This has come to be broadly accepted as a basic tenet of ecologically or environmentally sustainable development. See Marco Martuzzi and Roberto Bertollini, "The Precautionary Principle, Science and Human Health

Protection," *International Journal of Occupational Medicine and Environmental Health* (January 2004).

The Precautionary Principle also contributes to thinking in the areas of risk assessment and risk management in general. Human activities can damage health and the environment. Some people insist that action need not be taken against any particular activity until and unless there is solid, scientific proof that it is doing harm, and even then risks must be weighed against each other. Others insist that mere suspicion should be grounds enough for action. Sainath Suryanarayanan and Daniel Lee Kleinman, in "Disappearing Bees and Reluctant Regulators," *Issues in Science and Technology* (Summer 2011), argue that existing risk assessment, particularly in the Environmental Protection Agency (EPA), is biased toward avoiding type I or false positive errors (which incorrectly identify a safe substance or activity as dangerous) rather than type II or false negative errors (which incorrectly identify a dangerous substance or activity as safe). Their particular concern is colony collapse disorder in honey bees, which may be due to exposure to certain insecticides; however, the EPA refuses to call those insecticides dangerous and regulate them accordingly because of a lack of rigorous, experiment-based scientific evidence (as of 2012, the evidence is stronger; see Erik Stokstad, "Field Research on Bees Raises Concern About Low-Dose Pesticides," *Science*, March 30, 2012). They prefer a more precautionary approach, which would shift the balance toward avoiding type II errors. That is, they would rather err by calling safe substances or activities dangerous.

Since solid, scientific proof can be very difficult to obtain, the question of just how much proof is needed to justify action is vital. Not surprisingly, if action threatens an industry, that industry's advocates will argue against taking precautions, generally saying that more proof is needed. Those who feel threatened by an industry or a new technology are more likely to favor the Precautionary Principle; see John Dryzek, Robert E. Goodin, Aviezer Tucker, and Bernard Reber, "Promethean Elites Encounter Precautionary Publics: The Case of GM Foods," *Science, Technology & Human Values* (May 2009). The "Promethean Elites" are those who—like the Prometheus of myth—favor progress over the *status quo* and may argue that the Precautionary Principle holds back progress. Yet, says Charles Weiss in "Defining Precaution," *Environment* (October 2007), a review of *The Precautionary Principle*, UNESCO's World Commission on the Ethics of Scientific Knowledge and Technology Report, the principle "is an important corrective to the pressure from enthusiasts and vested interests to push technology in unnecessarily risky directions."

Not everyone agrees. Ronald Bailey, in "Precautionary Tale," *Reason* (April 1999), defines the Precautionary Principle as "precaution in the face of any actions that may affect people or the environment, no matter what science is able—or unable—to say about that action." "No matter what science says" is not quite the same thing

as "lack of full scientific certainty." Indeed, Bailey turns the Precautionary Principle into a straw man and thereby endangers whatever points he makes that are worth considering. One of those points is that widespread use of the Precautionary Principle would hamstring the development of the Third World. Roger Scruton, in "The Cult of Precaution," *National Interest* (Summer 2004), calls the Precautionary Principle "a meaningless nostrum" that is used to avoid risk and says it "clearly presents an obstacle to innovation and experiment," which are essential. Bernard D. Goldstein and Russellyn S. Carruth remind us in "Implications of the Precautionary Principle: Is It a Threat to Science?" *International Journal of Occupational Medicine and Environmental Health* (January 2004), that there is no substitute for proper assessment of risk. Jonathan H. Adler, in "The Precautionary Principle's Challenge to Progress," in Ronald Bailey, ed., *Global Warming and Other Eco-Myths* (Prima, 2002), argues that because the Precautionary Principle does not adequately balance risks and benefits, "The world would be safer without it." A. Benedictus, H. Hogeveen, and B. R. Berends, in "The Price of the Precautionary Principle: Cost-Effectiveness of BSE Intervention Strategies in the Netherlands," *Preventive Veterinary Medicine* (June 2009), found that measures taken to control the spread of BSE or mad cow disease were a very expensive way to protect human life. Peter M. Wiedemann and Holger Schutz, in "The Precautionary Principle and Risk Perception: Experimental Studies in the EMF Area," *Environmental Health Perspectives* (April 2005), report that "precautionary measures may trigger concerns, amplify . . . risk perceptions, and lower trust in public health protection." Cass R. Sunstein, in *Laws of Fear: Beyond the Precautionary Principle* (Cambridge, 2005), criticizes the Precautionary Principle in part because, he says, people overreact to tiny risks. John D. Graham, Dean of the Frederick S. Pardee RAND Graduate School, argues in "The Perils of the Precautionary Principle: Lessons from the American and European Experience," *Heritage Lecture* #818 (delivered January 15, 2004, at the Heritage Foundation, Washington, DC) that the Precautionary Principle is so subjective that it permits "precaution without principle" and threatens innovation and public and environmental health. It must therefore be used cautiously.

The 1992 Rio Declaration emphasized that the Precautionary Principle should be "applied by States according to their capabilities" and that it should be applied in a cost-effective way. These provisions would seem to preclude the draconian interpretations that most alarm the critics. Yet, say David Kriebel, et al., in "The Precautionary Principle in Environmental Science," *Environmental Health Perspectives* (September 2001), "environmental scientists should be aware of the policy uses of their work and of their social responsibility to do science that protects human health and the environment." Businesses are also conflicted, writes Arnold Brown in "Suitable Precautions," *Across the Board* (January/February 2002), because the

Precautionary Principle tends to slow decision-making, but he maintains that "we will all have to learn and practice anticipation."

Does the Precautionary Principle make us safer? The January 23, 2009 issue of *CQ Researcher* presents a debate, under that title, between Gary Marchant, who believes that the principle "fails to provide coherent or useful answers on how to deal with uncertain risks," and Wendy E. Wagner, who contends that the existing chemical regulatory system shows the consequences of not taking a precautionary approach. Many people agree with Wagner, and indeed in many parts of the world the Precautionary Principle is well accepted.

Do we actually need it? In the YES selection, Alexandros Khoury argues that the Precautionary Principle is valuable because it provides a framework for dealing with risk assessment and management. Used properly, it may highlight and address environmental and public health concerns without dictating courses of action. Precautionary *policies* should come only after citizen involvement in identifying concerns and desired levels of protection. In the NO selection, Jonathan H. Adler of the Center for Business Law and Regulation at Case Western Reserve University School of Law argues that although the basic idea of taking precautions is good, the regulatory system already takes many precautions. The Precautionary Principle is so vague, ill-defined, and value-ridden that it provides little guidance and holds the potential to do more harm than good. In addition, by focusing on the risks of action, it ignores the risks of inaction, and it is worth noting that despite the ill effects of novel technologies, the world is far better off because of those technologies.

YES

Alexandros Khoury

Is It Time for an EU Definition of the Precautionary Principle?

Introduction

In 'the complex and highly normative world of risk regulation', few notions have been as dominant but equally controversial as the precautionary principle (PP). With proponents supporting it enthusiastically and adversaries condemning it fervently, academic discourse has been uncharacteristically passionate. Indeed, the fundamental question of whether the principle adds value or whether its utility is less pronounced than its fame remains one of the hottest topics of (not only) academic debate in the area of risk governance. Amidst this polarity, the European Union (EU) has for quite some time formally adopted, but not defined, the PP and keenly promotes it in the international arena. This move, however, has come at a cost as the Union has been consistently criticised at various levels for succumbing to the alleged dangerous implications of the PP for reasons, *inter alia*, of political expediency. Against this background, the present contribution highlights the principle's potential as a democratising tool and argues that the articulation of an EU definition could assist in resolving some of the misunderstandings surrounding the Union's conception of precaution.

Before proceeding, a few observations are warranted. First, the PP—at least as it is understood in the EU—could be visualised as a risk regulation instrument that 'permit [s] [without prescribing] the adoption of preventative measures in the face of scientific uncertainty'. As such, it is more relevant to the 'process of decision-making [than] its outcome'. In general terms, and according to a baseline definition, it stipulates that 'where there are threats of serious or irreversible damage, lack of full scientific certainty shall not be used as a reason for postponing cost-effective measures to prevent environmental degradation'. Secondly, following its inauguration approximately four decades ago, the PP has appeared in several international agreements and has been mentioned or discussed by numerous courts and tribunals at the international as well as the national level. Thirdly, United States (US) law and policy is characterised by a parallel propensity towards precautionary regulation as compared to the EU despite American refusal to recognise the PP as a 'rule of international law'. The truth of this assertion is depicted in the fact that recent academic debate amongst leading scholars of comparative regulation does not revolve around the question of whether the two superpowers adopt precautionary measures; this is taken for granted. The real question involves the identification of transatlantic patterns of regulation in an effort to ascertain the extent to which European and American policies converge or diverge. And the secondary issue is whether the Old Continent has recently taken the lead in precautionary regulation, which is not to say, of course, that the US does not embrace it.

As regards the EU, it had discovered its precautionary consciousness long before formally recognising the PP in the early 1990s. What is more, it is also true that there are numerous instruments that employ it *explicitly* in the environmental/public health fields and, a few years ago, the General Court (ex Court of First Instance) promoted the PP to a general principle of EU law. Turning to the US, it has taken precautionary approaches throughout the years—that is, it has proceeded with regulation in the absence of conclusive evidence of risk—in a wide gamut of instances without, however, explicitly adopting the PP.

Ambiguity: Precaution's Achilles' Heel?

Wherever observers look in relation to precaution they might discover ambiguity. This feature is hard to deny even by its most dedicated advocates. And for critics, scrutinising the PP along these lines has become their bread and butter. Take the principle's status in international law for example. The range of academic opinion is very wide. Some scholars proclaim that precaution has become a principle of customary international law whilst others resist concurring with this assertion. On this note, the EU has argued at the World Trade Organization (WTO) level that the PP has attained the status of a general principle of international law. Notwithstanding the highlighted divergences, the truth of the matter is that every international court or tribunal that has been confronted, up to now, with the question relating to the status of precaution has treated the subject like a hot potato, refusing to recognise it explicitly as a principle of general or customary international law and politely declining to rule on this contentious issue. In summing up the reasons for this reality, Judge Laing of the International Tribunal for the Law of the Sea (ITLOS) stressed that 'Treaties and formal instruments use different language of obligation; the notion is stated variously . . . ; no authoritative judicial decision unequivocally supports the notion; doctrine is indecisive; and domestic juridical materials are uncertain or evolving'.

Another popular charge levelled against the PP concerns the multiplicity of definitions that aspire to express it. Characteristically, a decade ago a commentator managed to identify 19 different formulations of the principle, none of which could be claimed to be authoritative or decidedly superior to the rest. But, of course, it is not so much the wealth of definitions that generates critical commentary but the fact that they appear to be, by and large, irreconcilable with one another. At this point and in order to illuminate the ambiguity of the PP, sceptics often play a juxtaposition game. Various formulations of the principle are presented side by side, thus allowing the observer to trace the various differences and inconsistencies that exist between them. Some of these include the seriousness of the threat required to trigger regulation (which varies from definition to definition); the issue of which actor (regulators or industry) bears the burden of proving that no unacceptable risk exists; or the range and nature of possible responses to a given risk situation. The Rio Declaration, quoted earlier, is a usual point of departure and it is often employed as a benchmark from which to judge other, usually more aggressive, variants including the Wingspread Statement on the PP. The latter appears to have a wider ambit than the Rio Declaration as it applies to 'threats of harm' in general as opposed to 'threats of serious or irreversible damage'. Moreover, the Wingspread Statement seems to call for a reversal of the burden of proof to the effect that industry should be responsible for demonstrating the safety of the product/activity in question.

In the end, thoughtful and stimulating contributions, such as *Laws of Fear* offered by Cass Sunstein, declare that 'strong versions' of precaution—by this term, the author refers to situations when 'regulation is required whenever there is a possible risk to health, safety, or the environment, even if the supportive evidence remains speculative and even if the economic costs of regulation are high'—are 'paralysing. This assertion illuminates a simple reality, namely that risk regulation is essentially a balancing exercise and that it is certainly not free of charge. Consequently, addressing risks without taking into account the relevant financial repercussions and without paying attention to 'health-health tradeoffs' amounts to a recipe for irresponsible governance. And to be sure, there is ample logic to Sunstein's critique of 'strong precaution'—especially with reference to the way he conceptualises the concept—although he does not appear to be entirely certain as to who exactly stands in favour of it. Indeed, other commentators who have opted in favour of conceiving the PP along similarly extreme lines have characterised it, amongst other things, as 'deeply perverse'. This comes as no surprise.

On the other hand, Sunstein's suggestion that the EU flirts with the strong versions of the PP has proved somewhat controversial. This is so despite the acknowledgement that 'nations cannot plausibly be ranked along some continuum of precaution' and following an unequivocal denunciation of the simplistic axis that wants Europe to be the nest of protectionism and the US the heaven of liberalism. In any event, things are probably more straightforward than they appear upon first inspection. On this note, it would not be contentious to propose that the inconsistencies that characterise the PP's international presence are also partly reflected in EU policies and aspirations. But there is seemingly an unbridgeable chasm between pointing out the ambivalence of the PP in Europe and arguing that the Union stands in favour of paralysing precaution. Indeed, a combinative and systematic analysis of relevant sources reveals the presence of numerous mechanisms that aim to shield the principle from critiques, notably at the level of the WTO, combat arbitrariness and 'temper precaution' in the Old Continent. These include the clarification that the EU—whilst recognising that the absence of conclusive scientific evidence of harm should not necessarily bar regulatory action—is not searching in the quicksand of entirely hypothetical risks; and the prerequisite that a risk assessment ('as thorough . . . as possible') should precede the adoption of protective measures. The latter, if taken, should be consistent, *inter alia*, with the principles of proportionality and non-discrimination. Equally importantly, any regulatory action pursuant to the PP should pay respect to the relevant costs and benefits, 'including, where appropriate and feasible, an economic cost-benefit analysis'(CBA).

To sum up: despite occasionally sending out mixed signals, the EU does not, overall, appear to be subscribing to strong precaution. It has chosen a difficult path, namely, to organise its environmental/public health policies around specific notions (as opposed to adopting *ad hoc* practices) including the PP, which operate as compasses and help to generate dialogue about future directions. This enterprise is ongoing and cannot yet be considered an outright success. Ironing out, as far as possible, residual inconsistencies and applying precaution in a more meaningful way constitute issues that have to be addressed by the Union. This is where the challenge now lies.

The True Value of the PP

The next issue that requires exploration for the purposes of this analysis concerns the true value of the PP. Essentially, the question is this: what makes the principle so special to warrant the dedication of further EU grey matter to define it and continue to pursue its promotion internationally? On this note, one of the messages that runs through this contribution and is also highlighted by others has to do with the relationship of the PP with democracy. Indeed, if one distances oneself from all the rhetoric and literature regarding the principle and stands high above the condemnations and the dithyrambs, one would realise that the PP, beyond anything else, is a regulatory tool that, if used appropriately, grants the relevant authorities the flexibility to hearken to the demos and respect societal choices in the domain of risk regulation.

To appreciate the principle's dimension as a democratising device, it is essential to map out the leading and conflicting perspectives on risk regulation. On a general level, these could be conceptualised along a continuum that includes both intermediate and more extreme positions with some scholars giving primary weight to regulatory efficiency and others highlighting the crucial importance of democratic deliberation.

Despite the variations that undoubtedly exist, proponents of the former school of thought share a number of core ideas. Characteristically, they display a tendency to treat risk as something purely objective that should be measured and evaluated by insulated technocrats away from political/normative input. At the same time, objectivist scholars place emphasis on the limited ability of citizens to assess risk rationally and stress their vulnerability to heuristics, biases and media/political manipulation. They complain that regulatory bodies pay too much attention to people's false risk estimations and that the PP, by virtue of its alleged inattention to the whole picture of a given risk puzzle, becomes the alibi for inefficient regulatory policies. In addition, the notion that the scientific/technical component of risk analysis (risk assessment) should be hermetically sealed from the political component (risk management) is of central importance. Lastly, in terms of decision-making methodology, CBA is considered far superior to the PP and is proposed by some for reasons of economic efficiency and by others as an instrument to assist regulators in comparing risks and making tradeoffs between them.

At the other end of the spectrum there exists a wealth of social sciences literature that attempts to bring out risk's subjective dimension along with its relationship with contextual and cultural parameters contesting, by implication, the objectivist predisposition to consider risk assessment and CBA as value-free or neutral. Accordingly, the idea that there is a purely objective stage in risk analysis that should be left exclusively to experts is forcefully challenged. For example, it has been demonstrated that numerous decisions made at various stages of risk assessment including those of risk identification or dose-response assessment are infused with subjectivity. In a similar vein, 'every determination that the available scientific evidence warrants a finding of risk involves [value-laden and non-scientific] decisions about the acceptable degree of various types of uncertainty'. On the other hand, CBA seems to be facing important limitations as well. It has been routinely criticised for, *inter alia*, its inability to deal satisfactorily with scientific uncertainty, its flawed valuation protocols, its highly controversial practice of discounting', and its track record of consistently 'making regulation less stringent'.

To be sure, nothing mentioned thus far aims to diminish the fundamental importance of science or expertise in risk regulation. Instead, this analysis attempts to illuminate that a plethora of normative judgements and cultural influences permeate single-metric approaches. Consequently, polities that aspire to guarantee both the effectiveness and the legitimacy of risk governance have a duty to ensure that citizens' values are taken into account in the scientific as well as political stages of risk analysis. Within this framework, risk assessment is an important tool of public policy so long as it adopts an 'inclusive procedural approach' that incorporates 'public deliberation [which] can help establish priorities, define exactly what is at stake, and suggest crucial avenues of further scientific learning'. Analogously, some form of CBA could play a part in responsible governance provided that adequate safeguards are installed to prevent the aura of heavenly objectivity that often accompanies its numbers from hijacking value-choices. Devising a risk regulation system that simultaneously takes on board these concerns and guarantees efficiency involves delicate balancing exercises and represents a complex task of pivotal importance that is beyond the scope of this contribution.

Be that as it may, the preceding realisations entail important implications for the PP. In this vein, it should be emphasised that, for critics of objectivism, the adoption of the PP does not negate the significance of single-metric methods but places them in a framework where public opinion and deliberative politics are *also* important. In essence, then, by virtue of its attention to scientific uncertainties and value judgements in risk regulation, the PP provides a language for the latter. Conversely, for objectivists, risk assessment, CBA and 'sound science' *de facto* exhaust the range of available tools for managing risk governance, leaving, in effect, no room for the PP. Decisions should mainly be dictated by the normative evaluations of available scientific evidence in isolation from public opinion and, consequently, the role of uncertainties and values is rendered peripheral.

Defining the PP

The EU represents one of the biggest proponents of the PP on the international chessboard. And yet, in judgment after judgment international tribunals and courts point to the principle's ambiguity as the fundamental reason that leads them to decline to offer a verdict upon its status. A plethora of definitions and formulations is on offer as various actors are free to engage in 'cherry picking' depending on whether they wish to attack, defend or promote a specific version of the PP. With these realisations in mind, it is certainly regrettable that an EU definition does not exist. How can the Union push towards consolidation of the PP in the international sphere without first clarifying what it means by it? What are the alternatives that would allow the EU to shield its formal adoption of the principle from criticisms that are often solely based upon the (sometimes exaggerated) opinions of their authors in relation to the version that it supposedly endorses?

The EU's push for the institutionalisation of the PP in the international arena is a noble cause that might signal its commitment to sustainable development rather than incontrollable abuse. To go a step further, a potential universal acceptance of even a modest or unambitious definition of precaution could entail many benefits. First of all, it might help to structure, around the principle, a wider and overall dialogue about environmental strategies that aim to combat the challenges that the planet faces as a whole instead of focusing on *ad hoc* unilateral precautionary efforts that serve the designs of individual nations. Second, it could provide a working base upon which more effective policies may be built. And third, it might offer a defence for actors wishing to pursue high levels of environmental/public health protection within the framework of international organisations such as the WTO.

Moreover, the articulation of a European definition of the PP would have a significant advantage. It could provide a solid backing to the EU in view of its promotion of the principle in international fora by alleviating the fears of those who are concerned about its alleged irrational connotations. But for all this to happen, a modest version should be envisaged, possibly along the lines of the Rio Declaration. At this juncture, it is useful to emphasise the broad level of discretion enjoyed by EU institutions, not least in the domain of risk regulation. This discretion is, of course, watered down by virtue of the principle of proportionality and the requirement to consider the costs and benefits of regulatory actions. However, this does not mean that the relevant authorities do not have, within reason, ample room for manoeuvre in terms of achieving the desired level of protection, particularly if one considers the central importance that the Union attaches to public health protection. . . .

In any case, what the proposed version should avoid, to whatever extent possible, is addressing complex and controversial issues, including those relating to notions of harm or to the volume of scientific evidence necessary to trigger precautionary policies. These preoccupations are likely to hinder consensus and detract from the main focus, which should be to lead the international community towards a fundamentally different approach to tackling environmental/public health risks in situations of scientific uncertainty. Such approach will grant, for the first time, ecumenical recognition to the PP with all the implications that this reality will entail.

On this point, and without a doubt, the principle does represent a distinct worldview that aspires to emphasise the significance of environmental/public health concerns, but it amounts to more than that. It has the capacity to inspire deliberation and dismisses easy answers. Even an unassuming definition of precaution—for example the one found in the Rio Declaration—enhances the discretion of regulators to take into account scientific uncertainties without feeling pressed to arrive at decisions solely on the basis of available, often tentative, data. And in

famously inviting polities to take a 'step back' from 'sound science' as the only determinant of important decisions in the public health/environmental sphere, it encourages them to explore alternative avenues in search of authority for making tough, *albeit* legitimate, choices. Public opinion, through deliberation, could and should provide this authority because there is 'no safe depository of the ultimate powers of the society but the people themselves' as 'there is not now, and never will be, a class of empathetic, non self-interested elites who can be trusted to advance the common good'. In a catchphrase then, the PP stands in favour of inclusion as opposed to exclusion and this is precisely the feature that makes it special.

On the other hand, the limitations as well as the objections to the proposal articulated above can be envisaged and are certainly not without merit. Some might say that part of the PP's ingenuity lies in its ambiguity and flexibility. Others might point to the potential undesirability of tying the EU's hands into adopting an overly restrictive version of the PP that is in danger of appearing banal and unhelpful. But when, according to what has been concluded previously, one disentangles the capacity of polities to adopt even severe precautionary measures from the formal embrace of the PP, the picture changes considerably. Hence, the EU's hands will not be tied; they will be freed. It will establish a working base upon which it can build subsequently without giving away its right to choose, if the Union wants, more precautionary policies under a different heading. And a potential loss of the much-desired flexibility may not prove to be a high price to pay for the benefits it will bring.

Conclusion

The PP can provide a structured framework to assist in dealing with the difficult questions that risk regulation poses. If it is utilised properly, it may highlight and address environmental/public health concerns without dictating courses of action. In this sense, it should be seen as a framework provider and not as a fundamentalist bulldozer. But in order for it to produce the desired results it has to be liberated, as much as possible, from the elusiveness that accompanies it and the (sometimes justified) polemic against it.

On the other hand, precautionary policies cannot be adopted in a vacuum. There is, first and foremost, a need to ascertain through public deliberation what a given society considers important or sacred. These realisations, in turn, should inform the scope and intensity of scientific assessments as well as risk management measures. This is the point where the PP and democracy should meet and become indissoluble. Because the risks polities face are so many and so diverse, there is no point in them adopting precautionary measures without enquiring what is it exactly that they are protecting and for whose benefit. Consequently, citizens' values are important not only in terms of determining the desired level of protection but also in shaping the very subject matter of protection. In order to achieve this goal—and with particular reference to potential risks stemming from contentious technologies—people must be allowed and encouraged to deliberate not only with each other but also with scientists and political actors in the quest to decode their values from their (sometimes unfounded) fears. This process should precede and inform precautionary policies. Only then can the latter make sense because as Sunstein has rightly put it, 'in a democratic society, officials should respond to people's values, rather than to their blunders'.

Alexandros Khoury is a Lecturer in Law at the University of Leicester. His research focuses on the areas of EU law and risk regulation.

Jonathan H. Adler

→ **NO**

The Problems with Precaution:
A Principle without Principle

It's better to be safe than sorry. We all accept this as a commonsense maxim. But can it also guide public policy? Advocates of the precautionary principle think so, and argue that formalizing a more "precautionary" approach to public health and environmental protection will better safeguard human well-being and the world around us. If only it were that easy.

Simply put, the precautionary principle is not a sound basis for public policy. At the broadest level of generality, the principle is unobjectionable, but it provides no meaningful guidance to pressing policy questions. In a public policy context, "better safe than sorry" is a fairly vacuous instruction. Taken literally, the precautionary principle is either wholly arbitrary or incoherent. In its stronger formulations, the principle actually has the potential to do harm.

Efforts to operationalize the precautionary principle into public law will do little to enhance the protection of public health and the environment. The precautionary principle could even do more harm than good. Efforts to impose the principle through regulatory policy inevitably accommodate competing concerns or become a Trojan horse for other ideological crusades. When selectively applied to politically disfavored technologies and conduct, the precautionary principle is a barrier to technological development and economic growth.

It is often sound policy to adopt precautionary measures in the face of uncertain or not wholly known health and environmental risks. Many existing environmental regulations adopt such an approach. Yet a broader application of the precautionary principle is not warranted, and may actually undermine the goal its proponents claim to advance. In short, it could leave us more sorry and even less safe.

The Precautionary Principle Defined

According to its advocates, the precautionary principle traces its origins to the German principle of "foresight" or "forecaution"—*Vorsorgeprinzip*.[1] This principle formed the basis of social democratic environmental policies in West Germany, including measures to address the effects of acid precipitation on forests.[2] Germany was not alone, as other nations also adopted precautionary measures to address emerging environmental problems. So did various international bodies.[3]

The most common articulation of the precautionary principle is the Wingspread Statement on the Precautionary Principle, a consensus document drafted and adopted by a group of environmental activists and academics in January 1998.[4] The statement defined the precautionary principle thus:

> When an activity raises threats of harm to human health or the environment, precautionary measures should be taken even if some cause and effect relationships are not fully established scientifically.
>
> In this context the proponent of an activity, rather than the public, should bear the burden of proof.
>
> The process of applying the Precautionary Principle must be open, informed and democratic and must include potentially affected parties. It must also involve an examination of the full range of alternatives, including no action.

Contrary to what its advocates claim, this principle does not provide a particularly useful, let alone prudent, guide to developing or implementing environmental and public health measures. Taken literally, it does not even provide much guidance at all. Harvard law professor Cass Sunstein, who currently serves as administrator of the Office of Information and Regulatory Affairs in the Obama administration, is particularly harsh in assessing the precautionary principle. According to Sunstein, "The precautionary principle, for all its rhetorical appeal, is deeply incoherent. It is of course true that we should take precautions against some speculative dangers. But there are always risks on both sides of a decision; inaction can bring danger, but so can action. Precautions, in other words, themselves create risks—and hence the principle bans what it simultaneously requires."[5]

The Wingspread Statement itself embodies many of the problems with the precautionary principle. The initial portion of the principle calls for something already done in the United States and other developed countries. Regulatory measures are routinely adopted when "some cause and effect relationships are not fully established scientifically." There is rarely, if ever, perfect certainty about the nature and causes of health and environmental threats, so environmental and public health regulations are almost always adopted despite some residual uncertainty.

American environmental law is filled with regulatory programs that authorize, or even compel, regulatory action in the absence of scientific certainty about the nature and extent of potential risks to the environment or public health. The Clean Air Act, for example, requires that the administrator of the Environmental Protection Agency (EPA) adopt measures to control various types of air pollution when, in the administrator's "judgment," the emission of certain pollutants "may reasonably be anticipated to endanger public health or welfare."[6] Absolute proof or scientific certainty is not required. Rather, a "reasonable" belief in the possibility of future harm is sufficient—indeed, such a belief may *require* action.[7] Other portions of the Clean Air Act require the EPA to set air-quality standards at the level "requisite to protect the public health" with "an adequate margin of safety," irrespective of the cost."[8] Thus, the law requires the EPA to set the standard at a level more stringent than that known to threaten public health.

The Endangered Species Act also requires government action in the absence of scientific certainty. Under the Endangered Species Act, the Fish and Wildlife Service must list species as endangered or threatened on the basis of the "best scientific and commercial data available.[9] That the "best available" scientific evidence may be inconclusive or uncertain does not relieve the Fish and Wildlife Service of its obligation to list a species if the "best available" evidence suggests that it could be threatened with extinction.

Federal environmental law also includes requirements that federal agencies consider the "full range of alternatives" and their likely environmental effects before taking action. Consider the National Environmental Policy Act, which requires the federal government to consider not only the likely environmental consequences of major federal actions but also various alternatives, including forgoing the proposed action altogether.[10] This seems to line up nicely with the Wingspread Statement's call for "an examination of the full range of alternatives, including no action."

However precautionary these various regulatory measures may be, most precautionary principle advocates consider them insufficient. According to the Wingspread Statement, existing environmental regulations "have failed to protect adequately human health and the environment," and more is required. Those who signed the statement or point to it as a model for policy have not sought to defend existing laws so much as they have sought the adoption of more stringent environmental and other regulations. In the eyes of precautionary principle advocates, U.S. environmental regulations are too slow and inflexible, and insufficiently precautionary.[11]

The real teeth of the principle, as articulated in the Wingspread Statement, come from shifting the burden of proof to "the proponent of an activity." Here, "better safe than sorry" means that no activity which "raises threats of harm to human health or the environment" should proceed until proven "safe." Interpreted this way, the principle erects a potential barrier to any activity that could alter the status quo. Applied literally to all activities, it would be a recommendation for not doing anything of consequence, as all manner of activities "raise threats of harm to human health or the environment." As Sunstein observes, "Read for all its worth, it leads in no direction at all. The principle threatens to be paralyzing, forbidding regulation, inaction, and every step in between."[12]

Yet precautionary principle advocates rarely call for applying this principle neutrally across the board. Rather, they seek to burden private actors, most notably corporations, that propose altering the environmental landscape in some way or introducing a new product or technology into the stream of commerce. This creates a higher barrier to adopting and implementing new technologies, and justifies lengthy approval programs and restrictions on technological advance.

An obvious question: why is it safer or more "precautionary" to focus on the potential harms of new activities or technologies without reference to the activities or technologies they might displace? There is no a priori reason to assume that newer technologies or less-known risks are more dangerous than older technologies or familiar threats. In many cases, the exact opposite will be true. A new, targeted pesticide may pose fewer health and environmental risks than a pesticide developed ten, twenty, or thirty years ago. Shifting the burden of proof, as the Wingspread Statement calls for, is not a "precautionary" policy so much as a reactionary one. This myopic focus on the threats posed by new activities or technologies can actually do more harm than good.

The last portion of the precautionary principle, with its emphasis on open decision-making and information, does not appear to be focused much on "precaution," but on transparency, accountability, and democratic consent. These are all important values in public policy, but there is no clear connection between these values and a more precautionary or risk-averse approach to public health and environmental policy. Affected communities, when informed about the relevant risks and trade-offs, may choose to accept certain environmental risks in return for economic or other benefits.

How this part of the principle operates in practice is contingent on how one defines "democratic" decision-making and "potentially affected parties," and whether these values are allowed to conflict. Are the "potentially affected parties" those directly affected by the development of a technology or a given good or service? Or does any individual or interest group with a potential concern get to have their say, too? In some cases, "democratizing" decisions about the acceptability of given technologies or marketplace transactions involves supplanting the decisions of those most involved through a political process. There is no guarantee this ensures adequate representation of those most affected or produces a more precautionary or environmentally protective result.[13]

The Precautionary Principle Abroad

The precautionary principle is not simply an idea promoted by activists and academics. The principle has found its way into a variety of legal instruments, including several international treaties. Although the principle has not yet reached the status of customary international law, it is increasingly common in international documents and statements of principle.

A soft articulation of the precautionary principle can be found in the Rio Declaration, adopted at the 1992 Earth Summit in Rio de Janeiro, Brazil. It provides: "Where there are threats of serious or irreversible damage, lack of full scientific certainty shall not be used as a reason for postponing cost-effective measures to prevent environmental degradation."[14] This is a soft articulation because it does not apply to all innovations or all potential threats, but only those that pose "threats of *serious* or *irreversible* damage." It further embodies a measure of proportionality, as it only calls for the adoption of "cost-effective measures" to address such risks. Insofar as reasonable measures can reduce an uncertain but serious risk, the Rio Declaration calls for action. As a consequence, it embodies a degree of proportionality.

A somewhat stronger formulation of the principle can be found in the preamble to the Convention on Biological Diversity. It similarly declares that "where there is a threat of significant reduction or loss of biological diversity, lack of full scientific certainty should not be used as a reason for postponing measures to avoid or minimize such a threat."[15]

A stronger formulation still is contained in the World Charter for Nature: "Where potential adverse effects are not fully understood, the activities should not proceed.[16] This is a reactionary and completely unworkable standard. Applied to technology, it would bring technological progress to a halt, as there are always uncertain and unpredictable effects of new technologies. But this is not simply true of technology. All economic and social innovations have implications that are less than fully understood at the time they are adopted. The same goes for government policy. Indeed, were the World Charter for Nature's formulation of the precautionary principle applied to governmental interventions in the economy, regulatory bureaucrats would forever sit on their hands, as centralized government decision makers never have complete knowledge or a full understanding of the likely effects of their policies.[17]

The 1992 Maastricht Treaty creating the European Union explicitly calls for adopting the precautionary principle in European environmental policy, even as it also urges consideration of the likely costs and benefits of specific measures. Article 130R(2) of the treaty provides:

> Community policy on the environment shall aim at a high level of protection taking into account the diversity of situations in the various regions of the Community. It shall be based on the precautionary principle and on the principles that preventive action should be taken, that environmental damage should as a priority be rectified at source and that the polluter should pay. Environmental protection requirements must be integrated into the definition and implementation of other Community policies.

Article 130R(3) further calls for the consideration of various factors in the development of environmental policy, including "available scientific and technical data" and "the potential benefits and costs of action or lack of action."[18]

Since the signing of the Maastricht Treaty, the precautionary principle has been incorporated into various aspects of European policy. Some nations have cited the principle as justification for prohibiting the use of hormones in livestock production or importing genetically modified crops. The European Council of Ministers adopted a formal resolution in April 1999 calling the European Commission "to be in the future even more determined to be guided by the precautionary principle" in its legislative proposals.[19] This led to the European Commission's Communication from the *Commission on the Precautionary Principle*, which declared that the EU would apply the precautionary principle "where preliminary objective scientific evaluation indicates that there are reasonable grounds for concern that the potentially dangerous effects on the environment, human, animal, or plant health may be inconsistent with the high level of protection chosen for the community.[20] The resulting policies have created what some characterize as a "guilty until proven innocent" approach to approving new products.[21]

Precautionary principle advocates often point to EU environmental policies as better exemplars of precautionary policy than U.S. regulations. As Joel Tickner and Carolyn Raffensperger explain, "European decision-makers have no pressure to justify what are essentially political decisions in the artificially rational language of science or economics.[22] Critics likewise note that the EU appears to use precautionary rhetoric about environmental threats to defend decisions that appear more motivated by economic or cultural concerns, such as preserving small farms and rural communities across the European countryside. The precautionary principle seems to be invoked against some risks and not against others.[23] Even in Europe, it is difficult to maintain that the precautionary principle provides the foundation for safety or health-enhancing policies.

Is Precaution Safer Than the Alternatives?

The biggest problem with the precautionary principle is that it does not clearly enhance the protection of public health and the environment. As University of Texas law professor Frank Cross observes, "The truly fatal flaw of the precautionary principle, ignored by almost all the commentators, is the unsupported presumption that an action aimed at public health protection cannot possibly have negative effects on public health."[24] In any policy decision, policy makers can make two potential errors regarding risk. On the one hand, policy makers may err by failing to adopt measures to address a health or environmental risk that exists. On the other hand, policy makers may adopt regulatory measures to control a health or environmental risk that does not exist. Both types of error can increase risks to public health.

Regulatory drug approval, as conducted by the Food and Drug Administration, provides a good example of how both types of error can increase net risks to public health. The FDA must approve new drugs before they may be used or prescribed. FDA approval is fairly precautionary, as it will only approve drugs shown to be "safe and effective." This standard is designed to prevent the release of an unsafe drug. Delaying the availability of potentially lifesaving treatment, however, poses risks of its own. In the simplest terms, if a new drug or medical treatment will start saving lives once approved, then the longer it takes for the government to approve the drug, the more likely people will die awaiting treatment.[25]

The negative consequences of failing to quickly approve a lifesaving drug can be significant. Consider the example of Misoprostol, a drug that prevents gastric ulcers.[26] Misoprostol was developed in the early 1980s and first approved in some nations in 1985. The FDA, however, did not approve Misoprostol until 1988. Even though the drug was already available in several dozen foreign countries, the FDA subjected Misoprostol to a nine-and-one-half-month review. At the time, between 10,000 and 20,000 people died of gastric ulcers per year. Had Misoprostol been approved more rapidly, it could have saved as many as 8,000 to 15,000 lives. Thus, in seeking to prevent one risk—the risk of approving an unsafe drug—the FDA contributed to the risk of gastric ulcers by preventing the use of a potentially lifesaving drug.

If the goal is to have a drug approval process that maximizes protecting public health, the risks of premature drug approval need to be weighed against the risks of failing to approve new medications rapidly enough. In this context, the precautionary principle, and in particular the burden-shifting framework advocated by its proponents, cannot be assumed to result in greater protection of public health and the environment. In fact, as the experience with the FDA drug approval process in the United States shows, a more precautionary approach could do more harm than good. Whether or not one accepts this indictment of the FDA drug approval process, the underlying lesson is clear: Increased precaution does not come without potential costs to the very values precaution is supposed to protect.

Precaution and Pesticides

The same risk-balancing necessary in the drug approval context must be used in the case of agricultural chemicals if the goal is to maximize the protection of public health and the environment. Just as two types of errors can be made in the drug approval process, two types of errors can be made in regulating agricultural chemicals. On the one hand, policy makers may fail to control adequately the pesticides and other chemicals that pose a significant risk to public health or the environment. On the other hand, regulations may prevent or discourage the use of agricultural chemicals that could enhance human welfare and even enhance the protection of public health and the environment.

Pesticides and other agricultural chemicals may present many different risks. Exposure or consumption of agricultural chemicals may pose a direct threat to human health. Misusing such chemicals can also contaminate water supplies, disrupt ecosystems, and harm other animal and plant species.

The risks posed by improper or excessive use of agricultural chemicals are serious, but do not justify precautionary regulation, as there are countervailing risks from the excess regulation of agricultural chemicals. In particular, limits on the use of agricultural chemicals may reduce crop yields, which may require planting more acreage or a price increase for agricultural products. Making fresh fruits and vegetables more expensive may have negative human health impacts as consumers substitute other, less healthy foods. Reducing crop productivity can have serious environmental effects, including the loss of species habitat and consequent effects on biodiversity. If precautionary controls are imposed only on the adoption or introduction of new agricultural chemicals, this may result in the continued use of less safe or less effective compounds, thereby enhancing the very risks precautionary regulation is meant to avoid.

The fungicide ethylene dibromide (EDB) illustrates the risk-risk nature of agricultural chemical.[27] EDB was used to control mold on grains, but was restricted in 1983 due to fears that it was a potential carcinogen.[28] Yet molds themselves can be a source of carcinogens in the human diet. Perhaps more important, the agricultural chemicals that replaced EDB were less effective and used in higher quantities, increasing the risk for exposed workers. Thus, while regulating EDB reduced certain risks, it also increased others. The only way to know whether regulating EDB increased the protection of public health and the environment is to compare these risks and weigh them against each other.

The environmental risks of forgoing advances in agricultural technologies are particularly severe. Global food demand is increasing dramatically, in part due to increases in global population. By 2050, there could be nearly nine billion mouths to feed worldwide.

It will not be enough for food production and distribution to keep up with increases in population, however. Approximately 800 million people receive inadequate nutrition in their diets at present, and more than 6 million children die annually of malnutrition. If agricultural productivity were held at 1997 levels, the world would need to put more than 1.5 billion hectares under plow to meet increased demand. Yet if agricultural productivity increases by as little as 1 percent a year from 1997 through 2050, that figure drops to an estimated 325 million hectares. If agricultural productivity could increase at a rate of 1.5 percent, it is possible we could return nearly 100 million hectares from agriculture to nature.[29] Limiting the development and introduction of new agricultural chemicals will make this task only more difficult. Applying a precautionary regulatory regime to agricultural chemicals could have serious environmental consequences.

Wealthier Is Healthier and Richer Is Cleaner

Advocates of the precautionary principle tend to assume that economic growth and development threaten public health and environmental protection. The Wingspread Statement, for example, speaks of the "substantial unintended consequences" brought about by the industrial society.[30] An underlying premise of the precautionary principle is that modern industrial society is unsustainable and threatens human survival, if not much of the planet as well. This assumption is highly questionable.[31] Economic growth and technological progress have been a tremendous boon to both human health and environmental protection. Efforts to limit such progress will likely be counterproductive. Regulatory measures that stifle innovation and suppress economic growth will deprive individuals of the resources necessary to improve their quality of life and deny societies the ability to make investments that protect people and our environs.

Modern industrial society produced an explosion of potential health and environmental risks, but it also generated a vast degree of wealth and technological advance that led to unprecedented improvements in public health. For centuries, average life expectancy was scarcely more than a few decades. In 1900, U.S. life expectancy was less than 50 years.[32] Today, however, U.S. life expectancy is approaching 80 years.[33] Similar advances have been observed in other nations as they have developed.[34] Infant and maternal mortality rates plummeted over the past century, as have the incidence and mortality rates of typhoid, diphtheria, tuberculosis, and numerous other diseases.[35] These positive trends are largely the result of increased wealth and the benefits such wealth brings. Higher economic growth and aggregate wealth strongly correlate with reduced mortality and morbidity.[36] This should be no surprise, as accumulating wealth is necessary to fund medical research, support markets for advanced lifesaving technologies, and build infrastructure necessary for better food distribution, and so on. In a phrase, poorer is sicker, and wealthier is healthier."[37]

Economic progress is no less essential for environmental protection than for protecting public health. Environmental protection is a good, and, like all goods,

it must be purchased. Wealth is necessary to finance environmental improvements, from purifying drinking water and controlling raw sewage to developing cleaner combustion technologies and low-emission modes of transport. Wealthier societies have both the means and the desire to address a wider array of environmental concerns and improve public health.[38] Economic growth fuels technological advances and generates the resources necessary to deploy new methods for meeting human needs efficiently and effectively. Public support for environmental measures, both public and private, is correlated with changes in personal income.[39] . . .

New technologies pose risks, to be sure. This is as true for agricultural and industrial chemicals as for anything else. Without question, some of the chemicals and other technologies targeted by advocates of the precautionary principle can cause problems if misused. Yet it is notable that the proliferation of these technologies has coincided with the greatest explosion of prosperity and longevity in human history. If modern society were as risky as precautionary principle advocates suggest, this should not be the case. . . .

In Search of Safety

Can the precautionary principle provide any sort of useful guide for health and environmental policy? Indur Goklany has proposed a set of tiered criteria that could be used to implement the precautionary principle while enhancing health and safety.[40] In his framework, greater attention should be paid to identified risks to human mortality and morbidity, for example, than to less certain or immediate environmental threats. His framework may well provide a more effective way to ensure that precautionary policies have a truly precautionary effect, but it clearly does not produce the results most principle advocates defend. As applied by Goklany, for example, the precautionary principle would counsel against restrictions on using DDT and agricultural biotechnology, and it would not support drastic measures to address the threat of climate change.

Whether or not Goklany's framework presents the best way forward, one thing should be abundantly clear: True precaution requires recognizing that risks trade against other risks. The risks of new technologies must be weighed against the risks of doing without. The harms of environmental disruptions must be weighed against the consequences of merely preserving the status quo.

The stated aim of the precautionary principle is to enhance protection of public health and environmental concerns. In practice, however, the precautionary principle is only applied to the risks of technological change and industrial society, with little appreciation for the risks that wealth and technology prevent. New technologies can be risky. Some industrial chemicals may cause health problems, even if used carefully. But this does not justify adopting a blanket precautionary rule suppressing chemical use and technological development.

If the true aim is a safer world, and not merely to restrict industrial activity for its own sake or to retard technological progress, the risks of new chemicals or products must be weighed against the risks that they ameliorate or prevent. The risks of change must be weighed against the risk of stagnation. In every case, "the empirical question is whether the health [and environmental] gains from the regulation of the substances involved are greater or lesser than the health [and environmental] costs of the regulation."[41]

While the advocates of the precautionary principle rely on the rhetoric of prudence and public health protection, they encourage the exclusive focus on one set of risks while ignoring others. Contrary to what your mother may have told you, "better safe than sorry" isn't always safer. In fact, when it comes to policies to protect public health and the environment, this type of thinking could do us in.

Notes

1. Carolyn Raffensperger and Joel Tickner, "Introduction: To Foresee and Forestall," in Protecting Public Health and the Environment: Implementing the Precautionary Principle (Washington, DC: Island Press, 1999), 4.

2. Ibid.; Andrew Jordan and Timothy O'Riordan, "The Precautionary Principle in Contemporary Environmental Policy and Politics," in Protecting Public Health and the Environment, 19.

3. For an overview of the incorporation of the precautionary principle in various international instruments, see Julian Morris, "Defining the Precautionary Principle," in Rethinking Risk and the Precautionary Principle, ed. Julian Morris (Woburn, MA: Butterworth-Heinemann, 2000).

4. See "Wingspread Statement on the Precautionary Principle," Science and Environmental Health Network, http://www. sehn.org/wing.html.

5. Cass R. Sunstein, "Throwing Precaution to the Wind: Why the 'Safe' Choice Can Be Dangerous," Boston Globe, July 13, 2008. For a more extensive critique, see Cass R. Sunstein, The Laws of Fear: Beyond the Precautionary Principle (Cambridge: Cambridge University Press, 2005).

6. See, for example, 42 U.S.C. §7521(a)(1), which provides in relevant part: "The Administrator shall by regulation prescribe (and from time to time revise) in accordance with the provisions of this section, standards applicable to the emission of any air pollutant from any class or classes of new motor vehicles or new motor vehicle engines, which in his judgment cause, or contribute to, air pollution which may reasonably be anticipated to endanger public health or welfare." Other provisions in the Clean Air Act contain similar requirements.

7. See, for example, Natural Resources Defense Council, Inc. v. Train, 411 F.Supp. 864 (S.D.N.Y. 1976).

8. See 42 U.S.C. §7409(b)(1), which provides in relevant part: "National primary ambient air quality standards, prescribed under subsection (a) of this section shall be ambient air quality standards the attainment and maintenance of which in the judgment of the Administrator, based on such criteria and allowing an adequate margin of safety, are requisite to protect the public health. Such primary standards may be revised in the same manner as promulgated." As interpreted by the Supreme Court, this language precludes considering costs when setting the standards. Whitman v. American Trucking Associations, 531 U.S. 457 (2001).

9. See 16 U.S.C. §1533.

10. See 42 U.S.C. §4332.

11. See, for example, Joel Tickner and Carolyn Raffensperger, "The Politics of Precaution in the United States and the European Union," Global Environmental Change 11 (2001): 175, 177.

12. Sunstein, Laws of Fear, 14.

13. For an extended discussion of why expanding the range of potentially affected parties who have standing to challenge environmental harms may not improve environmental protection, see Jonathan H. Adler, "Stand or Deliver? Citizen Suits, Standing, and Environmental Protection," Duke Environmental Law and Public Policy Forum 12, no. 1 (2001): 39–83.

14. Rio Declaration on Environment and Development, UN Conference on Environment and Development, UN Doc. A/CONF.151/5/Rev.1 (1992).

15. United Nations Convention on Biological Diversity, June 5, 1992, preamble.

16. World Charter for Nature, G.A. Res. 37/7, UN GAOR, 37th Sess., Supp. No. 51, at section (II)(11)(b), UN Doc. A/Res/37/7 (1992).

17. As Nobel laureate economist F.A. Hayek observed, "The knowledge of the circumstances of which we must make use never exists in concentrated or integrated form but solely as the dispersed bits of incomplete and frequently contradictory knowledge which all the separate individuals possess." F. A. Hayek, "The Use of Knowledge in Society," American Economic Review 35, no. 4 (1945): 519–30.

18. Treaty on European Union, February 7, 1992, 1992 O.J. (C 191) 1, art. 130R(3).

19. European Commission, Communication from the Commission on the Precautionary Principle (Brussels: Commission of the European Communities, 2000), 7.

20. Ibid., 9–10.

21. See Tim Lougheed, "Understanding the Role of Science in Regulation," Environmental Health Perspectives 117, no. 3 (March 2009): A109.

22. Tickner and Raffensperger, "Politics of Precaution," 178.

23. See Jonathan Weiner and Michael Rogers, "Comparing Precaution in the United States and Europe," Journal of Risk Research 5 (2002): 317–49.

24. Frank B. Cross, "Paradoxical Perils of the Precautionary Principle," Washington and Lee Law Review 53, no. 3 (1996): 860.

25. Sam Kazman, "Deadly Overcaution: FDA's Drug Approval Process," Journal of Regulation and Social Costs (September 1990).

26. This discussion is based on the Kazman article; ibid., 47–48.

27. George M. Gray and John D. Graham, "Regulating Pesticides," in Risk versus Risk: Tradeoffs in Protecting Health and the Environment, ed. John D. Graham and Jonathan Baert Weiner (Cambridge, MA: Harvard University Press, 1995), 186–87.

28. Cross, "Paradoxical Perils," 875–76.

29. See Indur M. Goklany, The Precautionary Principle: A Critical Appraisal of Environmental Risk Assessment (Washington, DC: Cato Institute, 2001), 31–32.

30. Raffensperger and Tickner, Appendix A, Protecting Public Health and the Environment, 353–55.

31. For a contrary view, see Bjorn Lomborg, The Skeptical Environmentalist: Measuring the Real State of the World (Cambridge: Cambridge University Press, 2001). For critical consideration of Lomborg's thesis, see the symposium "The Virtues and Vices of Skeptical Environmentalism," Case Western Reserve Law Review 53, no. 2 (2002).

32. "United States Life Tables," National Vital Statistics Reports 56, no. 9 (December 28, 2007): table 11.

33. According to the U.S. Centers for Disease Control and Prevention, life expectancy at birth was 77.7 in 2006. Melonie Heron et al., "Deaths: Final Data for 2006," National Vital Statistics Reports 57, no. 14 (April 17, 2009): 1.

34. See Nicholas Eberstadt, "Population, Food, and Income: Global Trends in the Twentieth Century," in The True State of the Planet, ed. Ronald Bailey (New York: Free Press, 1995), 21–26.

35. Ibid.

36. See, for example, Susan L. Ettner, "New Evidence on the Relationship between Income and Health," Journal of Health Economics 15 (1996): 67; John D. Graham et al., "Poorer Is Riskier," Risk Analysis 12, no. 3 (1992): 333–37; Ralph L. Keeney, "Mortality Risks Induced by Economic Expenditures," Risk Analysis 10, no. 1 (1990): 147–59.

37. This phrasing is attributed to the late Aaron Wildavsky.

38. See Goklany, Precautionary Principle, 22–24, 75–79.

39. See Matthew E. Kahn and John G. Matsusaka, "Demand for Environmental Goods: Evidence from Voting Patterns on California Initiatives," Journal of Law and Economics 40, no. 1 (1997).

40. See Goklany, Precautionary Principle, 8–11.

41. Aaron Wildavsky, But Is It True? (Cambridge, MA: Harvard University Press, 1995), 428.

JONATHAN H. ADLER is a Professor and Director of the Center for Business Law and Regulation at Case Western Reserve University School of Law.

EXPLORING THE ISSUE

Do We Need the Precautionary Principle?

Critical Thinking and Reflection

1. What should the place of cost-benefit or risk-benefit analysis be in applying the Precautionary Principle?
2. Should the Precautionary Principle be applied to both action and inaction?
3. How much scientific evidence that harm is likely should be needed before invoking the Precautionary Principle?

Is There Common Ground?

No one involved in the debates over the value and application of the Precautionary Principle would argue that it is a bad idea to look before you leap. They are more likely to differ in their visions of where the benefits lie. For instance, proponents of genetically modified crops believe that genetically modified organisms benefit the food supply far more than they threaten human or environmental health. Critics see a need for precaution in the potential risk to human or environmental health. Similar polarizations can be seen with many other issues. Look briefly at other issues in this book and identify opposing visions of benefit or risk.

Additional Resources

David Kriebel, et al., "The Precautionary Principle in Environmental Science," *Environmental Health Perspectives* (September 2001)

Marco Martuzzi and Roberto Bertollini, "The Precautionary Principle, Science and Human Health Protection," *International Journal of Occupational Medicine and Environmental Health* (January 2004)

Cass R. Sunstein, *Laws of Fear: Beyond the Precautionary Principle* (Cambridge University Press, 2005)

Create Central

www.mhhe.com/createcentral

Internet References . . .

The European Union

europa.eu/legislation_summaries/consumers
/consumer_safety/l32042_en.htm

The Science and Environmental
Health Network

www.sehn.org/precaution.html

The United Nations Environmental
Programme

www.unep.org/

Selected, Edited, and with Issue Framing Material by:
Thomas Easton, *Thomas College*

ISSUE

Are There Limits to Growth?

YES: Graham M. Turner, from "On the Cusp of Global Collapse?: Updated Comparison of the *Limits to Growth* with Historical Data," *GAIA—Ecological Perspectives for Science and Society* (2012)

NO: Tsvi Bisk, from "No Limits to Growth," *World Future Review* (Spring 2012)

Learning Outcomes
After reading this issue, you will be able to:
• Explain what sustainable development is.
• Explain why the world needs prompt, decisive decision-making to deal with long-term problems such as sustainability.
• Explain the difference between making a forecast for the future and developing scenarios for the future.
• Explain what is meant by "ecological footprint."

ISSUE SUMMARY

YES: Graham M. Turner argues that the computer-modeled projections of the 1972 *Limits to Growth* study's "standard run" or "business as usual" scenario, which forecast a twenty-first century collapse in living standards and population due to resource constraints, continue to be matched very well by actual data. We have now reached the point where we can expect global collapse to begin in the current decade (it may have started already). In addition, it may now be too late to avoid the crisis; planning should focus on coping with it.

NO: Tsvi Bisk argues that the prophets of doom have consistently been proved wrong by the infinite resource of the human mind. Yes, we have problems, but new technologies now under development mean that there are no real limits to growth. By 2100, the Earth will be able to support 12 billion people (well above UN projections) with an American standard of living and a tenth of the present environmental impact.

Over the last 50 years, many people have expressed concerns that humanity cannot continue indefinitely to increase population, industrial development, and consumption of resources. The trends and their impacts on the environment are amply described in numerous books, including historian J. R. McNeill's *Something New Under the Sun: An Environmental History of the Twentieth-Century World* (W. W. Norton, 2000).

"Can we keep it up?" is the basic question behind the issue of sustainability. In the 1960s and 1970s, it was expressed as the "Spaceship Earth" metaphor, which said that we have limited supplies of energy, resources, and room, and we must limit population growth and industrial activity, conserve, and recycle if we are not to run into crucial shortages. In 1972, Dennis Meadows, an MIT social policy analyst, and his colleagues, published *The Limits to Growth: A Report for the Club of Rome's Project on the Predicament of Mankind* (Universe Books, 1974). The Club of Rome was an independent, international group of scientists, economists, humanists, and industrialists who shared the view that traditional institutions and policies could no longer properly evaluate the complex, major problems that the world was facing. The Meadows group used a computer model called World3 (primitive by today's standards) to study five factors that might limit global growth and development: population, agricultural production, natural resources, industrial production, and pollution. The conclusion was that if current policies of unrestrained growth were allowed to continue, without some conscious effort to achieve a state of developmental equilibrium, the next century would likely see a disastrous decline in population and industrial productivity, resulting from resource depletion and widespread pollution.

In 1992, the Meadows group published *Beyond the Limits*, a 20-year update of the original study using a much-improved World3 computer model. Their conclusion at that time was that future prospects had not improved and that, in fact, "humanity had already overshot the limits of Earth's support capacity." In 2002, the 30-year update did not alter that conclusion. Nor did the 40-year update in 2012, Jorgen Randers, *2052: A Global Forecast for the Next Forty Years (A Report to the Club of Rome Commemorating the 40th Anniversary of* The Limits to Growth) (Chelsea Green Publishing, 2012). Randers argues that humanity's ecological footprint is too large. That is, humanity uses more resources than one planet Earth can supply. Between now and 2052, there will be efforts to reduce the footprint, but they will be too little and too late, in large part because of short-term thinking by governments and businesses. As a result, population will peak about 2040 at just over 8 billion (well below UN projections), economic growth and consumption will slow, and the developed world will decline, but resource and climate problems will not become catastrophic until after 2052. He does not trust technology to save the situation.

In 2008, Dr. Graham Turner, a senior research scientist at Australia's CSIRO Sustainable Ecosystems program, published "A Comparison of *The Limits to Growth* with Thirty Years of Reality," Socio-Economics and the Environment in Discussion, CSIRO Working Paper Series 2008–2009 (*Global Environmental Change*, August 2008), and found that the projections and the data show a remarkable match. Indeed, lacking immediate action, global collapse before the middle of this century seems far too likely.

"Sustainability" entered the global debate in the early 1980s, when the United Nations' secretary general asked Gro Harlem Brundtland, a former prime minister and minister of environment in Norway, to organize and chair a World Commission on Environment and Development and produce a "global agenda for change." The resulting report, *Our Common Future* (Oxford University Press, 1987) defined "sustainable development" as "development that meets the needs of the present without compromising the ability of future generations to meet their own needs." It recognized that limits on population size and resource use cannot be known precisely, that problems may arise not suddenly but rather gradually, marked by rising costs, and that limits may be redefined by changes in technology. But it did also recognize that limits exist and must be taken into account when governments, corporations, and individuals plan for the future.

The *Brundtland Report* led to the UN Conference on Environment and Development held in Rio de Janeiro in 1992. The Rio Conference set sustainability firmly on the global agenda and made it an essential part of efforts to deal with global environmental issues and promote equitable economic development. In brief, sustainability means such things as cutting forests no faster than they can grow back, using groundwater no faster than it is recharged by precipitation, stressing renewable energy sources rather than exhaustible fossil fuels, and farming in such a way that soil fertility does not decline. In addition, economics must be revamped to take account of environmental costs as well as capital, labor, raw materials, and energy costs.

Many add that the distribution of the Earth's wealth must be made more equitable as well.

The first of the Rio Declaration's 22 principles states, "Human beings are at the center of concerns for sustainable development. They are entitled to a healthy and productive life in harmony with nature." Any solution to the sustainability problem therefore should not infringe human welfare. This makes any solution that involves limiting or reducing human population or blocking improvements in standard of living very difficult to sell. Yet solutions may be possible. David Malin Roodman suggests in *The Natural Wealth of Nations: Harnessing the Market for the Environment* (W. W. Norton, 1998) that taxing polluting activities instead of profit or income would stimulate corporations and individuals to reduce such activities or to discover nonpolluting alternatives. Arun Agrawal and Maria Carmen Lemos, in "A Greener Revolution in the Making? Environmental Governance in the 21st Century," *Environment* (June 2007), argue that budget cuts and globalization are eroding the power of the state in favor of international "hybrid" arrangements that stress public–private partnerships, markets, and community and local participation.

In June 2012, the UN World Summit on Sustainable Development (Rio + 20) was held in Rio de Janeiro, Brazil. Among its aims were to assess the progress to date and to secure new commitments to sustainable development. Among the results were $513 billion in pledges to work on sustainability, but there were no provisions for enforcing commitments. See Simon Romero and John M. Broder, "Progress on the Sidelines as Rio Conference Ends," *New York Times* (June 23, 2012) (www.nytimes.com/2012/06/24/world/americas/rio20-conference-ends-with-some-progress-on-the-sidelines.html). The outcome document agreed to at the conference, *The Future We Want* (www.uncsd2012.org/content/documents/727The%20Future%20We%20Want%2019%20June%201230pm.pdf) said (among other things): "We recognize that poverty eradication, changing unsustainable and promoting sustainable patterns of consumption and production, and protecting and managing the natural resource base of economic and social development are the overarching objectives of and essential requirements for sustainable development. We also reaffirm the need to achieve sustainable development by: promoting sustained, inclusive and equitable economic growth, creating greater opportunities for all, reducing inequalities, raising basic standards of living, fostering equitable social development and inclusion, and promoting integrated and sustainable management of natural resources and ecosystems that supports *inter alia* economic, social and human development while facilitating ecosystem conservation, regeneration and restoration and resilience in the face of new and emerging challenges."

These goals are laudable, and failing to meet them may have dire consequences. In February 2007, Sigma Xi and the United Nations Foundation released the Scientific Expert Group Report on Climate Change and Sustainable Development, *Confronting Climate Change: Avoiding the*

Unmanageable and Managing the Unavoidable (www.sigmaxi
.org/about/news/UNSEGReport.shtml) (Executive Summary,
American Scientist, February 2007). Among its many points
is that climate change from global warming is a huge
threat to sustainability. Even in spite of feasible attempts
at mitigation and adaptation, there is a serious risk of
"intolerable impacts on human well-being." Lester R.
Brown, "Could Food Shortages Bring Down Civilization?"
Scientific American (May 2009), warns that unsustainable
agricultural practices and use of groundwater, along with
global warming, threaten to diminish food supplies to the
point where society may actually break down. Underlining
this concern, in October 2008, the European Environment
and Sustainable Development Advisory Councils held a
conference on "Sustaining Europe for a Long Way Ahead."
The EEAC website (www.eeac.eu/) noted that "Addressing
the very long term through the lens of sustainable development is now a matter of urgency. The prospect of highly
damaging, and extremely costly, effects of global change
in climate, in natural hazards caused by human intervention, the loss of biodiversity and disruption of food
security, poses serious threats to personal and collective
human health and wellbeing. . . . The long term is indeed
here already."

Given continuing growth in population and demand
for resources, sustainable development is clearly a difficult proposition. Some think it can be done, but others
think that for sustainability to work, either population
or resource demand must be reduced. The World Business Council for Sustainable Development's (www.wbcsd
.org/home.aspx) *Business Solutions for a Sustainable World*
(February 2010) says it will require bringing the costs of
environmental impacts into the marketplace, shifting
away from fossil fuels, and improving efficiency. James
H. Brown, et al., "Energetic Limits to Economic Growth,"
BioScience (January 2011), find that there is a tight match
between rising energy use and rising per capita gross
domestic product (GDP); energy availability thus defines
a serious limit. See also Daniel N. Lipson, "Is the Great
Recession Only the Beginning? Economic Contraction in
an Age of Fossil Fuel Depletion and Ecological Limits to
Growth," *New Political Science* (December 2011). Anthony
R. Leiserowitz, Robert W. Kates, and Thomas M. Parris, in
"Do Global Attitudes and Behaviors Support Sustainable
Development?" *Environment* (November 2005), find that
though the world's people appear to support the component concepts of sustainable development, there is a mis-
match between that support and their behavior. In the
long term, they say, what is needed is a "shift from mate-
rialist to post-materialist values, from anthropocentric to
ecological worldview, and a redefinition of the good life."
Unfortunately, that shift remains in the future. See also
Joel Magnusson, "Cultural Lag and Limits to Growth,"
Interconnections (2011), who argues that people will have
to get by with less and "local institutions have to be made
anew and cut from very different cloth and that means
with a different consciousness and wholly different

expectations of what is meant by 'the good life'." P. Aarne
Vesilind, Lauren Heine, and Jamie Hendry, in "The Moral
Challenge of Green Technology," *TRAMES: A Journal of the
Humanities & Social Sciences* (vol. 1, 2006), "conclude that
the unregulated free market system is incompatible with
our search for sustainability. Experience has shown that if
green technology threatens profits, green technology loses
and profitability wins." See also Bill McKibben, "Breaking
the Growth Habit," *Scientific American* (April 2010). Not
surprisingly, many people see sustainable development as
in conflict with business and industrial activities, private
property rights, and such human freedoms as the freedoms to have many children, to accumulate wealth, and
to use the environment as one wishes.

The degrowth movement has taken shape to express
the environmentalist, anti-consumerist, and anti-capitalist
ideas noted above; see Baria Gencer Baykan, "From Limits
to Growth to *Degrowth* within French Green Politics,"
Environmental Politics (June 2007), and Takis Fotopoulos, "Is Degrowth Compatible with a Market Economy?"
International Journal of Inclusive Democracy (January 2007),
who argues that the root cause of our environmental and
social justice problems is the market-based growth economy. Sustainability requires a change to a non-growth
economy with local production and a focus on meeting
real human needs. The First International Degrowth Conference occurred in Paris, France, in 2008. The second
was held in Barcelona, Spain, in 2010. A related conference was held in Vancouver, Canada, later in 2010. John
Bellamy Foster, in "Capitalism and Degrowth—An Impos-
sibility Theorem," *Monthly Review* (January 2011), says:
"It is undeniable today that economic growth is the main
driver of planetary ecological degradation. But to pin one's
whole analysis on overturning an abstract 'growth society'
is to lose all historical perspective . . . . it can only take on
genuine meaning as part of a critique of capital accumulation and part of the transition to a sustainable, egalitarian,
communal order." Unlike most degrowth theorists, Robert
Engelman, in "Population and Sustainability," *Scientific
American Earth 3.0* (Summer 2009), notes that sustainability can only be achieved if population is controlled at the
same time as consumption.

Many people find the ideas that there are limits to the
growth of human civilization, that the way we organize
our political and economic systems is fundamentally
flawed, that we must reduce population and economic
activity, and even—in the name of social justice—
redistribute wealth, repugnant. This was already apparent
when the original *Limits to Growth* study was published in
1972. Critiques of the study came from every direction.
Conservatives insisted that the marketplace would work
to prevent disaster with no need to impose international
plans or controls. Liberals argued that restraints on growth
would hurt the poor more than the affluent. And radicals
contended that the results were only applicable to the
type of profit-motivated growth that occurs under capitalism. But later studies using more sophisticated models

have reinforced the basic conclusion that growth without limits will ultimately result in disaster. The 1980 *Global 2000 Report*, which reiterated the need for developmental controls; the 1992 United Nations Conference on Environment and Development in Rio de Janeiro; the report of the World Commission on Environment and Development, which focused attention on the need for sustainable development; and the UN World Summit on Sustainable Development (Rio + 20) all have roots that extend back to *The Limits to Growth.* It has also reshaped the nature of policy discussions in general. As Michael M. Crow wrote in "None Dare Call It Hubris: The Limits of Knowledge," *Issues in Science and Technology* (Winter 2007), "Our current approach to framing problems can be traced back to the 1972 publication of . . . *The Limits to Growth.* . . . Since that time, the way we think about human activity and the environment and the way we translate this thinking into our science policy and subsequent R&D, public debate, and political action have been framed by the idea of external limits—defining them, measuring them, seeking to overcome them, denying their existence, or insisting that they have already been exceeded."

In the YES selection, Graham M. Turner argues that the computer-modeled projections of the 1972 *Limits to Growth* study's "standard run" or "business as usual" scenario, which forecast a twenty-first century collapse in living standards and population due to resource constraints, continue to be matched very well by actual data. We have now reached the point where we can expect global collapse to begin in the current decade (it may have started already). In addition, it may now be too late to avoid the crisis; planning should focus on coping with it. In the NO selection, futurist Tsvi Bisk argues that the prophets of doom have consistently been proved wrong by the infinite resource of the human mind. Yes, we have problems, but new technologies now under development mean that there are no real limits to growth. By 2100, the Earth will be able to support 12 billion people (well above UN projections) with an American standard of living and a tenth of the present environmental impact.

YES ↵

Graham M. Turner

On the Cusp of Global Collapse?: Updated Comparison of the *Limits to Growth* with Historical Data

Although it has been about 40 years since *The Limits to Growth (LtG)* was first published, it is more pertinent than ever to review what this ground-breaking scenario and modelling study can tell us about the sustainability, or collapse, of the global economy and population. Through a dozen scenarios simulated in a global model *(World3)* of the environment and economy, Meadows et al. identified that "overshoot and collapse" was avoidable only if considerable change in social behaviour and technological progress was made early in advance of environmental or resource issues. When this was not achieved in the simulated scenarios, collapse of the economy and human population occurred in the 21st century, sometimes reducing living conditions to levels akin to the early 20th century.

Despite the *LtG* initially becoming a best-selling publication, the work was subsequently largely relegated to the "dustbin of history" by a variety of critics. These critics perpetuated the public myth that the *LtG* had been wrong, saying that it had forecast collapse to have occurred well before year 2000 when the *LtG* had not done this at all.

Over the last decade, however, there has been something of a revival in the awareness and understanding of the *LtG*. A thorough account of the *LtG* as well as associated debates and developments is provided by Bardi. Most recently, Randers a *LtG* co-author, has published his forecast of the global situation in 2052 and renewed the lessons from the original publication. A turning point in the debate occurred in 2000 with the energy analyst Simmons raising the possibility that the *LtG* modelling was more accurate than generally perceived. Others have made more comprehensive assessments of the model output indeed, we found that 30 years of historical data compared very well with the *LtG* baseline or *standard run* scenario. The *standard run* scenario embodies the business-as-usual social and economic practices of the historical period of the model calibration (1900 to 1970), with the scenario modelled from 1970 onwards.

The following paper presents an update on our data comparison, to coincide with the 40 years since the original *LtG* publication. In addition, an update is worthy because of questions raised about the economic downturn currently being experienced—commonly associated with the global financial crisis (GFC)—and the onset of collapse in the *LtG standard run* scenario. Is it possible that aspects leading to the collapse in the *LtG standard run* scenario have contributed to the GFC-related economic downturn? Could it be that this downturn is therefore a harbinger of global collapse as modelled in the *LtG*?

We begin with briefly reviewing the data that is available for our update, comparing it then with three key scenarios from the *LtG*, namely the *standard run comprehensive, technology* and *stabilized world* scenarios, the latter avoiding collapse. On the basis of the comparison, we discuss what the modelling might mean for a resource-constrained global economy. In particular, the paper examines the issue of peak oil and the link between energy return on investment (EROI) and the *LtG World3* model. The findings lead to a discussion of the role of oil constraints in the GFC, and a consideration of the link between these constraints and a general collapse depicted in the *LtG*.

Data Update

The data presented here follows that of our 30-year review. This data covers the variables, i.e., demographic variables and five sub-systems of the global economic system, displayed in the *LtG* output graphs:

- population (and crude birth and death rates),
- industrial output per capita,
- food supply per capita,
- services per capita,
- persistent global pollution, and
- fraction of non-renewable resources available.

Data sources are all publically available, many of them through the various United Nations (UN) organizations (and websites). In our review, we discuss details on these data sources and aspects such as interpretation, uncertainties and aggregation. However, some additional data and calculation were necessary since measured data to 2010 was not always available (and even when it is the data may be forecast estimates). A summary of the data is provided in the following.

Population data is readily available from the Population Division of the Department of Economic and Social Affairs of the UN Secretariat (obtained via the online *EarthTrends* database of the World Resources Institute); but data from 2006 onwards is a forecast. Given the short gap to 2010 and typical inertia in population dynamics, the 2010 estimate will be sufficiently accurate for the comparison made here.

Industrial output was available only to 2007 directly from the UN *Statistical Yearbooks* now accessible online. Industrial output per capita is used as a measure of material wealth in the *LtG* modelling, but the industrial output also supplies capital for use in other sectors, including agriculture and resource extraction.

Food supply was based on energy supply data (calories) from the Food and Agriculture Organization (FAO), with the extension to 2009/2010 generated from comparison with production data, which was scaled to the energy supply data for each corresponding food type in the production data.

Service provision has been measured by proxy indicators: electricity consumed per capita and literacy rates. In the former case, for the most recent data it was necessary to scale electricity generation data (from BP 2011) to consumption values, and hence account for electricity transmission losses. Literacy rates were updated from the United Nations Educational, Scientific and Cultural Organization (UNESCO) *Statistics database*, which is the source for the *Earth Trends* data. Values are provided for time ranges rather than single years.

Global persistent pollution was measured by the greenhouse gas CO_2 concentration, available to 2008 on the *Earth Trends* database, with latest measurements to 2010 from Pieter Tans, National Oceanic and Atmospheric Administration (NOAA) Earth System Research laboratory (ESRL), and Ralph Keeling, Scripps Institution of Oceanography. The 300 ppm CO_2 concentration at 1900 was subtracted from the measured data to represent an effective background of zero global pollution in 1900.

Finally, the ***fraction of non-renewable resources*** available is estimated from production data on energy resources, since other resources are conservatively assumed to be infinitely substitutable or there to be unlimited resources. Energy production data to 2010 was obtained from the BP *Statistical Review* which was subtracted from the ultimate resource originally available to obtain the remaining resources. To account for considerable uncertainty in the ultimate resource, upper and lower estimates were made based on optimistic and constrained assessments, respectively.

Comparison of Data with *LtG* Scenarios

This section presents a graphical comparison of the historical data with three scenarios from the original *LtG* modelling. The three scenarios effectively span the extremes of technological and social responses as investigated in the *LtG*.

The ***standard run*** represents a business-as-usual situation where parameters reflecting physical, economic and social relationships were maintained in the *World3* model at values consistent with the period 1900 to 1970.

The ***comprehensive technology*** approach attempts to solve sustainability issues with a broad range of purely technological solutions. This technology-based scenario incorporates levels of resources that are effectively unlimited, 75 percent of materials are recycled, pollution generation is reduced to 25 percent of its 1970 value, agricultural land yields are doubled, and birth control is available world-wide.

For the ***stabilized world*** scenario, both technological solutions and deliberate social policies are implemented to achieve equilibrium states for key factors including population, material wealth, food and services per capita. Examples of actions implemented in the *World3* model include: perfect birth control and desired family size of two children; preference for consumption of services and health facilities and less toward material goods; pollution control technology: maintenance of agricultural land through diversion of capital from industrial use; and increased lifetime of industrial capital. . . .

The statistical analysis undertaken in our 30-year review was not reproduced here as the changes would be minor, and add little further to the assessment. There are some other points of detail on the data comparison to be noted below.

It is evident that the data generally continues to align favourably to the *standard run* scenario (for most of the variables), and not to the other two scenarios. This comparison demonstrates that the original work cannot be dismissed as many critics have attempted, and increases confidence in the *LtG* scenario modelling. In contrast, there do not appear to be other economy-environment models that have demonstrated such comprehensive and long-term data agreement. Nevertheless, this agreement is not a complete validation of the model (partly due to the nonlinear nature of the *World3* model) or the *standard run* scenario. Achieving validation requires at least that key inputs and nonlinear (or threshold) assumptions also be verified. This verification is partially initiated in the *Discussion* section with an examination of the imposts of resource extraction. It is noteworthy that despite the nonlinearity of the *World3* model, the general outcomes of the scenarios are not sensitive to reasonable uncertainties in key parameters.

The demographic variables . . . continue to show the same comparisons as seen in our 30-year review, so that population would peak somewhat higher than the *standard run* by 2030 or later according to an extrapolation of the difference between the birth and death rates. It is more evident now, however, that the crude death rate has leveled off while the birth rate continues to fall, which are general trends seen in the three scenarios, albeit at different values. Notably, the death rate reverses its monotonic decline and begins to climb in all scenarios within a decade—significantly so in the *standard run* (and *comprehensive technology*) scenario by 2020.

Outputs of the economic system show trends mostly commensurate with the *LtG standard run*. Importantly, any downturn in industrial activity due to the GFC has not been captured in the historic data since these were only available to 2007. Nevertheless, the observed industrial output per capita illustrates a slowing rate of growth that is consistent with the *standard run* reaching a peak. In this scenario, the industrial output per capita begins a substantial reversal and decline at about 2015. Observed food per capita is broadly in keeping with the *LtG standard run*, with food supply increasing only marginally faster than population. Literacy rates show a saturating growth trend, while electricity generation per capita . . . grows more rapidly and in better agreement with the *LtG* model.

Global pollution measured by CO_2 concentration is most consistent with the *standard run* scenario, but this ten-year data update indicates that it is rising at a somewhat slower rate than that modelled. This could be due to a number of factors, which cannot be separately identified in this analysis. For instance, in comparison with the *standard run* model output, lower observed industrial output per capita is consistent with lower observed pollution generation, though this effect will be offset by the slightly higher observed population levels. It is also possible that the dynamics of persistent pollution generation by different economic activities or assimilation in the environment are not parameterized in the *World3* model precisely in terms of actual CO_2 dynamics (which is still a topic of active research). In this possibility, the recent data are consistent with a slightly higher assimilation rate, or alternatively, a lower pollution generation rate in the agriculture sector compared with the industrial sector (since the relative rate of food production is greater than industrial output). Regardless of the explanation, the level of global pollution is sufficiently low (in all scenarios, and the data) to not have a serious impact neither on the environment nor human life-expectancy. In the *World3* standard setting, current pollution levels decrease life expectancy by less than one percent.

In contrast, the [lower estimate] of non-renewable resources remaining [demonstrate] a closer alignment with the *standard run* while the upper estimate aligns well with the *comprehensive technology* scenario. The lower estimate also shows a significant fall toward the point (50 to 60 percent of the original resource) when the *World3* model incorporates a growing diversion of capital toward the resource sector in order to extract more difficult resources. This is the primary cause of collapse in the *standard run* scenario, as described below. The observed data is based on energy resources conservatively assuming full substitution potential among the different primary energy types. The assumption may not be entirely accurate, for instance, in the case of transport fuels essential for the smooth functioning of the economy; the following section reflects upon this question further.

Discussion—Is Collapse Imminent?

Based simply on the comparison of observed data and the *LtG* scenarios presented above, and given the significantly better alignment with the *standard run* scenario than the other two scenarios, it would appear that the global economy and population is on the cusp of collapse. This contrasts with other forecasts for the global future, which indicate a longer or indeterminate period before global collapse; Randers for example forecasts collapse after 2050, largely based around climate change impacts, with features akin to the *LtG comprehensive technology* scenario. This section therefore examines more closely the mechanisms behind the near-term *standard run* collapse and explores whether these resemble any real-world developments.

Essentially, the collapse in the *standard run* scenario is caused by resource constraints. The dynamics and interactions incorporated in the *World3* model that play out in this scenario are summarized in the following. During the 20th century, increasing population and demand for material wealth drives more industrial output, which grows at a faster rate than population. Pollution from increasing economic activity increases, but from a very low level, and does not seriously impact the population or environment.

However, the increased industrial activity requires ever increasing resource inputs (albeit offset by improvements in efficiency), and resource extraction requires capital (machinery) which is produced by the industrial sector (which also produces consumption goods). Until the nonrenewable resource base is reduced to about 50 percent of the original or ultimate level, the *World3* model assumed only a small fraction (five percent) of capital is allocated to the resource sector, simulating access to easily obtained or high quality resources, as well as improvements in discovery and extraction technology. However, as resources drop below the 50 percent level in the early part of the simulated 21st century and become harder to extract and process, the capital needed begins to increase. For instance, at 30 percent of the original resource base, the fraction of total capital that is allocated in the model to the resource sector reaches 50 percent, and continues to increase as the resource base is further depleted.

With significant capital subsequently going into resource extraction, there is insufficient available to fully replace degrading capital within the industrial sector itself. Consequently, despite heightened industrial activity attempting to satisfy multiple demands from all sectors and the population, actual industrial output per capita begins to fall precipitously, from about 2015, while pollution from the industrial activity continues to grow. The reduction of inputs to agriculture from industry, combined with pollution impacts on agricultural land, leads to a fall in agricultural yields and food produced per capita.

Similarly, services (e.g., health and education) are not maintained due to insufficient capital and inputs.

Diminishing per capita supply of services and food cause a rise in the death rate from about 2020 (and somewhat lower rise in the birth rate, due to reduced birth control options). The global population therefore falls, at about half a billion per decade, starting at about 2030. Following the collapse, the output of the *World3* model for the *standard run* shows that average living standards for the aggregate population (material wealth, food and services per capita) resemble those of the early 20th century.

The dynamics in the *World3* model leading to collapse resonate with aspects of other conceptual accounts of failed civilizations. Tainter's proposition of diminishing returns from growing complexity relates to the increasing inefficiency of extracting depleting resources in the *World3* response. It also aligns with a more general observation in the *LtG* that successive attempts to solve the sustainability challenges in the *World3* model, which lead to the *comprehensive technology* scenario, result in even more substantial collapse. The existence in *World3* of delays in recognizing and responding to environmental problems resonates with key elements in Diamond's characterization of societies that have failed. And Greer's mechanism of "catabolic collapse," i.e., increases in capital production outstripping maintenance, coupled with serious depletion of key resources, describes the core driver of breakdown in the *LtG standard run*.

The authors of the *LtG* caution that the dynamics in the *World3* model continue to operate throughout any breakdown. This could be realistic, or different dynamics might come to prominence that either exaggerate or ameliorate the collapse, such as wars or alternatively global leadership. Other researchers have contemplated how society might respond to serious resource constraints. Various degrees of hostility are foreshadowed, as well as lifestyles in developed countries that revert to greater self-reliance.

Instead, we now consider whether the key dynamics underlying the breakdown described above resemble actual developments. Since the collapse in the *standard run* scenario is predominantly associated with resource constraint and the diversion of capital to the resource sector, it is pertinent to examine peak oil, or other resource peaks. Peak oil refers to the peak in production of oil, as opposed to demand which is generally assumed to increase. Publications on peak oil have flourished in recent years as the possibility of a global peak has become more widely accepted (e.g., by the otherwise conservative International Energy Agency). These publications tend to focus on the question of when the peak will occur and what the oil supply volume will be. Sorrell et al. review many of these and find that independent researchers generally expect peaking to occur within about a decade, or to have occurred recently estimates of peaking made by oil industry representatives tend to be decades away. Unfortunately, these oil production profiles themselves say little analytically about the implications of reduced oil supply rates on the economy, though qualitatively a constrained supply of ubiquitous transport fuel is likely to be deleterious to global and national economies.

What is more relevant than the oil supply rates per se to our analysis of the *LtG* and collapse is the "opportunity cost" associated with extracting diminishing supplies of conventional oil or difficult extraction of non-conventional oil (e.g., tar sands, deep water, coal-to-liquids, etc.). In the *LtG*, the fraction of capital allocated to obtaining resources (FCAOR) represents this opportunity cost. In the peak oil literature, the relevant measure of opportunity cost is the energy return on investment (EROI), which is related to the net energy available after energy is used extracting the resource. The EROI is defined as the ratio of gross energy produced, TE_{Prod}, to energy invested to obtain the energy produced, E_{Res}. . . . .

The EROI can be related to the FCAOR used in the *LtG*. Since the capital (machinery, e.g., pumps, vehicles) operated in the resource sector, C_{Res}, is basically representative of the overall machinery stock, C_{Ttl}, the energy intensities will be similar and therefore the ratio of capital can be approximated by the ratio of energy used in the resource sector, E_{Res}, to total energy consumed, TE_{Cons}. . . .

Since the total energy consumed in any year will be approximately equal to the total energy produced (because stocks of energy stored are relatively small and don't change significantly from year to year), $TE_{Cons} \approx TE_{Prod}$ . . .

The collated data and model of EROI in Dale et al. can therefore be converted to FCAOR at corresponding values for the fraction of the oil resource remaining. This can be then be compared against the data used in the *LtG*. If the peak of conventional oil has occurred, or is about to occur, then approximately half the resource has been consumed, i.e., non-renewable resource fraction remaining, NRFR ≈ 0.5. Contemporary estimates of EROI are in the range 10 to 20 (or 1/EROI of 0.1–0.05). This agrees with the values and trends of the key parameter, FCAOR, used in the *LtG*.

Therefore, in addition to the data comparison made for modelled outputs, this data on oil resource extraction corroborates a key driver of dynamics in the *LtG standard run* scenario. In other words, the key mechanism driving the collapse in the *standard run* is observed in real world data. Further, there are other aspects of constraints in oil supply outlined below that also lend support to the mechanism of collapse.

Oil price rises have been linked to recent increases in food prices. There are direct and indirect links between oil and food associated with fuel for machinery and transport, both on-farm and in processing and distribution, as well as feedstock for inputs such as pesticides. Also, although nitrogen fertilizer is largely manufactured from natural gas, the price of these commodities is also linked to that of oil. More recently, production of biofuel as an alternative transport fuel, such as corn-based ethanol, has displaced food production and has been a factor in food price increases. These developments resemble the dynamics in the *LtG standard run* where agricultural production is negatively affected by reduced inputs. There may also be evidence of global pollution beginning to impact food production (which is a secondary factor in the *standard run* scenario) in the recent occurrence of major droughts, storms and fires (e.g., Russia, Australia) that are potentially early impacts of global climate change driven by anthropogenic greenhouse gas emissions.

The role of oil (and food) prices extends further, into more general economic shocks. For instance, other aggregate modelling of the role of energy in the economy finds that energy constraints cause a long-term economic downturn, as well as reducing greenhouse gas emissions, which are similar outcomes to those in the *LtG* collapse. Empirically, there is clear evidence of a connection between many oil price increases and economic recessions (just as there exists a strong correlation between energy consumption and growth in economic indicators). Hamilton's econometric analysis indicates that the latest (US) recession, associated with the GFC was different from previous oil-related shocks in that it appears caused by the combination of strong world demand confronting stagnating world production. His analysis downplays the role of financial speculation.

Nevertheless, the overriding proximate cause of the GFC is evidently financial: excessive levels of debt (relative to gross domestic product, GDP), or more accurately, the actual capacity of the real economy to pay back the debt). Such financial dynamics were not incorporated in the *LtG* modelling. Das highlights correlated defaults in high-risk debts, such as sub-prime housing mortgages, as a key trigger of the GFC. The financial models used did not properly account for a high number of defaults occurring simultaneously, being based on statistical analysis from earlier periods which suggest less correlation in defaults. Correlation may be caused by specific aspects of the financial instruments created recently, including for example, adjustments upward in interest rates of sub-prime mortgages after an initial "teaser" period of negligible interest rates. Even so, some spread in defaults would be expected in this case. Another potential factor could be the price increases in oil and related commodities, which would be experienced by all households simultaneously (but with a disproportionate impact on large numbers of households with low discretionary income).

Regardless of what role oil constraints and price increases played in the current GFC, a final consideration is whether there is scope of a successful transition to alternative transport fuel(s) and renewable energy more generally. Due to the GFC, there may be a lack of credit for funding any coordinated (or spontaneous) transition. And economic recovery may be interrupted, repeatedly, by increased oil prices associated with any recovery. Additionally, even if a transition is initiated it may take about two decades to properly implement the change over to a new vehicle fleet and distribution infrastructure. To transition requires introducing a new transport fuel to compensate for possible oil production depletion rates of four percent (or higher) while also satisfying any additional demand associated with economic growth. It is unclear that these various conditions required for a transition are possible.

Conclusion

Our previous comparison of global data with the *LtG* modelled scenarios has been updated here to cover the 40-year period 1970 to 2010, i.e., from when the scenario simulations begin. The data has been compared with the outputs of the *World3* model for three key *LtG* scenarios: *standard run, comprehensive technology,* and *stabilized world.* The data review continues to confirm that the *standard run* scenario represents real-world outcomes considerably well. This scenario results in collapse of the global economy and population in the near future. It begins in about 2015 with industrial output per capita falling precipitously, followed by food and services. Consequently, death rates increase from about 2020 and population falls from about 2030—as death rates overtake birth rates.

The collapse in the *standard run* is primarily caused by resource depletion and the model response of diverting capital away from other sectors in order to secure less accessible resources. Evidence for this mechanism operating in the real world is provided by comparison with data on the energy required to secure oil. Indeed, the EROI has decreased substantially in recent decades, and is quantitatively consistent with the relevant parameter in the *World3* model. The confirmation of the key model mechanism underlying the dynamics of the *standard run* strengthens the veracity of the *standard run* scenario. The issue of peak oil has also affected food supply and evidently played a role in the current global financial crisis. While the CFC does not directly reflect collapse in the *LtG standard run*, it may well be indirectly related.

The corroboration here of the *LtG standard run* implies that the scientific and public attention given to climate change, whilst important, is out of proportion with, and even deleteriously distracting from the issue of resource constraints, particularly oil. Indeed, if global collapse occurs as in this *LtG* scenario then pollution impacts will naturally be resolved, though not in any ideal "sense."

Another implication is the imminence of possible collapse. This contrasts with the general commentary on the *LtG* that describes collapse occurring sometime mid-century; and the *LtG* authors stressed not interpreting the time scale too precisely. However, the alignment of data trends with the model's dynamics indicates that the early stages of collapse could occur within a decade, or might even be underway. This suggests, from a rational risk-based perspective, that planning for a collapsing global system could be even more important than trying to avoid collapse.

Graham M. Turner is a Senior Researcher in Ecosystem Sciences with CSIRO, the Commonwealth Scientific and Industrial Research Organisation, in Canberra, Australia.

Tsvi Bisk

→ **NO**

No Limits to Growth

Prophets of doom are everywhere. We are headed for global catastrophe: climate change, population growth, shortages of water, deforestation, food shortages, and the end of oil are all converging to destroy the planet and modern human civilization as we have known it! We must stop economic growth in order to "save the planet" and survive as a species!

There are even calls to "de-develop." James Lovelock (of Gaia fame) has said, "The whole idea of sustainable development is wrongheaded. We should be thinking about sustainable retreat." Environmental activist George Monbiot says that the campaign against climate change "is a campaign not for abundance but for austerity; not for more freedom but for less . . . It is a campaign not just against other people, but against ourselves." We must forgo our growth fetish and addiction to unfettered consumerism and embrace the halcyon simplicity of years gone by, in his view. Monbiot says this policy must be achieved by "political restraint"—i.e., enforced by the police power of the state; in effect criminalizing innovation and creativity.

American consumer capitalism is claimed to be a luxury that the Earth's ecosystem cannot sustain. We are already at one-and-a-half times the carrying capacity of the planet, and the rate of environmental degradation is increasing. We must change the habits of the developed world and disabuse billions in the undeveloped world of their ambition to emulate the West.

Such calls for simplicity are often made by celebrities with seven-figure incomes, or by tenured academics and NGO groupies with six-figure incomes who supplement their earnings with books and speaking tours. However, the "wretched of the Earth" in the developing world already living in "halcyon simplicity," and the working and middle classes of the developed world struggling not to slide back into such "simplicity" are less enchanted by this eco-romanticism. The middle class will not give up its lifestyle and the non-middle class will not cease striving to become middle class.

Having lived and traveled widely in pre-consumerist societies, I can attest to how culturally barren and harsh the life of the vast majority of people in these societies is. The collapse of Communism, as well as the "Arab Spring," reflects the collective scream for escape from such a barren environment.

History matters! It teaches us that prophets of doom from Malthus to Ehrlich to the Club of Rome have been proven wrong time after time by that "infinite resource": the human mind. This essay will challenge the assumptions of the "catastrophe industry." It will show that growth in value (of GDP) is not synonymous with growth in the *volume* of raw materials—that we can grow value while using less of our natural resources; and that the wealth of nations can increase as the human burden on the carrying capacity of the planet decreases.

It has been estimated that close to 30% of world GDP is moved by air transport, but that this constitutes only 1% of the volume of goods and services transported around the globe. By this reckoning, air transport per unit of GDP is the least environmentally damaging form of transport, even when energy use is figured in. A 2009 U.S. Government Accountability Office report states the following:

> According to IPCC (Intergovernmental Panel on Climate Change [of the UN]), aviation currently accounts for about 2 percent of human-generated global carbon dioxide emissions, the most significant greenhouse gas—and about 3 percent of the potential warming effect of global emissions that can affect the earth's climate, including carbon dioxide.

If 95% of greenhouse gases are produced by nature and 5% by human activity, then air travel produces 0.0015% of total greenhouse gas. Compare this to greenhouse gases per unit of GDP produced globally by conventional field agriculture, and one discovers that air freight is greener per unit of value than agriculture. Globally, agriculture is responsible for 20% of manmade greenhouse gas emissions (or 0.01% of total greenhouse gases).

When one factors in the growth of the proportion of GDP of services, the picture becomes even less foreboding. Services have become the major drivers of global economic growth. They constitute over 63% of global GDP and well over 70% of GDP in the developed world, and the proportion is increasing. While services such as translation, consulting, planning, accounting, massage therapy, legal advice, etc., also consume some natural resources, the quantity is infinitesimal in relation to the economic value produced.

If I manufacture a car for $10,000 I have consumed a huge amount of natural resources and energy. But if I translate a 100,000-word book for $10,000 I have consumed merely the electricity necessary to run my computer, and the electricity used to send the translation as an email attachment, and little else that is even measurable.

Consumers and the consumer society are not the problem. The problem lies in the production methods *presently* used by manufacturing, mining, and agriculture. And the solution lies in the revolutions that are also *presently* taking place in material science, water engineering, energy harvesting, and food production.

We might speculate that the world economy can continue to grow *indefinitely* at 4%–5% a year while our negative footprint on the planet simultaneously declines. By the end of this century the planet will be able to carry a population of 12 billion people with an American standard of living and one-tenth the present negative environmental impact.

My Environmental Due Diligence

The energy/environment conundrum is the central issue of human civilization in the twenty-first century, but how are we to evaluate the *facts* while steering clear of fashionable catastrophism? After all, if it is already too late, and the earth is doomed, why even bother? Anxieties about global warming are so severe that there are reports of children requiring psychological treatment, and a sub-specialty of psychology that treats "Global Warming Anxiety Disorder" has arisen. . . .

Let us first define our terms precisely. *Ecology* is a science; *environmentalism* is an "ism"—i.e., an ideology, a set of beliefs about how human beings ought to live in their "environs." *Ecology* literally means *knowledge* of our house (our "house" in this instance being nature itself) as economy means *management* of our house (our house in this instance being commercial human society). *Environmentalism* is a value-laden philosophy and a social movement concerned with how human beings treat their environment (both natural and social). This ideology can either be based on the science of ecology (knowledge) or it can stray into theology (a dogmatic system of beliefs).

Environmentalists can be defined as:

1. *Human Centered (anthropocentric)*—advocating a clean environment because it is good for human beings. This is my viewpoint, which I call humanistic environmentalism.
2. *Nature Centered (biocentric)*—advocating the value of nature per se, and denigrating *anthropocentric* motivations. I call this *pagan environmentalism*—a latter-day version of nature worship.

There is a third minor category that I call *Marxist environmentalism;* ideologues who are resentful that human history has refuted Marxist historicism but who still wish to influence economic policy by maintaining that consumer capitalism is the root cause of environmental degradation. This, of course, ignores the environmental desolation left by the former Soviet Union—devastation several times greater than the capitalist West without the saving grace of having delivered a high standard of living.

I accept two axioms about global warming based on the *science* of ecology:

1. Global warming has been a *natural* phenomenon for the past 20,000 years. Consider that 20,000 years ago New York City was covered by a mile-thick sheet of ice, and ocean levels were 60 meters (200 feet) lower than today.
2. Human activity since the Industrial Revolution has increased the pace of global warming beyond what natural cycles can account for and might trigger a cascading effect that could be catastrophic for human civilization.

Since it is a natural phenomenon, I believe that "stopping global warming" and "saving the planet" are unscientific catchphrases. The globe will warm or cool as it pleases (we are in for another "ice age" in about 10,000 years no matter what we do) and the planet will be here for billions of years after that and will eventually be destroyed by our expanding sun. More accurate but less sexy slogans would be to "minimize the human contribution to the present natural cycle of global warming" in order to slow its rate and "save ourselves." It is not the planet that needs saving, or even life on this planet—it is us, the human race, that needs saving.

The planetary ecosystem has survived mass extinctions in the past (95% of species wiped out; 65% of species wiped out etc.). Evolution will certainly rise above our puny efforts to lay waste to the ecosystem. The real question is will the human race survive its own criminal negligence regarding the environment?

The view that reducing humanity's burden on the environment would wreck our economy is silly. If history is any judge, the opposite will likely be the case. It will create a more robust economic foundation for human civilization and expand economic opportunity for more individuals. The positive economic potential of environmentalism is a subject that has long been overlooked. The space program could serve as an analogy. At the time, no economist would have recommended it on purely economic grounds. Its great expense was considered justified for security and political reasons alone. Yet our entire modern economy is derived from it. . . .

There is a rich history of how other proactive national projects instituted for non-economic reasons have revolutionized the economy. These include the U.S. interstate highway system and the Internet. Might we not expect that a massive national or international project to liberate the planet from dependence on fossil fuels by 2050 could have much the same results? We would create economic sectors and products (and thus economic opportunity and growth) that we cannot even imagine.

Acknowledging the Limits to Growth Arguments

Before I make my case of "no limits to growth," I want to make the case of the "limits to growth" advocates. I do this out of fairness to, and respect for, their arguments but also to avoid possible future objections to my position.

I will never convince those whose entire intellectual careers are invested in gloom and doom about the future of the human race. Their historical analogy is the Renaissance university men who pressured the church to persecute Galileo because his proof of the Copernican universe threatened their academic careers, which consisted of teaching Aristotelian/Ptolemaic cosmology. I hope, however, to convince those who have no personal interest in the gloom and doom prophecies but have bought into them because they appear internally logical and because, heretofore, no coherent alternative has been presented to them. The following are my "limits to growth" arguments.

Resources, Food, and Water

While two billion people today are without clean water, 10 billion people will inhabit the planet by 2060. How will these people be fed? We will have to add a farmland area the size of Brazil. Eighty percent of the Earth's arable land is *already* in use, and 15% of that is degraded. There are not enough mineral resources on the planet even to satisfy the basic needs of 10 billion people, much less provide them all with a middle-class lifestyle.

The average age of U.S. farmers in 2007 was 57.1, up from 55.3 in 2002. More than one out of every four farmers is now more than 65 years old and near retirement. Less than 6% of all American farmers are younger than 35. Moreover, the majority of farmers are reluctant to see their children become farmers, and their children even more reluctant to follow in their footsteps.

Infatuation with "the good earth" has always been the fantasy of urban romantics or rural gentry that did not have to do the real work. Real farmers, subject to the unpredictability of nature and market and who actually perform the back-breaking labor, have usually been less enamored. Dissatisfaction with life on the farm has become a global phenomenon. People are voting with their feet. In 2008, more than half the world's population, 3.3 billion people, was living in urban areas. By 2030, this will reach 5 billion.

Thinking conventionally and extrapolating from our present methods of doing things, the claims of the anti-growth crowd are accurate. But there are already new practices in the pipeline regarding water, food, and material science that can obviate these arguments and enable us to expand the prosperity of the entire human race to American levels while reducing strain on the environment.

Sense and Nonsense about Energy

Energy is central to the arguments of "limits to growth" proponents, and their critique of the present energy paradigm is essentially correct. But first, let us be clear about terminology. Energy independence and energy self-sufficiency are not the same thing. Self-sufficiency, or *autarchy,* is the attribute of being completely self-contained and not relying on the external world for any resources. Modern examples of *attempts* at autarchy include communist Albania, the Khmer Rouge in Cambodia, North Korea, Burma, and Maoist China. Not only did none of these attempts succeed, but the policy itself guaranteed grinding poverty for their people.

Energy independence, on the other hand, simply means not being dependent on any one supplier or group of suppliers or on any one resource for your energy supplies. For example, the United States is energy independent and almost self-sufficient in regard to its electricity supply. It draws on a variety of sources: coal, natural gas, nuclear, hydro-electric (some of which it imports from Canada) and increasingly on alternative energy sources, as well. Regarding transportation, however, the planet is heavily dependent on a single substance: petroleum, and thus on 14 major oil exporting nations, most of which are unstable/unpredictable non-democratic countries that have created a de facto cartel that precludes true competition.

Petroleum also dominates consumer products. It is estimated that 300,000 products are made out of materials whose feedstock is oil. Worldwide, 40% of the petroleum consumed is used to manufacture products (25% in the U.S.). Oil-derived products include plastics, chemicals, herbicides, pesticides, fertilizers, medicines, etc. Petroleum is the most versatile, multipurpose natural resource that Mother Nature has provided us with, and we burn it to move mass from A to B. Future historians will probably look back at us and conclude that human society must have had a collective psychotic breakdown to use this resource in this way. We might as well burn Louis XIV's furniture to move mass. . . .

The Middle-Class Revolution

According to the Brookings Institute and other sources, the global middle class has been growing by 80 million people per year, and this rate is increasing. By 2015 the number of Asian middle-class consumers will equal the number in Europe and North America combined. By 2021, on present trends, there could be more than 2 billion Asians in middle-class households. By 2025 China alone could have more than 600 million middle-class consumers compared with 150 million today. According to the McKinsey Global Institute, India's middle class will be 20 times its present size by 2030—somewhere around a half a billion people. By 2020, the global middle class will have grown to 52% of the global population, up from 30% in 2008, and heading toward 80% by 2060.

The energy needs of such a population are tremendous. In 2005, China added as much electricity generation as Britain produces in a year. In 2006, it added as much as France's total supply. In India, more than 400 million people still don't have electric power. The demand for electricity in India will grow fivefold in the next 25 years.

The middle class desire for cars will negate the effects of increased petroleum production from non-conventional sources. The U.S. has 750 cars for every 1,000 people while

China only has 12 cars per 1,000 people, but it is *already* the world's largest car market. There will be 200 million cars in China by 2020—its market is growing by 20% a year. India and the rest of Asia have begun to follow suit (some projections have India passing China in car ownership by mid-century).

The middle class also consumes "stuff"—including those 300,000 products based on petroleum. Even if the developed world increases energy productivity and cuts energy consumption, global energy consumption will still rise by 1.5% a year.

The Case for No Limits to Growth

Notwithstanding all of the above, I want to reassert that by imagineering an alternative future—based on solid science and technology—we can create a situation in which there are "no limits to growth." It begins with a new paradigm for food production now under development: the urban vertical farm.

This is a concept popularized by Prof. Dickson Despommier of Columbia University. A 30-story urban vertical farm located on five square acres could yield food for fifty thousand people. We are talking about high-tech installations that would multiply productivity by a factor of 480: four growing seasons, times twice the density of crops, times two growing levels on each floor, times 30 floors = 480. This means that five acres of land can produce the equivalent of 2,600 acres of conventionally planted and tended crops. Just 160 such buildings occupying only 800 acres could feed the entire city of New York. Given this calculus, an area the size of Denmark could feed the entire human race.

Vertical farms would be self-sustaining. Located contiguous to or inside urban centers, they could also contribute to urban renewal. They would be urban lungs, improving the air quality of cities. They would produce a varied food supply year-round. They would use 90% less water. Since agriculture consumes two-thirds of the water worldwide, mass adoption of this technology would solve humanity's water problem.

Food would no longer need to be transported to market; it would be produced at the market and would not require use of petroleum intensive agricultural equipment. This, along with lessened use of pesticides, herbicides and fertilizers, would not only be better for the environment but would eliminate agriculture's dependence on petroleum and significantly reduce petroleum demand. Despite increased efficiencies, direct (energy) and indirect (fertilizers, etc.) energy use represented over 13% of farm expenses in 2005–2008 and have been increasing as the price of oil rises.

Many of the world's damaged ecosystems would be repaired by the consequent abandonment of farmland. A "rewilding" of our planet would take place. Forests, jungles, and savannas would reconquer nature, increasing habitat and becoming giant CO_2 "sinks," sucking up the excess CO_2 that the industrial revolution has pumped into the atmosphere. Countries already investigating the adoption of such technology include Abu Dhabi, Saudi Arabia, South Korea, and China—countries that are water starved or highly populated.

Material Science, Resources, and Energy

The embryonic revolution in material science now taking place is the key to "no limits to growth." I refer to "smart" and superlight materials. Smart materials "are materials that have one or more properties that can be significantly changed in a controlled fashion by external stimuli." They can produce energy by exploiting differences in temperature (thermoelectric materials) or by being stressed (piezoelectric materials). Other smart materials save energy in the manufacturing process by changing shape or repairing themselves as a consequence of various external stimuli. These materials have all passed the "proof of concept" phase (i.e., are scientifically sound) and many are in the prototype phase. Some are already commercialized and penetrating the market.

For example, the Israeli company Innowattech has underlain a one-kilometer stretch of local highway with piezoelectric material to "harvest" the wasted stress energy of vehicles passing over and convert it to electricity. They reckon that Israel has stretches of road that can efficiently produce 250 megawatts. If this is verified, consider the tremendous electricity potential of the New Jersey Turnpike or the thruways of Los Angeles and elsewhere. Consider the potential of railway and subway tracks. We are talking about tens of thousands of potential megawatts produced without any fossil fuels.

Additional energy is derivable from thermoelectric materials, which can transform wasted heat into electricity. As Christopher Steiner notes, capturing waste heat from manufacturing alone in the United States would provide an additional 65,000 megawatts: "enough for 50 million homes." Smart glass is already commercialized and can save significant energy in heating, air-conditioning and lighting—up to 50% saving in energy has been achieved in retrofitted legacy buildings (such as the former Sears Tower in Chicago).

New buildings, designed to take maximum advantage of this and other technologies could save even more. Buildings consume 39% of America's energy and 68% of its electricity. They emit 38% of the carbon dioxide, 49% of the sulfur dioxide, and 25% of the nitrogen oxides found in the air. Even greater savings in electricity could be realized by replacing incandescent and fluorescent light bulbs with LEDS which use l/10th the electricity of incandescent and half the electricity of fluorescents.

These three steps: transforming waste heat into electricity, retrofitting buildings with smart glass, and LED lighting, could cut America's electricity consumption and its CO_2 emissions by 50% within 10 years. They would also generate hundreds of thousands of jobs in construction and home improvements. Coal driven electricity generation would become a thing of the past. The coal released could be liquefied or gasified (by new *environmentally friendly* technologies) into the energy equivalent of 3.5 million barrels of oil a day. This is equivalent to the amount of oil the United States imports from the Persian Gulf and Venezuela together.

Conservation of energy and parasitic energy harvesting, as well as urban agriculture would cut the planet's energy consumption and air and water pollution significantly. Waste-to-energy technologies could begin to replace fossil fuels. Garbage, sewage, organic trash, and agricultural and food processing waste are essentially hydrocarbon resources that can be transformed into ethanol, methanol, and biobutanol or biodiesel. These can be used for transportation, electricity generation or as feedstock for plastics and other materials. Waste-to-energy is essentially a recycling of CO_2 from the environment instead of introducing new CO_2 into the environment. . . .

Superlight Materials

But it is superlight materials that have the greatest potential to transform civilization and, in conjunction with the above, to usher in the "no limits to growth" era. I refer, in particular, to carbon nanotubes—alternatively referred to as Buckyballs or Buckypaper (in honor of Buckminster Fuller). Carbon nanotubes are between 1/10,000th and 1/50,000th the width of a human hair, more flexible than rubber and 100–500 times stronger than steel per unit of weight.

Imagine the energy savings if planes, cars, trucks, trains, elevators—everything that needs energy to move—were made of this material and weighed 1/100th what they weigh now. Imagine the types of alternative energy that would become practical. Imagine the positive impact on the environment: replacing many industrial processes and mining, and thus lessening air and groundwater pollution.

Present costs and production methods make this impractical but that infinite resource—the human mind—has confronted and solved many problems like this before. Let us take the example of aluminum. A hundred fifty years ago, aluminum was more expensive than gold or platinum. When Napoleon III held a banquet, he provided his most honored guests with aluminum plates. Less-distinguished guests had to make do with gold!

When the Washington Monument was completed in 1884, it was fitted with an aluminum cap—the most expensive metal in the world at the time—as a sign of respect to George Washington. It weighed 2.85 kilograms, or 2,850 grams. Aluminum at the time cost $1 a gram (or $1,000 a kilogram). A typical day laborer working on the monument was paid $1 a day for 10–12 hours a day. In other words, today's common soft-drink can, which weighs 14 grams, could have bought 14 ten-hour days of labor in 1884.

Today's U.S. minimum wage is $7.50 an hour. Using labor as the measure of value, a soft drink can would cost $1,125 today (or $80,000 a kilogram), were it not for a new method of processing aluminum ore. The Hall-Héroult process turned aluminum into one of the cheapest commodities on earth *only two years after* the Washington Monument was capped with aluminum. Today aluminum costs $3 a kilogram, or $3000 a metric ton. The soft drink can that would have cost $1,125 today without the process now costs $0.04.

Today the average cost of industrial grade carbon nanotubes is about $50–$60 a kilogram. This is already far cheaper in *real* cost than aluminum was in 1884. Yet revolutionary methods of production are now being developed that will drive costs down even more radically. At Cambridge University they are working on a new electrochemical production method that could produce 600 kilograms of carbon nanotubes per day at a projected cost of around $10 a kilogram, or $10,000 a metric ton.

This will do for carbon nanotubes what the Hall-Héroult process did for aluminum. Nanotubes will become the universal raw material of choice, displacing steel, aluminum, copper, and other metals and materials. Steel presently costs about $750 per metric ton. Nanotubes of equivalent strength to a metric ton of steel would cost $100 if this Cambridge process (or others being pursued in research labs around the world) proves successful.

Ben Wang, director of Florida State's High-Performance Materials Institute claims that: "If you take just one gram of nanotubes, and you unfold every tube into a graphite sheet, you can cover about two-thirds of a football field." Since other research has indicated that carbon nanotubes would be more suitable than silicon for producing photovoltaic energy, consider the implications. Several grams of this material could be the energy-producing skin for new generations of superlight dirigibles—making these airships energy autonomous. They could replace airplanes as the primary means to transport air freight.

Modern American history has shown that anything human beings decide they want done can be done in 20 years if it does not violate the laws of nature. The atom bomb was developed in four years; putting a man on the moon took eight years. It is a reasonable conjecture that by 2020 or earlier, an industrial process for the inexpensive production of carbon nanotubes will be developed, and that this would be the key to solving our energy, raw materials, and environmental problems all at once.

Mitigating Anthropic Greenhouse Gases

Another vital component of a "no limits to growth" world is to formulate a rational environmental policy that saves money; one that would gain wide grassroots support because it would benefit taxpayers and businesses, and would not endanger livelihoods. For example, what do sewage treatment, garbage disposal, and fuel costs amount to as a percentage of municipal budgets? What are the costs of waste disposal and fuel costs in stockyards, on poultry farms, throughout the food processing industry, and in restaurants? How much aggregate energy could be saved from all of the above?

Some experts claim that we could obtain enough liquid fuel from recycling these hydrocarbon resources to satisfy all the transportation needs of the United States. Turning the above waste into energy by various means would be a huge cost saver and value generator, *in addition to* being a blessing to the environment.

The U.S. army has developed a portable field apparatus that turns a combat unit's human waste and garbage into bio-diesel to fuel their vehicles and generators. It is called TGER—the Tactical Garbage to Energy Refinery. It eliminates the need to transport fuel to the field, thus saving lives, time, and equipment expenses. The cost per barrel must still be very high.

However, the history of military technology being civilianized and revolutionizing accepted norms is long. We might expect that within 5–10 years, economically competitive units using similar technologies will appear in restaurants, on farms, and perhaps even in individual households, turning organic waste into usable and economical fuel. We might conjecture that within several decades, centralized sewage disposal and garbage collection will be things of the past and that even the Edison Grid (unchanged for over one hundred years) will be deconstructed. . . .

The Promise of the Electric Car

There are 250 million cars in the United States. Let's assume that they were all fully electric vehicles (EVs) equipped with 25-kWh batteries. Each kWh takes a car two to three miles, and if the average driver charges the car twice a week, this would come to about 100 charge cycles per year. All told, Americans would use 600 billion kWh per year, which is only 15% of the current total U.S. production of 4 trillion kWh per year. If supplied during low demand times, this would not even require additional power plants. If cars were made primarily out of Buckypaper, one kWh might take a car 40–50 miles. If the surface of the car was utilized as a photovoltaic, the car of the future might conceivably become energy autonomous (or at least semi-autonomous).

A kWh produced by a coal-fired power plant creates two pounds of CO_2, so our car-related CO_2 footprint would be 1.2 trillion pounds if all electricity were produced by coal. However, burning one gallon of gas produces 20 pounds of CO_2. In 2008, the U.S. used 3.3 billion barrels of gasoline, thereby creating about 3 trillion pounds of CO_2. Therefore, a switch to electric vehicles would cut CO_2 emissions by 60% (from 3 trillion to 1.2 trillion pounds), even if we burned coal exclusively to generate that power.

Actually, replacing a gas car with an electric car will cause zero increase in electric draw because refineries use seven kWh of power to refine crude oil into a gallon of gasoline. A Tesla Roadster can go 25 miles on that 7 KWh of power. So the electric car can go 25 miles using the same electricity needed to refine the gallon of gas that a combustion engine car would use to go the same distance.

Additional Strategies

The goal of mitigating global warming/climate change without changing our lifestyles is not naïve. Using proven Israeli expertise, planting forests on just 12% of the world's semi-arid areas would offset the annual CO_2 output of one thousand 500-megawatt coal plants (a gigaton a year). A global program of foresting 60% of the world's semi-arid areas would offset five thousand 500-megawatt coal plants (five gigatons a year). Since mitigation goals for global warming include reducing our CO_2 emissions by eight gigatons by 2050, this project alone would have a tremendous ameliorating effect.

Given that large swaths of semi-arid land areas contain or border on some of the poorest populations on the planet, we could put millions of the world's poorest citizens to work in forestation, thus accomplishing two positives (fighting poverty and environmental degradation) with one project. Moving agriculture from its current field-based paradigm to vertical urban agriculture would eliminate two gigatons of CO_2. The subsequent re-wilding of vast areas of the earth's surface could help sequester up to 50 gigatons of CO_2 a year, completely reversing the trend.

The revolution underway in material science will help us to become "self-sufficient" in energy. It will also enable us to create superlight vehicles and structures that will produce their own energy. Over time, carbon nanotubes will replace steel, copper and aluminum in a myriad of functions.

Converting waste to energy will eliminate most of the methane gas humanity releases into the atmosphere. Meanwhile, *artificial photosynthesis* will suck CO_2 out of the air at 1,000 times the rate of natural photosynthesis. This trapped CO_2 could then be combined with hydrogen to create much of the petroleum we will continue to need. As hemp and other fast-growing plants replace wood for making paper, the logging industry will largely cease to exist. Self-contained fish farms will provide a major share of our protein needs with far less environmental damage to the oceans.

Population Explosion or Population Implosion

One constant refrain of anti-growth advocates is that we are heading towards 12 billion people by the end of the century, that this is unsustainable, and thus that we must proactively reduce the human population to 3 billion–4 billion in order to "save the planet" and human civilization from catastrophe. But recent data indicates that a *demographic winter* will engulf humanity by the middle of this century. More than 60 countries (containing over *half* the world's population) already do not have replacement birth rates of 2.1 children per woman. This includes the entire EU, China, Russia, and half a dozen Muslim countries, including Turkey, Algeria, and Iran.

If present trends continue, India, Mexico, and Indonesia will join this group before 2030. The human population will peak at 9–10 billion by 2060, after which, for the first time since the Black Death, it will begin to shrink. By the end of the century, the human population might be as low as 6 billion–7 billion.

The real danger is not a population explosion; but the consequences of the impending population implosion. This demographic process is not being driven by famine or disease as has been the case in all previous history. Instead, it is being driven by the greatest Cultural Revolution in the history of the human race: the liberation and empowerment of women. The fact is that even with *present* technology, we would still be able to sustain a global population of 12 billion by the end of the century if needed. The evidence for this is cited above.

Conclusion

The current crisis of human civilization is not a consequence of the consumer society or of hyper-industrialization or of the spread of vulgar Americanization. It is a consequence of the incompetence and dearth of imagination of our politicians and intellectuals coupled with the inertia of outmoded ideas. The "limits to growth" advocates are right: If we continue to run our societies and economies as we have been doing, we will head toward environmental and civilizational catastrophe.

But there is no reason to continue to run our societies and our economies as we have. Nor is there any reason to deprive ourselves and our progeny of the material comforts that the industrial revolution has provided us in the name of "saving the planet." Instead we can envision and then construct a planetary civilization totally committed to enabling the self-actualization of every single human being on the planet; a civilization that will have banished want and hunger for all time; a civilization dedicated to realizing the god-like potential of the human spirit. And in that day we will finally be free, Amen!

Tsvi Bisk is an American Israeli futurist and Director of the Center for Strategic Futurist Thinking. His latest book is *The Optimistic Jew: A Positive Vision for the Jewish People in the 21st Century* (Maxanna, 2007).

EXPLORING THE ISSUE

Are There Limits to Growth?

Critical Thinking and Reflection

1. Should we be more concerned about future human well-being than about future income levels?
2. Can a market-based economy exist without continuous growth in production and consumption?
3. How does population relate to the sustainability issue?
4. In what ways is the precautionary principle an essential component of sustainable development?

Is There Common Ground?

It is hard to avoid the conclusion that sustainability—meaning meeting present needs while leaving enough resources unused to permit our children and grandchildren to meet their needs—is important. Yet many people do manage to avoid the corollary, the idea that because the Earth is finite, there are limits to how much population and economic activity can grow. Some, like Tsvi Bisk, insist that technology can solve all potential problems of sustainability.

1. In what ways does population control help us achieve both sustainability and environmental protection?
2. Tsvi Bisk does not mention one interesting technology currently gaining ground. This is 3D printing, meaning machines that can make small objects of many kinds. In the future they may be found in community centers or even in homes. Look this up, and consider whether it would help defeat potential limits to growth.
3. Graham M. Turner and Tsvi Bisk agree that our use of resources can pose problems for future well-being. Explain why Tsvi Bisk believes that these problems are not insuperable.

Create Central

www.mhhe.com/createcentral

Additional Resources

James H. Brown, et al., "Energetic Limits to Economic Growth," *BioScience* (January 2011)

Lester R. Brown, "Could Food Shortages Bring Down Civilization?" *Scientific American* (May 2009)

Gro Harlem Brundtland, *Our Common Future* (Oxford University Press, 1987)

Takis Fotopoulos, "Is Degrowth Compatible with a Market Economy?" *International Journal of Inclusive Democracy* (January 2007)

Dennis Meadows, et al., *The Limits to Growth: A Report for the Club of Rome's Project on the Predicament of Mankind* (Universe Books, 1972)

Donella Meadows, et al., *Beyond the Limits: Confronting Global Collapse, Envisioning a Sustainable Future* (Chelsea Green, 1993)

Donella Meadows, et al., *Limits to Growth: The Thirty Year Update* (Chelsea Green, 2004)

Jorgen Randers, *2052: A Global Forecast for the Next Forty Years (A Report to the Club of Rome Commemorating the 40th Anniversary of* The Limits to Growth) (Chelsea Green Publishing, 2012)

Internet References . . .

Global Sustainability Institute

www.anglia.ac.uk/ruskin/en/home/microsites/global
_sustainability_institute.html

Limits to Growth: Exploring the Physical Limits to Economic Growth

www.limits-to-growth.org/

The Club of Rome

www.clubofrome.org/

The Earth Day Network

www.earthday.org/

The International Institute for Sustainable Development

www.iisd.org/

The Natural Resources Defense Council

www.nrdc.org/

Selected, Edited, and with Issue Framing Material by:
Thomas Easton, *Thomas College*

ISSUE

Should We Be Pricing Ecosystem Services?

YES: David C. Holzman, from "Accounting for Nature's Benefits: The Dollar Value of Ecosystem Services," *Environmental Health Perspectives* (April 2012)

NO: Marino Gatto and Giulio A. De Leo, from "Pricing Biodiversity and Ecosystem Services: The Never-Ending Story," *BioScience* (April 2000)

Learning Outcomes

After reading this issue, you will able to:

- Explain what ecosystem services are and why they are important to people.
- Discuss whether it makes sense to assess the economic value of ecosystem services.
- Describe how payments for ecosystem services can help in the management of natural ecosystems.

ISSUE SUMMARY

YES: David C. Holzman argues that ecosystems provide valuable services without which humanity would perish. Efforts to assess the value of these services are difficult but necessary to help policy makers and resource managers make rational decisions that factor important environmental and human health outcomes into the bottom line.

NO: Professors of Applied Ecology Marino Gatto and Giulio A. De Leo contend that the pricing approach to valuing nature's services is misleading because it falsely implies that only economic values matter.

Human activities frequently involve trading a swamp or forest or mountainside for a parking lot or housing development or farm (among other things). People generally agree that these developments are worthwhile projects, for they have obvious benefits to people. But are there costs as well? Construction costs, labor costs, and material costs can easily be calculated, but what about the value of the swamp itself, of the forest, and of the species living there?

How much is a species worth? One approach to answering this question is to ask people how much they would be willing to pay to keep a species alive. If the question is asked when there are a million species in existence, few people will likely be willing to pay much. But if the species is the last one remaining, they might be willing to pay a great deal. Most people would agree that both answers fail to get at the true value of a species, for nature is not expressible solely in terms of cash values. Yet some way must be found to weigh the effects of human activities on nature against the benefits gained from those activities. If it is not, we will continue to degrade the world's ecosystems and threaten our own continued well-being. Indeed, human dependence is growing more acute;

see Zhongwei Guo, Lin Zhang, and Yiming Li, "Increased Dependence of Humans on Ecosystem Services and Biodiversity," *PLoS One* (www.plosone.org) (October 2010). See also Peter Kareiva, Heather Tallis, Taylor H. Ricketts, Gretchen C. Daily, and Stephen Polasky, eds., *Natural Capital: Theory and Practice of Mapping Ecosystem Services* (Oxford University Press, 2011).

Traditional economics views nature as a "free good." That is, forests generate oxygen and wood, clouds bring rain, and the Sun provides warmth, all without charge to the humans who benefit. At the same time, nature has provided ways for people to dispose of wastes—such as dumping raw sewage into rivers or emitting smoke into the air—without paying for the privilege. This "free" waste disposal has turned out to have hidden costs in the form of the health effects of pollution (among other things), but it has been up to individuals and governments to bear the costs associated with those effects. The costs are real, but in general they have not been borne by the businesses and other organizations that caused them. They have thus come to be known as "external" costs or externalities.

Environmental economists have recognized the problem of external costs, and government regulators have devised a number of ways to make those who are responsible accept the bill, such as instituting requirements for pollution control and fining those who exceed permitted emissions. Yet some would say that this approach does not help enough.

The ecosystem services approach recognizes that undisturbed ecosystems do many things that benefit human beings. A forest, for instance, slows the movement of rain and snowmelt into streams and rivers; if the forest is removed, floods may follow (a connection that a few years ago forced China to deemphasize forest exploitation). Swamps filter the water that seeps through them. Food chains cycle nutrients necessary for the production of wood and fish and other harvests. Bees pollinate crops and thus make the production of many food crops possible. These services are valuable—even essential—to us, and anything that interferes with them must be seen as imposing costs just as significant as those associated with the illnesses caused by pollution. Yet standard economics encourages the conversion of "free" natural resources into marketable commodities without considering those costs; see Christopher L. Lant, J. B. Ruhl, and Steven E. Kraft, "The Tragedy of Ecosystem Services," *Bioscience* (November 2008).

How can the costs of interfering with ecosystem services be assessed? In 1997, Robert Costanza and his colleagues published an influential paper entitled "The Value of the World's Ecosystem Services and Natural Capital," *Nature* (May 15, 1997). In it, the authors listed a variety of ecosystem services and attempted to estimate what it would cost to replace those services if they were somehow lost. The total bill for the entire biosphere came to $33 trillion (the middle of a $16–54 trillion range), compared to a global gross national product of $25 trillion. Costanza, et al. stated that this was surely an underestimate. Janet N. Abramovitz, "Putting a Value on Nature's 'Free' Services," *WorldWatch* (January/February 1998), argues that nature's services are responsible for the vast bulk of the value in the world's economy and that attaching economic value to those services may encourage their protection. Rebecca L. Goldman, "Ecosystem Services: How People Benefit from Nature," *Environment* (September/October 2010), argues that ecosystem services are crucial to human well-being, both now and for the sustainable future. They are also affected by human behavior, both at the individual and at the national level. Assessing their economic value is difficult but essential to public decision making. Shuang Liu, Robert Costanza, Stephen Farber, and Austin Troy, "Valuing Ecosystem Services," *Annals of the New York Academy of Sciences* (January 2010), argue that valuing ecosystem services is needed to provoke society to acknowledge the value of natural capital.

In "Can We Put a Price on Nature's Services?" *Report from the Institute for Philosophy and Public Policy* (Summer 1997), Mark Sagoff objects that trying to attach a price to ecosystem services is futile because it legitimizes the accepted cost–benefit approach and thereby undermines efforts to protect the environment from exploitation. The

March 1998 issue of *Environment* contains Environmental Economics Professor David Pearce's detailed critique of the 1997 Costanza et al.'s study. Pearce objects chiefly to the methodology, not the overall goal of attaching economic value to ecosystem services. Costanza et al. reply to Pearce's objections in the same issue. Pearce and Edward B. Barbier have published *Blueprint for a Sustainable Economy* (Earthscan, 2000), in which they discuss how governments worldwide are now applying economics to environmental policy. Nikolas Kosoy and Esteve Corbera, "Payments for Ecosystem Services as Commodity Fetishism," *Ecological Economics* (April 2010), argue that payments for ecosystem services are a symptom and consequence of commodity fetishism (very simply, excessive market thinking), which disregards ecosystem complexity, fails to account for value in the broader sense, and preserves inequality in society. A better approach to valuating ecosystem services would rely on environmental ethics, not market economics. Kerry R. Turner, Sian Morse-Jones, and Brendan Fisher, "Ecosystem Valuation," *Annals of the New York Academy of Sciences* (January 2010), favor a more pluralistic approach that recognizes the complexity of ecosystem value.

Despite the controversy over the worth of assigning economic value to various aspects of nature, researchers continue the effort. Gretchen C. Daily et al., in "The Value of Nature and the Nature of Value," *Science* (July 21, 2000), discuss valuation as an essential step in all decision making and argue that efforts "to capture the value of ecosystem assets . . . can lead to profoundly favorable effects." Daily and Katherine Ellison continue the theme in *The New Economy of Nature: The Quest to Make Conservation Profitable* (Island Press, 2002). In "What Price Biodiversity?" *Ecos* (January 2000), Steve Davidson describes an ambitious program funded by Australia's Commonwealth Scientific and Industrial Research Organization (CSIRO) and the Myer Foundation that is aimed at developing principles and methods for objectively valuing "ecosystem services—the conditions and processes by which natural ecosystems sustain and fulfill human life—and which we too often take for granted. These include such services as flood and erosion control, purification of air and water, pest control, nutrient cycling, climate regulation, pollination, and waste disposal." Jim Morrison, "How Much Is Clean Water Worth?" *National Wildlife* (February/March 2005), argues that such ecosystem services have sufficient economic value to make it profitable to spend millions of dollars to protect natural systems. Shahid Naeem, J. Emmett Duffy, and Erika Zavaleta, "The Functions of Biological Diversity in an Age of Extinction," *Science* (June 15, 2012), see in the worldwide loss of biodiversity threats to "food, water, energy, and biosecurity." Anthony D. Barnosky et al., "Approaching a State Shift in Earth's Biosphere," *Nature* (June 7, 2012), warn that the human impact on the environment is bringing the global ecosystem to a "tipping point" after which the loss of biodiversity and ecosystem services will threaten human well-being.

Stephen Farber et al., "Linking Ecology and Economics for Ecosystem Management," *Bioscience* (February 2006), find "the valuation of ecosystem services . . . necessary for the accurate assessment of the trade-offs involved in different management options." Ecosystem valuation is currently being used to justify restoration efforts, "linking the science to human welfare," as shown (for example) in Chungfu Tong et al., "Ecosystem Service Values and Restoration in the Urban Sanyang Wetland of Wenzhou, China," *Ecological Engineering* (March 2007). See also J. R. Rouquette et al., "Valuing Nature-Conservation Interests on Agricultural Floodplains," *Journal of Applied Ecology* (April 2009), and Z. M. Chen et al., "Net Ecosystem Services Value of Wetland: Environmental Economic Account," *Communications in Nonlinear Science & Numerical Simulation* (June 2009). Joshua Farley and Robert Costanza, "Payments for Ecosystem Services: From Local to Global," *Ecological Economics* (September 2010), say that "Payment for Ecosystem Services (PES) is becoming increasingly popular as a way to manage ecosystems using economic incentives." Just as important is the role that assessing the economic value of the environment can play in weaving the environment into "the fabric of economic thinking"; see "Price Fixing: Why It Is Important to Put a Price on Nature," *The Economist* (January 18, 2010) (www.economist.com/node/15321193?story_id=15321193).

In 2011, researchers released *The UK National Ecosystem Assessment* (UNEP-WCMC, Cambridge, UK, 2011). Among the key findings were that about 30 percent of ecosystems were declining in their ability to deliver services. They suggest that ecosystem management would benefit if the government "took into account the economic value of a broader array of natural benefits"; see Erik Stokstad,

"Appraising U.K. Ecosystems, Report Envisions Greener Horizon," *Science* (June 3, 2011). A. P. Kinzig et al., "Paying for Ecosystem Services—Promise and Peril," *Science* (November 4, 2011), note that because "existing markets seldom reflect the full social cost of production, we have incorrect measures" (or none) of the value of most ecosystem services, but devising effective pricing mechanisms is difficult. John W. Day, Jr. et al., "Ecology in Times of Scarcity," *Bioscience* (April 2009), report that "In an energy-scarce future, services from natural ecosystems will assume relatively greater importance in supporting the human economy." As a result, the practice of ecology will become both more expensive and more valuable, and ecological engineering (ecoengineering) and restoration will become more common and necessary. Perhaps the connection between ecosystem services and economic value will become less debatable. Already, the CEO of a major corporation, Andrew Liveris of Dow Chemical, can say, "protecting nature can be a profitable corporate priority and a smart global business strategy. . . . The economy and the environment are interdependent"; see Bryan Walso, "Paying for Nature," *Time* (February 21, 2011).

In the YES selection, David C. Holzman argues that ecosystems provide valuable services without which humanity would perish. Efforts to assess the value of these services are difficult but necessary to help policy makers and resource managers make rational decisions that factor important environmental and human health outcomes into the bottom line. In the NO selection, professors of applied ecology Marino Gatto and Giulio A. De Leo contend that the pricing approach to valuing nature's services is misleading because it falsely implies that only economic values matter.

YES ↵

David C. Holzman

Accounting for Nature's Benefits: The Dollar Value of Ecosystem Services

Healthy ecosystems provide us with fertile soil, clean water, timber, and food. They reduce the spread of diseases. They protect against flooding. Worldwide, they regulate atmospheric concentrations of oxygen and carbon dioxide. They moderate climate. Without these and other "ecosystem services," we'd all perish.

One hallmark of the history of civilization is an ever-increasing exploitation of ecosystem services coupled with substitution of technology for these services, particularly where ecosystems have been exploited beyond their ability to provide. Agriculture is a hybrid of exploitation and substitution that enabled people to live in greater, denser populations that drove further exploitation and substitution. Modern plumbing made close quarters far less noxious but led to exploitation of ecosystems' ability to break down sewage, and to substitution with expensive sewage treatment technologies. Exploitation of fossil fuels led to a slew of modern conveniences, including fishing fleets that are so effective at catching their prey that they threaten fisheries globally. All this exploitation strained ecosystems, but in the past, when the population was a fraction of what it is now, these strains were local rather than global phenomena.

In 2005 the Millennium Ecosystem Assessment (MA), a sweeping survey conducted under the auspices of the United Nations, found that approximately 60% of 24 ecosystem services examined were being degraded or used unsustainably. "Every year we lose three to five trillion dollars' worth of natural capital, roughly equivalent to the amount of money we lost in the financial crisis of 2008–2009," says Dolf de Groot, leader of the Research Program on Integrated Ecosystem Assessment and Management at Wageningen University, the Netherlands.

The value of ecosystem services typically goes unaccounted for in business and policy decisions and in market prices. For commercial purposes, if ecosystem services are recognized at all, they are perceived as free goods, like clean air and water. So it's not surprising that much of the degradation of ecosystems is rooted in what the President's Council of Advisors on Science and Technology (PCAST), an independent group of U.S. scientists and engineers, describes as "widespread underappreciation of the importance of environmental capital for human well-being and . . . the absence of the value of its services from the economic balance sheets of producers and consumers." PCAST and other groups are working to build recognition of ecosystem services and, importantly, to valuate them— that is, calculate values for these services to help policy makers and resource managers make rational decisions that factor important environmental and human health outcomes into the bottom line.

An Idea Whose Time Has Come?

In July 2011 PCAST called upon the federal government to assess quadrennially the condition of the nation's ecosystems and the social and economic value of services they provide. The goal was to improve methods for evaluating those services and to establish an ecoinformatics initiative to pull together existing knowledge and gather new information in a format that interested parties can easily use.

But the concept of valuating ecosystem services is not new. John P. Holdren, now science advisor to President Barack Obama, introduced it to students in his class "Quantitative Aspects of Global Environmental Problems" at the University of California, Berkeley, in the 1970s. He emphasized that technological substitutions for ecosystem services are often costly, sometimes to the point of impracticality, and that sometimes an incomplete understanding of how they function makes such substitutions impossible. Geoengineering to mitigate global climate disruption in the face of increasing emissions, for example, is widely viewed as extremely risky, because the climate is so complex.

In 1997 Robert Costanza, Distinguished University Professor of sustainability at Portland State University, Oregon, and colleagues first estimated that ecosystem services worldwide are worth an average $33 trillion annually ($44 trillion in today's dollars), nearly twice the global GNP of around $18 trillion ($24 trillion in today's dollars). Although the $33 trillion has been difficult to substantiate, this study was widely praised for drawing attention to the value of ecosystem services, says Rick Linthurst, national program director of the Ecosystem Services Research Program at the U.S. Environmental Protection Agency (EPA).

Payments to preserve ecosystem services date to at least the early 1980s, when the United States implemented

Holzman, David C. Adapted from the original article that appeared in *Environmental Health Perspectives*, April 2012. National Institute of Environmental Health Sciences.

wetland and stream credit banking, but the idea really took off in the mid 1990s. For instance, in 1996 Costa Rica began paying landowners $42 per hectare per year to preserve forest. At the time, that country had the highest deforestation rate in the world; now it has among the lowest, says Gretchen C. Daily, Bing Professor of Environmental Science at Stanford University.

China responded to a devastating drought in 1997, followed by massive floods in 1998, by inaugurating various payments for ecosystem services and a policy for conserving areas that are important sources of ecosystem services. These are known as Ecosystem Function Conservation Areas. Among the benefits: soil erosion fell sufficiently to cut sediment in the Yellow River by 38% over the period 2000 through 2007, and carbon sequestration rose by an estimated 1.3 billion tons between 1998 and 2010, says Jianguo Liu, the Rachel Carson Chair in Sustainability and University Distinguished Professor of fisheries and wildlife at Michigan State University. But he adds that some benefits probably came at the expense of natural capital elsewhere in the world, as declines in forest cutting coincided with a rise in imported timber.

In 2010 the World Bank launched a program to help countries incorporate the value of ecosystem services into their accounting systems with an eye toward managing ecosystems to maximize economic benefit. Colombia, beset for several years by unusually persistent and damaging rains, is one of five pilot countries working with the bank.

Elsewhere, Norway is paying Indonesia $1 billion to preserve rainforest for carbon storage and sequestration to limit the impacts of climate change. And in Vietnam, an investment of $1.1 million in mangroves, which protect coastal regions from flooding, saved $7.3 million annually that would have gone to maintaining dikes.

A number of agencies of the U.S. federal government are conducting research on ecosystem services. The EPA is assembling a national atlas that can overlay visual information, like that used in Google Earth, with ecological and economic analyses to reveal variability in ecosystem service provision. The agency also has pilot programs in four regions of the United States enabling interested parties to project different resource-use scenarios into the future to help guide decision making now.

And in 2007 environment ministers from the G8+5 countries agreed to begin analyzing the global economic benefits that derive from ecosystems and biodiversity, and to compare the costs of failure to protect these resources with the costs of conserving them. In the ensuing years, the resulting initiative, The Economics of Ecosystems and Biodiversity (TEEB), has produced a series of reports for decision makers at the international, national and local levels aimed at enabling practical responses.

Another leader in guiding decision makers on payments for ecosystem services is the Natural Capital (NatCap) Project, cofounded by Stanford's Daily in 2006. NatCap has created software called InVEST (Integrated Valuation of Environmental Services and Tradeoffs) to model tradeoffs among environmental, economic, and social benefits, so that decision makers can explore the implications of alternative land-use scenarios. For any given piece of real estate, InVEST can take existing data on various ecosystem services—each of which may fall under a different field of study—and provide one consistent platform for assessing all of them together, to determine the optimal use(s) of that land, says Heather Tallis, lead scientist of NatCap. For example, "trade-off curves" can reveal how much timber can be harvested before causing major profit loss to hydropower, flood damage, or loss of biodiversity. The tools are available free through NatCap.

NatCap's consulting group is currently working on numerous projects within the United States and with 15 other countries in Africa, Latin America, the Pacific, North America, and Asia. Foremost among those countries is China, which is spending a total of around $100 billion— more than any other country—to preserve forestlands through logging bans, to buy farms that are perched unsustainably on steep slopes for conversion to forests, and to restore wetlands. The Chinese government will shift farmers either to more sustainable locations or to other occupations, says Daily. NatCap is using InVEST to assess how many resource-intensive livelihoods—in farming, forestry, herding, and other fields—could be supported sustainably in a certain area under given practices, and to evaluate how shifting inhabitants to an alternative mix of livelihoods would impact natural capital and ecosystem services. This helps inform the investments needed to enable desired shifts, as well as to ascertain who will benefit and who will be hurt by the shifts, and to determine appropriate compensation, says Daily.

Assigning a Dollar Value

Ecosystem services are valued, ideally, by how much human welfare they can provide. The most convenient measure of welfare is dollars, although at this early stage of development of the science, that is not always a practical measure.

Values for provisioning services [see sidebar, "What Are Ecosystem Services?"] are relatively easy to determine. The simplest and least controversial methods to assess value draw on existing prices in the marketplace, says Emily McKenzie, manager of the NatCap Project at the World Wildlife Fund U.S. office. For example, coastal and marine ecosystems support the production of fish. The value of this service can be assessed based on revenues, a function of the price and quantity of harvested fish.

Thus, the value of the provisioning service is equal to how much all of its current and future production is worth today—what economists call its "present value." The further into the future the production lies, the lower the present value of the service. That's because money invested today in a safe investment, such as a Treasury bill, almost certainly will grow. If Treasury bills are earning 3%, $100 invested today will become $103 a year from now, $106.09 two years from now, and so on. That means that $106.09 two years from now is no more valuable than $100 today.

Many ecosystem services, such as scenery, recreational value, and most regulatory services, including those moderating infectious disease, lack a market price. One way to address this problem involves asking people what they would pay for a particular service, says Stephen Polasky Fesler-Lampert Professor of ecological/environmental economics at the University of Minnesota; this is called "stated preference." Another method, "revealed preference," involves determining values from related actual purchases, such as the money people spend to travel to bucolic tourist destinations, or the extra cost of a house with a water view over a similar nearby house without the view.

Another valuation technique is estimating "replacement cost." This is the cost of the least expensive technical fix as a replacement for an ecosystem service. For example, New York City recently paid landowners in its watershed more than $1 billion to change their farm management practices to prevent animal waste and fertilizer from washing into the waterways. In doing so, the city avoided spending $6–8 billion on a new water filtration plant and $300–500 million annually to run it—the replacement cost of the natural filtration provided by waterways. "Protecting the watershed [along with the ecosystem service it provides] can be said to be worth at least six to eight billion dollars because that is the cost of replacing the service," says Polasky who notes that the value of clean water is far higher still.

The value of an ecosystem service depends on local and/or regional socioeconomic conditions as well as supply and demand. Thus, the value of clean water is much higher in New York City's watershed, where it serves 19 million, people, than it would be in, say, Alaska, says Polasky.

When it comes to valuating ecosystem services, the economics is the easy part—easy being a relative term. The major difficulties have more to do with the fact that ecology is a relatively young science, and there is much that we don't yet understand about it, says Polasky, echoing colleagues. "Nature is probably the most complex system we know of," Daily explains.

Huge Error Bars and Heroic Assumptions

Part of the problem, generally speaking, is that the multiple uncertainties about how ecosystems do what they do add up to "huge error bars," says Polasky. Dollar values are often based on "heroic assumptions" that don't stand on much data, says Lisa Wainger, a research associate professor at the University of Maryland Center for Environmental Science.

For instance, scientific understanding of feedback among the many ecosystem services remains wanting—

"If you have more carbon in soil, are plants better able to take up nitrogen?" asks Polasky, as one example. A good deal of ecological uncertainty stems from a lack of information about basic natural history. The PCAST report notes that "groups of organisms likely to be most important in ecological terms, such as species that determine soil fertility, promote nutrient cycling, or consume wastes . . . are among the least familiar and least visible—e.g., fungi, nematodes, mites, insects, and bacteria. Populations of ecologically dominant marine organisms, most of which are either invertebrates or microbes, are just as poorly understood."

Climate change magnifies all these ecological uncertainties. It's "the mother of all externalities," writes Richard S.J. Tol, a professor of economics at the University of Sussex, "larger, more complex, and more uncertain than any other environmental problem." Over the rest of the century, global climate shifts are likely to be the biggest driver of ecosystem change and may greatly reduce Earth's carrying capacity, according to PCAST.

It also remains difficult to link changes in the delivery of ecosystem services to changes in human welfare. "There are many mysteries about which species confer what dynamics to ecosystems or what benefits to people," says Daily. "We really don't know how much biodiversity is needed to sustain and fulfill human life."

But more precise knowledge of the economic value of those ecosystem services that can easily be valuated "would not, in itself provide insight into what fraction of the benefit would be lost in consequence of a given type or degree of ecosystem disruption," according to the PCAST report. There are thresholds in ecosystem function beyond which carrying capacity plummets. History and prehistory are littered with thresholds breached, from the degeneration of the Fertile Crescent into today's desertified Middle East (probably due to mismanagement of irrigation, says Daily) to the deforestation, extinction of all wild land birds, and human population collapse on Easter Island. One of the biggest fears about the impact of climate change is that global thresholds will be breached, but the ability to predict such with anything approaching precision is currently beyond ecological science.

Progress

Despite the challenges, considerable progress has been made over the last decade toward improved techniques for linking changes in ecosystem services to changes in human welfare. Part of that improvement is due to modeling methods, including In VEST, as well as to greater numbers of ecological studies, and part is due to improvements in the data, says Polasky. The field has been boosted by the revolution in GIS (geographic information system) technology and so-called spatially explicit data: "We now have very good images that enable us to know the heights of plants and elevations of terrain, and really good sensors that show us what's on the ground," Polasky says.

"You can combine that with monitoring. If we increase the deforestation upriver, we can monitor the sediment downriver. That's been a huge help."

Health & Ecosystems: Analysis of Linkages (HEAL), a consortium of more than 25 conservation and public health institutions, has embarked on the first rigorous, systematic attempt to measure the human health impacts of changes in a variety of natural systems. HEAL's projects are designed to evaluate what are thought to be key connections between the environment and health. Examples include the relationships between subsistence hunters' sustainable access to wildlife and their children's nutritional needs (particularly as related to iron and key micronutrient deficiencies); between upland deforestation on islands such as Fiji, erosion and waterborne diarrheal diseases in children, and downstream coral reef health and productivity; between deforestation patterns and malaria in the Amazon and other major forest systems; between landscape fires in Sumatra and smoke-related cardiopulmonary illness in the broader region downwind; and between fishers' access to Marine Protected Areas, food security, income to purchase health services, and the psychological dimensions of having a "sense of place" related to coastal resource security.

Perhaps most importantly, says Steve Osofsky, HEAL coordinator and director of health policy at the Wildlife Conservation Society, the project seeks to quantify all these types of relationships related to communicable diseases, noncommunicable diseases, nutrition, and the social and psychological dimensions of health. In Osofsky's words, "If it cannot be measured, it cannot be managed."

The ultimate goal of valuating ecosystem services "is to improve human well-being overall," says Daily. She cautions that there will always be people who lose out in any policy decision. However, she says, "The aim is to design these investments in natural capital so as to advance human development and alleviate poverty at the same time. This is the Holy Grail."

What Are Ecosystem Services?

Ecosystem services are highly interdependent and often overlap. These services are typically categorized under 4 types: provisioning, regulatory, supporting, and cultural.

Like factories, provisioning services maintain the supply of natural products: food, timber, fuel, fibers for textiles, water, soil, medicinal plants, and more.

Regulatory services keep different elements of the natural world running smoothly. They filter pollutants to maintain air and water quality, moderate the climate, sequester and store carbon, recycle waste and dead organic matter, and serve as natural controls for agricultural pests and disease vectors.

Supporting services can be thought of as the services that maintain the provisioning and regulatory services. These services include soil formation, photosynthesis, and provision of habitat. Healthy habitats preserve both species diversity and genetic diversity, which are critical underpinnings of all provisioning and regulatory services.

Finally, cultural services are defined as the intangible benefits obtained from contact with nature—the aesthetic, spiritual, and psychological benefits that accrue from culturally important or recreational activities such as hiking, bird watching, fishing, hunting, rafting, gardening, and even scenic road trips. Increasingly, these services are being tied to tangible health benefits, especially those related to stress reduction.

Quantifying the Impact of Ecosystem Disturbances on . . .

The idea that ecosystem services influence human health has been around for quite a while, says Rick Ostfeld, a disease ecologist at the Cary Institute of Ecosystem Studies in Millbrook, New York. Only more recently have researchers begun investigating the hypothesis that services provided by healthy ecosystems include moderating infectious disease.

There is growing evidence—some experimental and some correlational—that a decline in biodiversity can boost disease transmission and that "preserving intact ecosystems and their endemic biodiversity should generally reduce the prevalence of infectious diseases," Ostfelt and colleagues wrote in a recent review on the topic in Nature. When ecosystems are simplified or fragmented, as human development is wont to do, the changes often favor proliferation of more efficient vectors and wildlife reservoirs of infectious disease—chiefly arthropods and rodents, respectively—partly by reducing the population of predators that keep these creatures in check.

The most efficient natural reservoirs of disease "tend to be the weedy, resilient species with a 'live fast and die young' life history," says Ostfeld. "Those are the species left standing when we disturb or degrade the ecosystem." He explains that predators that feed on the natural reservoirs tend to be more sensitive and disappear first when ecosystems are disturbed.

There is direct evidence supporting an inverse relationship between biodiversity and infectious disease. In South America, for instance, converting forest to cereal production increases rodent populations, contributing to epidemics of viral hemorrhagic fever, says Samuel Myers, a research scientist in the Department of Environmental Health of the Harvard School of Public Health. Dams, irrigation systems, and deforestation have been linked to increases in malaria and schistosomiasis, diarrheal diseases are associated with road building, and dengue has been tied to urbanization. Nonetheless, "[t]here is a big gap between the research showing associations between changes in natural systems and health outcomes, and actually being able to quantify the specific health benefits or costs of incremental changes in the system," Myers and colleague Jonathan Patz of the University of

Wisconsin-Madison wrote in the 2009 edition of Annual Review of Environment and Resources.

A Complex Undertaking

The science that underlies ecosystem services is cumbersome. Taylor Ricketts, director of the Gund Institute for Ecological Economics at the University of Vermont, measured the productivity of bees as pollinators of coffee plantations in Costa Rica. The bees live in forests near the plantations.

In a series of carefully controlled and lengthy experiments, he found that coffee plants within a kilometer of the forest were fully pollinated, whereas those beyond a kilometer received insufficient pollination, producing 20% fewer coffee beans. Using these results, Ricketts calculated that two forest patches were contributing $62,000 worth of pollination services annually to a single coffee farm. If those forests were destroyed, that farm would suffer a 7% drop in productivity, he estimated. Thus, the present value of the annual $62,000 would be the value of the service provided by those forest patches.

However, pollination is just one of many ecosystem services performed by that forest, with others including carbon sequestration, support of biodiversity, and water purification, to name just a few, says Ricketts. At this early stage of the science of valuating ecosystem services, often it is impractical to determine the value of more than one or a few services. Part of the difficulty of determining the value of all the services within an ecosystem is that the methods for obtaining the necessary information are often so different for each service.

In some cases, the studies required may be enormously time-consuming or otherwise difficult. For example, Ricketts' pollination studies involved comparing coffee production among five similar trees at each of three different distances from the forest, after hand pollinating flowers on some branches at each location to simulate maximum pollination activity, allowing bees to do the job on other flowers, and covering some flowers to simulate no pollination. Measuring a forest's capacity to purify water might involve determining the purity of a stream that runs through the forest after passing a pollution source by assaying pollutants at regular distances to determine how quickly they are declining. Assaying biodiversity involves taking a census of all plants, animals, and invertebrates on a plot of land. Ricketts says a thorough determination of ecosystem services from a single piece of land might involve 20 different studies for as many services.

DAVID C. HOLZMAN is a writer whose work has appeared in *Smithsonian, The Atlantic Monthly,* and *The Journal of the National Cancer Institute.*

Marino Gatto and Giulio A. De Leo    **NO**

Pricing Biodiversity and Ecosystem Services: The Never-Ending Story

In 1844, the French engineer Jules Juvénal Dupuit introduced cost–benefit analysis to evaluate investment projects. . . . The application of cost–benefit analysis to ecological issues fell out of favor three decades ago, and it was gradually replaced by multicriteria analysis in the decision-making process for projects that have an impact on the environment. Although multicriteria analysis is currently used for environmental impact assessments [EIA] in many nations, [recently] the concept of cost–benefit analysis has again become fashionable, along with the various pricing techniques associated with it, such as contingent valuation methods, hedonic prices, and costs of replacement of ecological services. . . . Economists have generated a wealth of virtuosic variations on the theme of assessing the societal value of biodiversity, but most of these techniques are invariably based on price—that is, on a single scale of values, that of goods currently traded on world markets.

Perhaps the most famous recent study on the issue of pricing biodiversity and ecological services is that by Costanza et al., who argued that if the importance of nature's free benefits could be adequately quantified in economic terms, then policy decisions would better reflect the value of ecosystem services and natural capital. Drawing on earlier studies aimed at estimating the value of a wide variety of ecosystem goods and services, Costanza et al. estimated the current economic value of the entire biosphere at $16–54 trillion per year, with an average value of approximately $33 trillion per year. By contrast, the gross national product of the United States totals approximately $18 trillion per year. The paper, as its authors intended, stimulated much discussion, media attention, and debate. A special issue of *Ecological Economics* (April 1998) was devoted to commentaries on the paper, which, with few exceptions, were laudatory. Some economists have questioned the actual numbers, but many scientists have praised the attempt to value biodiversity and ecosystem functions.

Although Costanza et al. acknowledged that their estimates were crude and imperfect, they also pointed the way to improved assessments. In particular, they noted the need to develop comprehensive ecological economic models that could adequately incorporate the complex interdependencies between ecosystems and economic

systems, as well as the complex individual dynamics of both types of systems. Despite the authors' caveats and the fact that many economists have been circumspect in applying their own tools to decisions regarding natural systems, the monetary approach is perceived by scientists, policymakers, and the general public as extremely appealing; a number of biologists are also of the opinion that attaching economic values to ecological services is of paramount importance for preserving the biosphere and for effective decision-making in all cases where the environment is concerned.

In this article, we espouse a contrary view, stressing that, for most of the values that humans attach to biodiversity and ecosystem services, the pricing approach is inadequate—if not misleading and obsolete—because it implies erroneously that complex decisions with important environmental impacts can be based on a single scale of values. We contend that the use of cost–benefit analysis as the exclusive tool for decision-making about environmental policy represents a setback relative to the existing legislation of the United States, Canada, the European Union, and Australia on environmental impact assessment, which explicitly incorporates multiple criteria (technical, economic, environmental, and social) in the process of evaluating different alternatives. We show that there are sound methodologies, mainly developed in business and administration schools by regional economists and by urban planners, that can assist decision-makers in evaluating projects and drafting policies while accounting for the nonmarket values of environmental services.

The Limitations of Cost–Benefit Analysis and Contingent Valuation Methods

Historically, the first important implementation of cost–benefit analysis at the political level came in 1936, with passage of the US Flood Control Act. This legislation stated that a public project can be given a green light if the benefits, to whomsoever they accrue, are in excess of estimated costs. This concept implies that all benefits and costs are to be considered, not just actual cash flows from and to government coffers. However, public agencies (e.g., the US Army Corps of Engineers) quickly ran into a problem: They were not able to give a monetary value to many environmental effects, even those that were predictable in quantitative terms. For instance, engineers could calculate the reduction of downstream water flow resulting from construction of a dam, and biologists could predict the river species most likely to become extinct as a consequence of this flow reduction. However, public agencies were not able to calculate the cost of each lost species. Therefore, many ingenious techniques for the monetary valuation of environmental goods and services have been devised since the 1940s. These techniques fall into four basic categories.

- **Conventional market approaches.** These approaches, such as the replacement cost technique, use market prices for the environmental service that is affected. For example, degradation of vegetation in developing countries leads to a decrease in available fuelwood. Consequently, animal dung has to be used as a fuel instead of a fertilizer, and farmers must therefore replace dung with chemical fertilizers. By computing the cost of these chemical fertilizers, a monetary value for the degradation of vegetation can then be calculated.
- **Household production functions.** These approaches, such as the travel cost method, use expenditures on commodities that are substitutes or complements for the environmental service that is affected. The travel cost method was first proposed in 1947 by the economist Harold Hotelling, who, in a letter to the director of the US National Park Service, suggested that the actual traveling costs incurred by visitors could be used to develop a measure of the recreation value of the sites visited.
- **Hedonic pricing.** This form of pricing occurs when a price is imputed for an environmental good by examining the effect that its presence has on a relevant market-priced good. For instance, the cost of air and noise pollution is reflected in the price of plots of land that are characterized by different levels of pollution, because people are willing to pay more to build their houses in places with good air quality and little noise. . . .
- **Experimental methods.** These methods include contingent valuation methods, which were devised by the resource economist Siegfried V. Ciriacy-Wantrup. Contingent valuation methods require that individuals express their preferences for some environmental resources by answering questions about hypothetical choices. In particular, respondents to a contingent valuation methods questionnaire will be asked how much they would be willing to pay to ensure a welfare gain from a change in the provision of a nonmarket environmental commodity, or how much they would be willing to accept in compensation to endure a welfare loss from a reduced provision of the commodity.

Among these pricing techniques, the contingent valuation methods approach is the only one that is capable of providing an estimate of existence values, in which biologists have a special interest. Existence value was first defined by Krutilla as the value that individuals may attach to the mere knowledge that rare and diverse species, unique natural environments, or other "goods" exist, even if these individuals do not contemplate ever making active use of or benefiting in a more direct way from them. The name "contingent valuation" comes from the fact that the procedure is contingent on a constructed or simulated market, in which people are asked to manifest, through questionnaires and interviews, their demand function for a certain environmental good (i.e., the price they would pay for one extra unit of the good versus the availability of the good). . . .

The limits of cost–benefit analysis were discussed in the 1960s, after more than two decades of experimentation. In particular, many authors pointed out that cost–benefit analysis encouraged policymakers to focus on things that can be measured and quantified, especially in cash terms, and to disregard problems that are too large to be assessed easily. Therefore, the associated price might not reflect the "true" value of social equity, environmental services, natural capital, or human health. In particular, economists themselves recognize that the increasingly popular contingent valuation methods are undermined by several conceptual problems, such as free-riding, overbidding, and preference reversal.

When it comes to monetary valuation of the goods and services provided by natural ecosystems and landscapes specifically, a number of additional problems undermine the effectiveness of pricing techniques and cost–benefit analysis. These problems include the very definition of "existence" value, the dependence of pricing techniques on the composition of the reference group, and the significance of the simulated market used in contingent valuation.

The definition of "existence" value A classic example of contingent valuation methods is to ask for the amount of money individuals are willing to pay to ensure the continued existence of a species such as the blue whale. However, the existence value of whales does not take into account potential indirect services and benefits provided by these mammals. It is just the value of the existence of whales for humans, that is, the satisfaction that the existence of blue whales provides to people who want them to continue to exist. Therefore, there is a real risk that species with very low or no aesthetic appeal or whose biological role has not been properly advertised will be given a low value, even if they play a fundamental ecological function. Without adequate information, most people do not understand the extent, importance, and gravity of most environmental problems. As a consequence, people may react emotionally and either underestimate or overestimate risks and effects.

Therefore, it is not surprising that five of the seven guidelines issued by the National Oceanic and Atmospheric

Administration [NOAA] about how to conduct contingent valuation discuss how to properly inform and question respondents to produce reliable estimates (e.g., in-person interviews are preferred to telephone surveys to elicit values). Of course, acquisition of reliable and complete information is always possible in theory, but in practice strict adherence to NOAA guidelines makes contingent valuation methods expensive and time consuming.

Difficulties with the reference group for pricing   Pricing techniques such as contingent valuation methods provide information about individual willingness to pay or willingness to accept, which must be summed up in the final balance of cost–benefit analysis. Therefore, the outcome of cost–benefit analysis depends strongly on the group of people that is taken as a reference for valuation—particularly on their income. Van der Straaten noted that the Exxon *Valdez* oil spill in 1989 provides a good example of this dependence. The population of the United States was used as a reference group to calculate the damage to the existence value of the affected species and ecosystems using contingent valuation methods. Exxon was ultimately ordered to pay $5 billion to compensate the people of Alaska for their losses. This huge figure was a consequence of the high income of the US population. If the same accident had occurred in Siberia, where salaries are lower, the outcome would certainly have been different.

This example shows that contingent valuation methods simply provide information about the preferences of a particular group of people but do not necessarily reflect the ecological importance of ecosystem goods and services. Moreover, the outcome of cost–benefit analysis depends on which individual willingness to pay or willingness to accept are included in the cost–benefit analysis. If the quality of the Mississippi River is at issue, should the analysis be restricted to US citizens living close to the river, or should the willingness to pay of Californians and New Yorkers be included too? According to Krutilla's definition of existence value, for many environmental goods and ecological services that may ultimately affect ecosystem integrity at the global level, the preferences of the entire human population should potentially be considered in the analysis. Because practical reasons obviously preclude doing so, contingent valuation methods will inevitably only provide information about the preferences of specific groups of people. For many of the ecological services that may be considered the heritage of humanity, contingent valuation methods analyses performed locally in a particular economic situation should be extrapolated only with great caution to other areas. The process of placing a monetary value on biodiversity and ecosystem functioning through nonuser willingness to pay is performed in the same way as for user willingness to pay, but the identification of people who do not use an environmental good directly and still have a legitimate interest in its preservation is problematic.

Significance of the simulated market   Contingent valuation methods are contingent on a market that is constructed or simulated, not real. It is difficult to believe in the efficiency of what Adam Smith called the "invisible hand" of the market for a process that is the artificial production of economic advisors and does not possess the dynamic feedback that characterizes real competitive markets. Is it even possible to simulate a market where units of biodiversity are bought and sold? As Friend stated, "these contingency evaluation methods (CVM) tend to create an illusion of choice based on psychology (willingness) and ideology (the need to pay) which is supposed, somewhat mysteriously, to reflect an equilibrium between the consumer demand for and producer supply of environmental goods and services."

Many additional criticisms of pricing ecological services are more familiar to biologists. For many ecological services, there is simply no possibility of technological substitution. Moreover, the precise contribution of many species is not known, and it may not be known until the species is close to extinction. . . . In addition, specific ecosystem services, as evaluated by Costanza et al., should not be separated from one another and valued individually because the importance of any piece of biodiversity cannot be determined without considering the value of biodiversity in the aggregate. And finally, the use of marginal value theory may be invalidated by the erratic and catastrophic behavior of many ecological systems, resulting in potentially detrimental effects on the health of humans, the productivity of renewable resources, and the vitality and stability of societies themselves.

Despite the efforts of many economists, we believe that some goods and services, especially those related to ecosystems, cannot reasonably be given a monetary value, although they are of great value to humans. Economists coined the term "intangibles" to define these goods. Cost–benefit analysis cannot easily deal with intangibles. As Nijkamp wrote, more than 20 years ago, "the only reasonable way to take account of intangibles in the traditional cost–benefit analysis seems to be the use of a balance with a debit and a credit side in which all intangible project effects (both positive and negative) are represented in their own (qualitative or quantitative) dimensions" as secondary information. In other words, the result of cost–benefit analysis is primarily a single number, the net monetary benefit that comprises all the effects that can be sensibly converted into monetary returns and costs.

Commensurability of Different Objectives and Multicriteria Analysis

Cost–benefit analysis includes intangibles in the decision-making process only as ancillary information, with the main focus being on those effects that can be converted to monetary value. This approach is not a balanced solution to the problem of making political decisions that are acceptable to a wide number of social groups with a range of legitimate interests. . . .

However, even if the attempt to put a price on everything is abandoned, it is not necessary to give up the attempt to reconcile economic issues with social and environmental ones. Social scientists long ago developed multicriteria techniques to reach a decision in the face of multiple different and structurally incommensurable goals. The most important concept in multicriteria analysis was actually conceived by an Italian economist, Vilfredo Pareto, at the end of the nineteenth century. It is best explained by a simple example. Suppose that a natural area hosting several rare species is a target for the development of a mining activity. Alternative mining projects can have different effects in terms of profits from mining (measured in dollars) and in terms of sustained biodiversity (measured in suitable units, for instance, through the Shannon index). Profit from mining can be corrected using welfare economics to include those environmental and social effects that can be priced (e.g., the benefit of providing jobs to otherwise unemployed people, the cost of treating lung disease of miners, and the cost of the loss of the tourists who used to visit the natural area). . . .

The methods of multicriteria analysis are intended to assist the decision-maker in choosing among . . . alternatives . . . (a task that is particularly difficult when there are several incommensurable objectives, not just two). Nevertheless, the initial step of determining [these] alternatives is of enormous importance, for three reasons. First, [doing so] makes perfect sense even if there is no way of pricing a certain environmental good because each objective can be expressed in its own proper units without reduction to a common scale. Second, the determination of all the feasible alternatives . . . requires the joint effort of a multidisciplinary team that includes, for example, economists, engineers, and biologists and that must predict the effects of alternative decisions on all of the different environmental and social components to which humans are sensitive and which, therefore, deserve consideration. Third, the determination of [feasible alternatives] allows the objective elimination of inadequate alternatives because [they are] independent of the subjective perception of welfare . . . [and] in essence describe the tradeoff between the various incommensurable objectives when every effort is made to achieve the best results in all respects; the attention of the authority that must make the final decision is thus directed toward genuine potential solutions because nonoptimal decisions have already been discarded.

It should be noted that a cost–benefit analysis does not elicit tradeoffs between incommensurable goods because it also gives a green light to projects . . . , provided that the benefits that can be converted into a monetary scale exceed the costs. . . . Cost–benefit analysis, however, is not useful for eliciting the tradeoffs between two incommensurable goods, neither of which is monetary. For instance, there might be a conflict between the goals of preserving wildlife within a populated area and minimizing the risk that wild animals are vectors of dangerous diseases. A multicriteria analysis can describe this tradeoff, whereas a cost–benefit analysis cannot.

Another philosophical point concerning the issue of commensurability is the question of implicit pricing. Economists often argue that to make a decision is to put an implicit price on such intangibles as human life or aesthetics and, therefore, to reduce their value to a common scale (as pointed out also by Costanza et al.). . . .

Environmental Impact Assessment and Multiattribute Decision-Making

Because of the flaws of cost–benefit analysis, many countries have taken a different approach to decision-making through the use of environmental impact assessment legislation (e.g., the United States in 1970, with the signing of the National Environmental Policy Act, NEPA; France in 1976, with the act 76/629; the European Union in 1985, with the directive 85/337). Environmental impact assessment procedures, if properly carried out, represent a wiser approach than setting an a priori value of biodiversity and ecosystem services because these procedures explicitly recognize that each situation, and every regulatory decision, responds to different ethical, economic, political, historical, and other conditions and that the final decision must be reached by giving appropriate consideration to several different objectives. As Canter noted, all projects, plans, and policies that are expected to have a significant environmental impact would ideally be subject to environmental impact assessment.

The breadth of goals embraced by environmental impact assessment is much wider than that of cost–benefit analysis. Environmental impact assessment provides a conceptual framework and formal procedures for comparing different alternatives to a proposed project (including the possibilities of not developing a site, employing different management rules, or using mitigation measures); for fostering interdisciplinary team formation to investigate all possible environmental, social, and economic consequences of a proposed activity; for enhancing administrative review procedures and coordination among the agencies involved in the process; for producing the necessary documentation to enhance transparency in the decision-making process and the possibility of reviewing all the objective and subjective steps that resulted in a given conclusion; for encouraging broad public participation and the input of different interest groups; and for including monitoring and feedback procedures. Classical multiattribute analysis can be used to rank different alternatives. . . . Ranking usually requires the use of value functions to transform environmental and other indicators (e.g., biological oxygen demand or animal density) to levels of satisfaction on a normalized scale, and the weighting of factors to combine value functions and to rank the alternatives. These weights explicitly reflect the relative importance of the different environmental, social, and economic compartments and indicators.

A wide range of software packages for decision support can assist experts in organizing the collected information; in documenting the various phases of EIA; in guiding the assignment of importance weights; in scaling, rating, and ranking alternatives; and in conducting sensitivity analysis for the overall decision-making process. This last step, of testing the robustness and consistency of multiattribute analysis results, is especially important because it shows how sensitive the final ranking is to small or large changes in the set of weights and value functions, which often reflect different and subjective perspectives. It is important to stress that, although the majority of environmental impact assessments have been conducted on specific projects, such as road construction or the location of chemical plants, there is no conceptual barrier to extending the procedure to evaluation of plans, programs, policies, and regulations. In fact, according to NEPA, the procedure is mandatory for any federal action with an important impact on the environment. The extension of environmental impact assessment to a level higher than a single project is termed "strategic environmental assessment" and has received considerable attention.

Conclusions

An impressive literature is available on environmental impact assessment and multiattribute analysis that documents the experience gained through 30 years of study and application. Nevertheless, these studies seem to be confined to the area of urban planning and are almost completely ignored by present-day economists as well as by many ecologists. Somewhere between the assignment of a zero value to biodiversity (the old-fashioned but still used practice, in which environmental impacts are viewed as externalities to be discarded from the balance sheet) and the assignment of an infinite value (as advocated by some radical environmentalists), lie more sensible methods to assign value to biodiversity than the price tag techniques suggested by the new wave of environmental economists. Rather than collapsing every measure of social and environmental value onto a monetary axis, environmental impact assessment and multiattribute analysis allow for explicit consideration of intangible nonmonetary values along with classical economic assessment, which, of course, remains important. It is, in fact, possible to assess ecosystem values and the ecological impact of human activity without using prices. Concepts such as Odum's eMergy [the available energy of one kind previously required to be used up directly and indirectly to make the product or service] and Rees' ecological footprint [the area of land and water required to support a defined economy or population at a specified standard of living], although perceived by some as naive, may aid both ecologists and economists in addressing this important need.

To summarize our viewpoint, economists should recognize that cost–benefit analysis is only part of the decision-making process and that it lies at the same level as other considerations. Ecologists should accept that monetary valuation of biodiversity and ecosystem services is possible (and even helpful) for part of its value, typically its use value. We contend that the realistic substitute for markets, when they fail, is a transparent decision-making process, not old-style cost–benefit analysis. The idea that, if one could get the price right, the best and most effective decisions at both the individual and public levels would automatically follow is, for many scientists, a sort of Panglossian obsession. In reality, there is no simple solution to complex problems. We fear that putting an a priori monetary value on biodiversity and ecosystem services will prevent humans from valuing the environment other than as a commodity to be exploited, thus reinvigoraing the old economic paradigm that assumes a perfect substitution between natural and human-made capital. As Rees wrote, "for all its theoretical attractiveness, ascribing money values to nature's services is only a partial solution to the present dilemma and, if relied on exclusively, may actually be counterproductive."

Marino Gatto is a Professor of Applied Ecology in the Dipartimento di Elettronica e Informazione at Politecnico di Milano in Milan, Italy. He is Associate Editor of *Theoretical Population Biology*.

Giulio A. De Leo is a Professor of Ecology at the Universit degli Studi di Parma in Parma, Italy.

EXPLORING THE ISSUE

Should We Be Pricing Ecosystem Services?

Critical Thinking and Reflection

1. In what ways are natural ecosystems valuable?
2. Is it right to destroy natural ecosystems in order to generate money?
3. What is the best way to convince people to protect nature?
4. What is the best way to protect nature?

Is There Common Ground?

No one disagrees that nature provides benefits to human life. The disagreement arises when discussing whether we have a right to use nature in any way we like, even to the point of destroying it. Indeed the idea that we have such a right has a long history, for nature was long considered worthless wilderness or wasteland, gaining value only when it had been "improved" (built on, farmed, mined, logged, etc.) by human effort. In recent decades, the idea that nature has intrinsic value has been gaining ground, but most people agree that "intrinsic value" is not easily used to convince people to protect nature. Value in terms of benefits to humans works better.

1. Research "carbon offsets." In what way do they provide an example of payment for ecosystem services?
2. Do regulations that require mitigation of undesirable environmental impacts (such as building artificial wetlands to replace natural wetlands destroyed by construction projects) amount to another form of payment for ecosystem services? Can they motivate protection of the original, natural environment?
3. Are there situations where the ecosystem services approach to protection seems unlikely to work?

Create Central

www.mhhe.com/createcentral

Additional Resources

Gretchen Daily and Katherine Ellison, *The New Economy of Nature: The Quest to Make Conservation Profitable* (Island Press, 2002)

Rebecca L. Goldman, "Ecosystem Services: How People Benefit from Nature," *Environment* (September/October 2010)

Zhongwei Guo, Lin Zhang, and Yiming Li, "Increased Dependence of Humans on Ecosystem Services and Biodiversity," *PLoS One* (www.plosone.org) (October 2010)

Peter Kareiva, Heather Tallis, Taylor H. Ricketts, Gretchen C. Daily, and Stephen Polasky, eds., *Natural Capital: Theory and Practice of Mapping Ecosystem Services* (Oxford University Press, 2011)

Nikolas Kosoy and Esteve Corbera, "Payments for Ecosystem Services as Commodity Fetishism," *Ecological Economics* (April 2010)

Christopher L. Lant, J. B. Ruhl, and Steven E. Kraft, "The Tragedy of Ecosystem Services," *Bioscience* (November 2008)

Shuang Liu, Robert Costanza, Stephen Farber, and Austin Troy, "Valuing Ecosystem Services," *Annals of the New York Academy of Sciences* (January 2010)

Internet References . . .

Mainstreaming Ecosystem Services Initiative

www.wri.org/project/mainstreaming-ecosystem-services

Millennium Ecosystem Assessment

www.unep.org/maweb/en/index.aspx

The Ecosystem Services Project

www.ecosystemservicesproject.org/

Unit 2

UNIT

Principles versus Politics

*I*n many environmental issues, it is easy to tell what basic principles apply and therefore determine what is the right thing to do. Ecology is clear on the value of species to ecosystem health and the harmful effects of removing or replacing species. Medicine makes no bones about the ill effects of pollution. But are the environmental problems so bad that we must act immediately? Should we go slow on environmental regulations for fear of damaging the economy? Are lawsuits really needed to make sure regulations are enforced and businesses obey them? When businesses say they are acting in a green way, can we believe them?

Selected, Edited, and with Issue Framing Material by:
Thomas Easton, *Thomas College*

ISSUE

Does Designating "Wild Lands" Harm Rural Economies?

YES: Mike McKee, from Testimony before the House Natural Resources Committee Hearing on "The Impact of the Administration's Wild Lands Order on Jobs and Economic Growth," *Illinois House State Government Administration Committee* (March 1, 2011)

NO: Robert Abbey, from Testimony before the House Natural Resources Committee Hearing on "The Impact of the Administration's Wild Lands Order on Jobs and Economic Growth," *Illinois House State Government Administration Committee* (March 1, 2011)

Learning Outcomes
After reading this issue, you will be able to: • Explain how declaring public lands to be "wild lands" or wilderness can affect rural livelihoods.   • Explain why public involvement in land use planning is essential.   • Describe the various ways in which public lands are valuable to the nation.   • Describe how "wild lands" are designated.

ISSUE SUMMARY

YES: Mike McKee, Uintah County Commissioner, argues that the government's new "Wild Lands" policy is illegal, contradicts previously approved land use plans for public lands, and will have dire effects on rural economies based on extractive industries. It should be repealed.

NO: Robert Abbey, Director of the Bureau of Land Management, argues that the government's new "Wild Lands" policy is legal, restores balance and clarity to multiple-use public land management, and will be implemented in collaboration with the public. Destruction of local extractive economies is by no means a foregone conclusion.

On March 30, 2009, President Obama signed the Omnibus Public Lands Management Act, protecting more than two million acres of wilderness in nine states. Stephen Trimble, "Wilderness: 'The System' Delivers: Hope Abounds for Wilderness Bills," *Wilderness* (2009/2010), notes that "As we search for a resilient ethic of sustainability that balances traditional American vitality with the stark demands of the twenty-first century our solutions must balance wilderness and restoration, preservation and reconciliation, collaboration and core values, environmental and social justice." See also his book, *Bargaining for Eden: The Fight for the Last Open Spaces in America* (University of California Press, 2009). However, not everyone is interested in such balance. On December 23, 2010, Ken Salazar, Secretary of the United States Department of the Interior, issued an order proposing to designate public lands with wilderness characteristics as "Wild Lands" and then to manage them

to protect their wilderness value. The order can be seen at www.blm.gov/pgdata/etc/medialib/blm/wo/Communications_ Directorate/public_affairs/news_release_attachments.Par.26564. File.dat/sec_order_3310.pdf. Doc Hastings (R-WA), chairman of the House Natural Resources committee, called the order "a clear attempt to allow the Administration to create *de facto* Wilderness areas without Congressional approval." He added:

> "Designating land as Wilderness imposes the most restrictive land-use policies. Lands that are currently used for multiple-use—including recreation activities, agriculture, ranching, American energy production and other economic activities—are in danger of being placed off-limits."
>
> "This Secretarial Order will disproportionately impact rural communities, who depend on public lands for their livelihoods. These communities have already been hit hard by onerous

existing federal restrictions and by the current economic crisis. They suffer from some of the highest unemployment rates in the country. The 'wild lands' order threatens to inflict further economic pain. This is just one more example of the onslaught of harmful actions that the Obama Administration is imposing on rural America."

See also Brad Knickerbocker, "Republicans Fear 'war on the West' in New Wild Lands Protection," *Christian Science Monitor* (March 2, 2011).

This debate is not new. A great deal of the American West is public land, and a great many farmers, ranchers, and others have become accustomed to using (and misusing) it as if it were their own. When the federal government has attempted to rein in overgrazing and other misuses, protests have been loud and Western politicians have fought to protect their constituents' prior uses of public lands. The conflict is also seen in connection with water rights, as in California where the debate is over whether the waters of the San Joaquin River should be routed through the river to protect endangered species of fish or through pipes and canals to supply farmers with irrigation water and protect agricultural production and jobs. In that context, Representative Hastings has accused the Environmental Protection Agency and "the radical policies of the environmental left" of waging "an assault on rural America." For the history of the wilderness idea and the controversies surrounding it, see Michael P. Nelson and J. Baird Caldicott, eds., *The Wilderness Debate Rages on: Continuing the Great New Wilderness Debate* (University of Georgia Press, 2008).

As is so often the case in environmental issues, the basic question is one of priorities. Do human needs (jobs, incomes, etc.) come before the needs of wildlife? Do vested interests—farmers, ranchers, miners, the oil industry, etc.—have a higher claim to environmental resources than do forests, rivers, and wildlife? Are short-term benefits (jobs, incomes, etc.) more important than long-term benefits (sustainability, and hence survival for future generations)? To people like Representative Hastings, the choice is obvious. So too is it for environmentalists. It is the job of institutions such as the Environmental Protection Agency and the Department of the Interior to balance the claims to the environment, and it is a safe bet that no such balancing will satisfy everyone. David N. Laband, "Regulating Biodiversity: Tragedy in the Political Commons," *Ideas on Liberty* (September 2001), argues that one basic problem is that those who demand protection of endangered species do not bear the costs (in jobs and economic activity) of that protection. On the other hand, Howard Youth, "Silenced Springs: Disappearing Birds," *Futurist* (July/August 2003), argues that the actions needed to protect

biodiversity not only have economic benefits, but also are the same actions needed to ensure a sustainable future for humanity. According to Martin Jenkins, "Prospects for Biodiversity," *Science* (November 14, 2003), the consequences for human life of failing to protect the natural world are "unforeseeable but probably catastrophic." In the San Francisco Bay area, local land managers recognize this and are working to conserve wild lands; see Glen Martin, "Big Plans for Wild Lands," *Bay Nature* (April–June 2011). New York State recently bought a "conservation easement" on a tract of paper company land, committing it to sustainable use by timber interests, recreational users, and others; see Connie Prickett, "More than a Working Forest," *New York State Conservationist* (June 2011). However, Nick Salafsky, "Integrating Development with Conservation: A Means to a Conservation End, or a Mean End to Conservation?" *Biological Conservation* (March 2011), warns that trying to combine conservation and development often fails. If one cannot pursue conservation by itself, one must be aware of the inevitable trade-offs. See also Johan A. Oldekop, Anthony J. Bebbington, Dan Brockington, and Richard F. Preziosi, "Understanding the Lessons and Limitations of Conservation and Development," *Conservation Biology* (April 2010); and Thomas O. McShane, et al., "Hard Choices: Making Trade-Offs Between Biodiversity Conservation and Human Well-Being," *Biological Conservation* (March 2011).

In the YES selection, Mike McKee, Uintah County Commissioner, argues that the government's new "Wild Lands" policy is illegal, contradicts previously approved land use plans for public lands, and will have dire effects on rural economies based on extractive industries. It should be repealed. In the NO selection, Robert Abbey, Director of the Bureau of Land Management, argues that the government's new "Wild Lands" policy is legal, restores balance and clarity to multiple-use public land management, and will be implemented in collaboration with the public. Destruction of local extractive economies is by no means a foregone conclusion.

GUIDE TO ACRONYMS

APD   Application for permit to drill
BLM   Bureau of Land Management
CEO   Chief Executive Officer
DOI   Department of the Interior
FLPMA Federal Land Policy and Management Act
LWC   Lands with Wilderness Characteristics
NEPA National Environmental Policy Act
RMP   Resource Management Plan
WSA   Wilderness Study Area

YES ↵

Mike McKee

Testimony before the House Natural Resources Committee Hearing on "The Impact of the Administration's Wild Lands Order on Jobs and Economic Growth"

Thank you for holding this hearing on the Wild Lands Policy and its negative impacts on my constituents. In Uintah County we are proud of our history, our heritage, and the multiple uses on our public lands from recreation to development of our natural resources.

Uintah County is the largest producer of natural gas in the state of Utah, with 63% of the state's natural gas coming from our County. Oil and gas have been produced in Uintah County since the early 1900's. We remain committed to responsible development of our public lands in an environmentally safe manner.

In Uintah County, only 15% of our land is privately owned. Policy changes during the past two years have had a chilling and detrimental effect on the economy of our County. In 2009, Uintah County lost 3,200 jobs in the mining and extraction industry. Many of our citizens are relocating to other states in order to retain employment and family members are left behind with the hope that the jobs will return. Jobs and the economy are not the only consequences of this administration's policy actions. Uintah County is concerned about homelessness, drug abuse, domestic violence, crime, and other social impacts. Jobs and economy are important to the citizens of Utah and Uintah County. In Uintah County, 50% of our jobs and 60% of our economy are tied to the extractive industry. This fact underscores the importance of sound policy and procedure on our public lands. The Wild Lands Policy issued by the Secretary will make all of these lands off limits in the predictable future for natural gas production, oil production, and shale oil, which are in such rich abundance.

Our community is suffering, and this suffering can be directly tied to policies of the Department of Interior.

Wild Lands Policy which the Interior Secretary signed on December 23, 2010 directly repudiates a Settlement Agreement signed by the state of Utah, the Utah School and the Institutional Trust Lands Administration (SITLA), the Utah Association of Counties and Department of the Interior. The Interior Department incorrectly describes Wild Lands Policy as a revocation of the Norton no-more wilderness policy. The fact is that BLM adopted an instruction memorandum to implement an out-of-court settlement that resolved litigation between the state of Utah and the Department of the Interior.

Interior officials continue to say that there is no violation of this Settlement Agreement, presumably based on the incorrect premise that "Wild Lands" are different from "Wilderness Study Areas" or WSAs. But aside from the name, they are identical and are treated the same.

In the Settlement Agreement, the Department of the Interior committed to not manage public lands outside of WSAs as if they were WSAs. The Wild Lands Policy in fact manages non-WSA public lands under the same protective framework that DOI has applied to WSAs for more than 30 years. The Wild Lands Policy clearly violates the Utah Wilderness Settlement Agreement.

In the Settlement Agreement, the Department of the Interior also pledged not to create new WSAs. The Wild Lands Policy does just exactly that and changing the name does not make it any less of a violation.

No federal law gives the Interior Secretary the authority to implement Secretarial Order 3310, the Wild Lands Policy.

In addition to being poor policy, the Wild Lands Policy is illegal. Under the U.S. Constitution, Congress has the sole authority to regulate federal lands. For public lands, Congress delegates that authority to the Interior Secretary in a series of federal laws, including the Bureau of Land Management Organic Act or the Federal Land Policy and Management Act (FLPMA). For wilderness designation, Congress chose to retain the sole power to designate wilderness.

The Wild Lands Policy attempts to override the laws that apply to public lands in several key respects:

> The Wild Lands Policy declares protection of lands with wilderness character a management priority.
>
> FLPMA dedicates the public lands to multiple use, with principal emphasis on six multiple uses: including domestic livestock grazing, fish and wildlife development and utilization, mineral exploration and production, rights-of-way [including transmission lines and pipelines], outdoor recreation, and timber production.

U.S. House of Representatives, March 1, 2011.

FLPMA does not include the word 'wilderness' in its definition of multiple use. It defines 'wilderness' only with respect to the now-expired wilderness review program in Section 603.

The Wild Lands Policy attempts to revise federal law by changing land management priorities to promote wilderness protection over all of the other uses that, by federal law, apply to public lands. This contradicts FLPMA, which dedicates the public lands to other uses, several of which, like mineral exploration and development, conflict with wilderness management. It also contradicts the Wilderness Act, which reserves to the sole authority to designate wilderness only by Congress.

The Wild Lands Policy assumes that the Secretary can manage public lands to protect wilderness, although FLPMA provided for a single and limited wilderness review program. FLPMA defines wilderness solely in terms of Section 603, which prescribed a 15-year wilderness review period. It is widely accepted that the authority to study public lands for wilderness expired in 1991, 15 years after FLPMA was enacted. There is no new authority to manage public lands for wilderness protection without attempting to rewrite FLPMA, and only Congress can do so.

It is also worth pointing out that federal agencies must involve the public and local governments when making a significant public land management change. These procedures ensure that there is a robust discussion of the effects of a proposal, and in the case of federal lands, there is coordination with state and local governments. In his haste to issue this policy right before the Christmas holiday, the Interior Secretary ignored these procedural steps.

The Interior Department also ignored the significant adverse environmental impacts that will come from the Wild Lands Policy. Proponents of this policy forget that the Wild Lands Policy will also prohibit wind turbines and transmission lines that are necessary for the green energy promoted by the Interior Secretary. For two years we have heard how the Administration will fund and subsidize green energy for wind turbines, solar energy farms, and the transmission lines necessary to put these alternative energy projects into the electrical power grid. Many energy projects are proposed for public lands, without considering the fact that these structures will violate the Wild Lands Policy. The structures associated with wind and solar energy are prohibited as permanent development and cannot be said to conform to the visual standards applied to wild lands. These important impacts are entirely ignored in the discussion by the Interior Department. It also appears that the Energy Department, which is issuing millions of dollars in incentive grants and loans, is not coordinating with the Interior Department which has adopted a policy that will prohibit or certainly delay implementation of any project.

Since early 2009, DOI has imposed a *de facto* moratorium on drilling and leasing on these lands. Uintah County initiated litigation in October of 2010 because the management policies violated the Settlement Agreement, contradicted the approved land use plans for public lands, and also were harming the local economy.

The Wild Lands Policy could potentially close millions of acres to oil and gas leasing in the state of Utah. BLM previously studied the lands that were said to have wilderness character when it revised the land use plans between 2000 and 2008, so we know the scope of the lands which may be impacted in Utah. These lands do not meet the actual definition of wilderness but are being called wilderness even with dirt roads, livestock developments, oil and gas rigs, pipelines and transmission lines.

We are concerned that the Wild Lands Policy now creates defacto wilderness. In our County, this policy is already negatively affecting areas that were open for multiple use activity. Recently signed Resource Management Plans are being turned upside down by this policy. For example, current road improvement requests, oil and gas leases, and permits to drill are being affected based on Wild Lands Policy.

Historically, Uintah County, on behalf of its citizens, has fully participated in federal land management forums in numerous land management issues, including resource management plans, oil and gas leasing decisions, transportation corridors on Federal lands, and wilderness issues. The County has expended a tremendous amount of resources over the past 20 years to engage in these processes in a responsible manner and representing our constituents. When Secretary Salazar announced the Wild Lands Policy just two days before Christmas in 2010, it was not only a shock to our constituents but was clearly an effort to circumvent established public processes that have governed our federal lands. In an economy and energy situation that is already at rock bottom, this action is further proof that Secretary Salazar has little regard for jobs or energy security in the West.

Over the past decade, the BLM began a revision of the Resource Management Plan for Utah and the Uintah Basin. This process, governed by NEPA, was open to the public and Uintah County participated as a cooperating agency. Thousands of hours and well over a million dollars of tax payer funds were expended by Uintah County. Other entities participated to bring to fruition a management plan that takes a comprehensive look at all uses of public lands in Uintah County. Although long, sometimes painful, and certainly no one group liked everything in the plan; this is what NEPA contemplated. Concessions were made on all sides. Uintah County supports open, public processes where all views are heard and considered, and then the hard working professionals of the BLM make informed decisions. All of the issues the Secretary claims to address under the new Wild Lands Policy are addressed in the Resource Management Plan—the only difference is that the Secretary clearly disagrees with the outcome of this Plan. Instead of attempting to short circuit the NEPA process, we urge the Secretary to vigilantly defend the BLM's Resource Management Plans. We need to end the practice of settling claims with litigants for the sole purpose of setting new policy outside the bright light of public input. Simply, the Wild Lands Policy undermines the Resource Management Plans.

We also note that toward the conclusion of the Vernal Resource Management Plan process, alternative "E" was added. This alternative's sole purpose was to evaluate the full spectrum of potential wilderness and the management thereof. This process required an additional two years to complete. Director Bob Abbey, in a meeting recently held in Salt Lake City, Utah, stated that the reason for reanalyzing work that was already complete was because not enough wilderness was found. This continual upheaval, unrest, change of direction, and philosophy, is discouraging. Either the land has wilderness quality or it does not. Why, with the huge deficits of spending that the Government is going through, do we have the BLM redo that which they have already completed?

In real terms, this policy will make it economically less viable for natural resource developers to operate on federal lands in the West. The state of Utah processes applications for permit to drill (APD's) in 35 days, while BLM takes an average of one and a half years. The Wild Lands Policy will add years to the permitting process and effectively further reduce access to natural resource production. It will yet create another layer of unnecessary bureaucracy that will only result in the further loss of jobs in my County and in other public lands counties throughout the West. Moreover, Uintah County will be forced to spend precious tax payer dollars to fight our own government to try to force the Department of Interior to live by the law of the land.

The combination of regressive gas leasing policies and the new Wild Lands Policy will result in further job losses and economic impact in Uintah County and throughout the west. Recently, I visited with a local CEO whose business has a cutting edge technology in the natural gas industry, yet, he can see the writing on the wall with the current policies. He will likely move his headquarters. He just returned from Dhabi as an option. Why would a business owner even consider such an option with all the unrest in the Middle East? What is wrong with this picture? Is the business environment better in the Middle East than on our own public lands in Uintah County? Planned and balanced development of these resources takes years to move into production. Driving these companies overseas is detrimental to our economy and to our energy security.

Unfortunately, today's policies are stopping responsible development and endangering America's energy security. This is not a spigot you can simply turn on and off on a whim.

Many companies stand ready to invest large sums of money in our County over the next ten years. All told, these investments would exceed two billion dollars over a ten year period. However, the regulatory uncertainty and the adverse policies of the Department of Interior is keeping these companies from investing, and in many cases, driving them overseas where U.S. dollars are being invested in foreign economies.

Eastern Utah is a treasure chest of natural resources. Uintah County has a great opportunity to help America become energy independent. Utah has 6.7 trillion cubic feet of proven natural gas reserves, conventional oil reserves of 286 million barrels, much of these are found in Uintah County. According to a Rand Report, the Uintah Basin has a staggering amount of shale oil ranging from 56 billion barrels to 321 billion barrels.

Each morning our newspapers carry disturbing pictures of governmental unrest in the Middle East and news of more and larger oil supply disruptions. In less than a month, previously stable countries in northern Africa and the Middle East have erupted in violent demonstrations. The governmental overthrow of Tunisia and Egypt has gone viral in Yemen, Libya, Saudi Arabia, and Bahrain with new calls for changes in the governments of the region. These shifts in power will have profound changes for the future, especially for the United States that produces and transports oil from those regions to the United States.

The Wild Lands Policy threatens national security by sharply reducing the nation's energy independence. It applies equally to all sources of energy from public lands such that the country is made weaker at a time when it needs to be stronger and more self-sufficient.

In addition to this, the Wild Lands Policy will impact the education of our children. The state of Utah was granted upon statehood, school trust lands, which by State Constitution are mandated to generate income to fund schools in the state of Utah. These lands are interspersed with federal lands throughout the State of Utah and Uintah County. It is commercially unviable to develop these lands for natural resources without access to the surrounding lands. If the federal lands become off limits to development, state lands go undeveloped as well, and education suffers directly from the federal policies.

To sum it up, the Wild Lands Policy is a short-sighted initiative that undermines the interests of this country and its people. The Wild Lands Policy overreaches by revising federal law when only Congress can do so. We urge this Committee to take every action possible to repeal it.

Our natural resources should be responsibly developed pursuant to the laws of the land. We have a responsibility to carefully develop our resources for America, for energy security, for our economy, and jobs for our citizens. I commend the House for choosing to de-fund the Wild Lands Policy for this current fiscal year and I urge the Senate to follow your lead. The role of Congress is clear in terms of wilderness policy, and I urge this Congress to preserve its authority and reverse this policy to save my county and our country from further economic harm.

MIKE MCKEE is Commissioner of Uintah County, Utah. He also co-chairs the Western Legacy Homestead Alliance.

Robert Abbey

 **NO**

Testimony before the House Natural Resources Committee Hearing on "The Impact of the Administration's Wild Lands Order on Jobs and Economic Growth"

The Wild Lands policy, established by Secretarial Order 3310, restores balance and clarity to the management of our public lands and follows clear legal direction. This order directs the BLM to work collaboratively with the public and local communities to determine how best to manage the public lands, taking into account all of their potential uses, including uses associated with the wilderness characteristics of certain public lands. It does not dictate the results of that planning process.

Section 102 of the Federal Land Policy and Management Act (FLPMA) declares that preservation and protection of public lands in their natural condition are part of the BLM's mission. Just as conventional and renewable energy production, grazing, mining, off-highway vehicle use, and hunting are considered in the development of the BLM's Resource Management Plans (RMPs), so too must the protection of wilderness characteristics be considered in the agency's land use plans.

Lands with wilderness characteristics are valued for their outstanding recreational opportunities (such as hunting, fishing, hiking, photography, or just getting outdoors) as well as for their important scientific, cultural, and historic contributions. Failing to consider protecting these wild places would undermine the careful balance in management mandated by law, a balance that we need on our public lands. Public lands provide billions of dollars in local economic benefits and they should be managed for multiple uses and many values, including energy production, recreation, and conservation.

The BLM's Multiple-Use Mission/Economic Contributions

I have worked for over 30 years in public service, 25 of those years as a career BLMer. I believe in, and am dedicated to, the BLM's multiple-use mission. This multiple-use mission is what makes the agency unique among Federal land management agencies, and it is what makes us welcome members of every community in which we work and live. However, multiple-use does not mean every use on every acre.

The BLM strives to be a good neighbor and a vital part of communities across America. Public lands managed by the BLM contribute significantly to the nation's economy and, in turn, often have a positive impact on nearby communities. The BLM's management of public lands contributes more than $100 billion annually to the national economy, and supports more than 500,000 American jobs.

A key component of these economic benefits is the BLM's contribution to America's energy portfolio. The BLM expects its onshore mineral leasing activities to contribute $4.3 billion to the Treasury in Fiscal Year 2012. The BLM currently manages more than 41 million acres of oil and gas leases, although less than 30 percent of that acreage is currently in production. More than 114 million barrels of oil were produced from BLM-managed mineral estate in Fiscal Year 2010 (the most since Fiscal Year 1997), and the almost 3 billion MCF (thousand cubic feet) of natural gas produced made 2010 the second-most productive year of natural gas production on record. The coal produced from nearly a half million acres of federal leases powers more than one-fifth of all electricity generated in the United States.

The BLM is also leading the nation toward the new energy frontier with active solar, wind, and geothermal energy programs. The BLM has proposed 24 Solar Energy Zones within 22 million acres of public lands identified for solar development, and in 2010 approved nine large-scale solar energy projects. These projects will generate more than 3,600 megawatts of electricity, enough to power close to 1 million homes, and could create thousands of construction and operations jobs. Development of wind power is also a key part of our nation's energy strategy for the future. The BLM manages 20 million acres of public lands with wind potential; currently, there is 437 MW of installed wind power capacity on the public lands. Geothermal energy development on the public lands, meanwhile, accounts for nearly half of U.S. geothermal energy capacity and supplies the electrical needs of about 1.2 million homes.

Energy production is not the only way in which the BLM contributes to local communities and the national economy. The combined economic impacts of timber-related activities on BLM-managed lands, grazing-related activities, and activity attributable to non-energy mineral

U.S. House of Representatives, March 1, 2011.

production from BLM-managed mineral estate total more than $5 billion each year. Recreation on public lands also provides major economic benefits to local economies and communities. In 2010, more than 58 million recreational visits took place on BLM-managed lands and waters, contributing billions of dollars to the U.S. economy. The diverse recreational opportunities on BLM-managed lands draw crowds of backpackers, hunters, off-road vehicle enthusiasts, mountain bikers, anglers, and photographers. In an increasingly urbanized West, these recreational opportunities are vital to the quality of life enjoyed by residents of western states, as well as national and international visitors. It should be noted that many of these recreationists are seeking the primitive experience available in BLM's wilder places.

The BLM's multiple-use mission is all about balancing public land management, and balancing all of the myriad resource values of this nation's great public lands. Wilderness character is one of these many resource values, and the BLM's new Wild Lands policy is a rational approach to ensuring that balance.

Secretarial Order 3310—Wild Lands Policy

The BLM's authority to designate new Wilderness Study Areas (WSAs) under section 603 of the FLPMA expired after President George H.W. Bush completed his recommendations for wilderness designation to Congress in January 1993. However, the BLM was still required to inventory and consider wilderness characteristics in the land use planning process.

Secretary of the Interior Gale Norton and the State of Utah entered into an out-of-court settlement agreement (the "Norton-Leavitt settlement") in 2003 that resulted in BLM rescinding the agency's then existing guidance on wilderness inventory. Since that time, the BLM has been without long-term national guidance on how to meet the FLPMA requirements to inventory and manage lands with wilderness characteristics. In 2008, the Ninth Circuit Court of Appeals in *Oregon Natural Desert Association v. BLM* stated that FLPMA's requirement that BLM maintain an inventory of public lands and their resources and other values includes inventory of wilderness values and that BLM must consider those values in its land use planning when they are present in the planning area. Secretarial Order 3310 and the related BLM manuals address that previous lack of direction on inventorying and managing lands with wilderness characteristics.

On December 23, 2010, I joined Secretary Salazar in announcing clear direction for implementing the BLM's mandate under FLPMA to conduct wilderness characteristics inventories and decide how best to manage those lands. The BLM also issued draft manuals that were recently finalized. This Wild Lands policy restores balance to the BLM's multiple-use management of the public lands in accordance with applicable law. It also provides the

field with clear guidance on how to comply with FLPMA and more specifically how to take into account wilderness characteristics in the agency's planning process.

With this consistent guidance, we believe that the BLM will enhance its ability to sustain its land use plan and project level decisions. In the past, some of these decisions have been invalidated because the courts in the Ninth and Tenth Circuits have found the analysis of wilderness characteristics lacking.

Policy Implementation/ BLM's Manuals

There has been a great deal of confusion about what this new policy does, and perhaps more importantly, what it does not do. Be assured that the new policy itself does not immediately change the management or status of the public lands. I would like to outline for you the facts about the new policy and its implementation. The BLM's new manuals set out a two-step process for inventorying and managing lands that may have wilderness characteristics. The first step is to maintain an inventory of Lands with Wilderness Characteristics (LWCs) as required by section 201 of FLPMA. The BLM's new manual on Wilderness Characteristics Inventory provides guidance on both updating existing inventory information and inventorying lands not previously assessed.

The manual carefully spells out the process for making these determinations, based on size, naturalness, and outstanding opportunities for solitude or a primitive and unconfined type of recreation—using the same Wilderness Act criteria the agency has always used. This process makes no determination about how the lands should be managed; it simply documents the current state of the lands.

Step two of the process, deciding how LWCs should be managed, is an open, public process undertaken through the BLM's land use planning process. Through this public process, a decision may be made to protect LWCs as "Wild Lands" or to manage them for other uses. For example, the BLM may determine that impairment of LWCs is appropriate for some areas due to other resource considerations, such as energy development. Other areas may be managed as Wild Lands with restrictions on surface disturbance and the construction of new structures. In addition, Wild Lands designations must be consistent with other applicable requirements of law. The BLM must consider these additional statutory requirements, where appropriate, in determining whether LWCs can be managed to protect their wilderness characteristics.

It is important to emphasize that if lands are designated as Wild Lands they are not wilderness and they are not WSAs. First, Wild Lands may only be designated administratively through an open, public planning process. The designation of Wild Lands may be revisited, as the need arises, through a subsequent public planning process. Second, allowed uses in Wild Lands may include some forms of motorized and mechanized travel. Allowed uses in each specific Wild Land will be determined by the land use plan governing those lands and will be accomplished through a process that allows the public and local communities full access to that decision-making. These decisions will be made locally, not in Washington, D.C. This policy doesn't change the delegation of authority for land use planning decisions. The BLM's state and field offices will continue to be responsible for those planning decisions.

The BLM regularly makes project-level decisions for activities on public lands. These decisions can involve a wide range of proposals such as locating roads and power lines, filming commercials and movies, and permitting mineral extraction activities. When considering these proposals, the BLM relies on existing land use plans, as well as any new information, to make a determination of how and if these projects can be accommodated within the BLM's multiple-use mission. This determination is necessarily a balancing act, taking into account all of the resources for which the BLM is responsible—including wilderness characteristics—as mandated by FLPMA.

A Wild Lands designation will be made and modified through an open public process, and therefore these designations differ from designated wilderness areas and WSAs. Wilderness areas can only be designated through an act of Congress and modified through subsequent legislation. The BLM manages WSAs to protect their wilderness characteristics until Congress designates them as wilderness or releases them from WSA status.

I have heard concerns that the new Wild Lands policy has put a halt to new projects and will prevent important economic activity in local communities. This claim is, simply put, false. A recent example involves a potash lease proposal in Utah that the BLM has approved through this new process. Through the NEPA process, the BLM has undertaken a review of a proposal to offer a competitive lease sale for potash on Sevier Lake, a dry lake bed in southwestern Utah. Following the issuance of the Secretarial Order roughly two months ago, the BLM completed an inventory of the lands involved and determined that the area does not meet the criteria for LWCs. The project is moving forward and it has been reported that it may result in as many as 300 permanent jobs in the local community.

Conclusion

The BLM is committed, and I am personally committed, to working with Congress and other key stakeholders to ensure that the Wild Lands policy works. My staff and I have spoken with many of you directly about the policy. In January, I traveled to Utah at the request of Governor Herbert, and participated in several meetings and forums on the policy. We have heard your concerns, and we are listening.

The BLM's Wild Lands policy affirms the agency's responsibility to take into account all of the public land resources for which the BLM is responsible. The policy provides local communities and the public with a strong voice in the decisions affecting the nation's public lands. Working cooperatively with our stakeholders, and being sensitive to local needs, we will ensure that all of the potential uses of the public lands and the BLM's multiple-use mission are taken into account when determining how best to manage the nation's public lands.

ROBERT ABBEY is Director of the Bureau of Land Management, United States Department of the Interior.

EXPLORING THE ISSUE

Does Designating "Wild Lands" Harm Rural Economies?

Critical Thinking and Reflection

1. Do states with different amounts of publicly owned land warrant different land-management policies?
2. Why do many business interests oppose the concept of wilderness?
3. To what degree should public land management policy be left to the federal government?

Is There Common Ground?

The users of public lands fear the BLM's Wild Lands policy because of its potential effect on jobs and local economies. The BLM agrees that jobs and local economies are important but "preservation and protection of public lands in their natural condition are part of the BLM's mission." A crucial question is how to satisfy both the economic and protective demands. Public involvement in policy-making is one way to reduce conflicts; Stephen E. Decker and Alistair J. Bath, "Public versus Expert Opinions Regarding Public Involvement Processes Used in Resource and Wildlife Management," *Conservation Letters* (December 2010), note that "Successful public involvement efforts can reduce conflict and build trust between resource managers and the public."

1. Where else does public involvement in decision making play a part?
2. Find a local example (perhaps on campus). Has it been successful at reducing conflict and building trust?
3. Sometimes "public involvement" is replaced by "stakeholder involvement" in decision making. Who are the stakeholders in the Wild Lands debate?

Create Central

www.mhhe.com/createcentral

Additional Resources

Thomas O. McShane, et al., "Hard Choices: Making Trade-Offs Between Biodiversity Conservation and Human Well-Being," *Biological Conservation* (March 2011)

Michael P. Nelson and J. Baird Caldicott, eds., *The Wilderness Debate Rages On: Continuing the Great New Wilderness Debate* (University of Georgia Press, 2008)

Johan A. Oldekop, Anthony J. Bebbington, Dan Brockington, and Richard F. Preziosi, "Understanding the Lessons and Limitations of Conservation and Development," *Conservation Biology* (April 2010)

Stephen Trimble, *Bargaining for Eden: The Fight for the Last Open Spaces in America* (University of California Press, 2009)

Internet References . . .

SourceWatch

www.sourcewatch.org/index.php/SourceWatch

The Wilderness Society

wilderness.org/

The Wildlands Conservancy

www.wildlandsconservancy.org/

Selected, Edited, and with Issue Framing Material by:
Thomas Easton, *Thomas College*

ISSUE

Does Excessive Endangered Species Act Litigation Threaten Species Recovery, Job Creation, and Economic Growth?

YES: Brandon M. Middleton, from Testimony before the House Committee on Natural Resources Oversight Hearing on "The Endangered Species Act: How Litigation Is Costing Jobs and Impeding True Recovery Efforts" (December 6, 2011)

NO: James J. Tutchton from Testimony before the House Committee on Natural Resources Oversight Hearing on "The Endangered Species Act: How Litigation Is Costing Jobs and Impeding True Recovery Efforts" (December 6, 2011)

Learning Outcomes

After reading this issue, you will be able to:

- Explain the purpose of the Endangered Species Act.
- Explain how efforts to protect endangered species may conflict with various human concerns.
- Explain why endangered species cases often wind up in court.

ISSUE SUMMARY

YES: Brandon M. Middleton argues that the Endangered Species Act (ESA) elevates species protection above human well-being, punishes landowners, fails to bring species back from the brink of extinction, frustrates local conservation efforts, and unfairly favors environmental groups in the courtroom. As written, the ESA encourages litigation. A more balanced approach is needed.

NO: James J. Tutchton of WildEarth Guardians argues that litigation is a tool to enforce the law. It is not true that litigation costs jobs or impedes true recovery efforts, and there is no need to make litigation more difficult.

Extinction is normal. Indeed, 99.9 percent of all the species that have ever lived are extinct, according to some estimates. But the process is normally spread out over time, with the formation of new species by mutation and selection, balancing out the loss of old ones to disease, new predators, climate change, habitat loss, and other factors. Today, human activities are an important cause of species loss mostly because humans destroy or alter habitat but also because of hunting (including commercial fishing), the introduction of foreign species as novel competitors, and the introduction of diseases. According to Martin Jenkins, "Prospects for Biodiversity," *Science* (November 14, 2003), some 350 (3.5 percent) of the world's bird species may vanish by 2050. Other categories of living things may suffer greater losses, leading to a "biologically impoverished" world. Jenkins states that the consequences for human life are "unforeseeable but potentially catastrophic." Amphibians are particularly at risk, partly because frogs, toads, and salamanders do not get as much attention as mammals and birds; see Brian Gratwicke, Thomas E. Lovejoy, and David E. Wildt, "Will Amphibians Croak under the Endangered Species Act?" *Bioscience* (February 2012). According to Sarah DeWeerdt, "Bye Bye, Birdie," *World Watch* (July/August 2006), over a third of all species may be on their way to extinction by 2050. Julia Whitty says "By the End of the Century Half of All Species Will Be Gone. Who Will Survive?" *Mother Jones* (May/June 2007). Yet in the face of limited funds and other resources, it seems impossible to protect all species; see Madeleine C. Bottrill, et al., "Is Conservation Triage Just Smart Decision Making?" *Trends in Ecology and Evolution* (vol. 23, no. 12, 2008), and Michelle Nijhuis, "Which Species Will Live?" *Scientific American* (August 2012). A. D. Barnosky, et al., "Has the Earth's Sixth Mass Extinction Already Arrived?" *Nature* (March 3, 2011), contend that current high extinction rates argue for strong and effective conservation measures. More recently, A. D. Barnosky, et al.,

"Approaching a State Shift in Earth's Biosphere," *Nature* (June 7, 2012), argue that population growth, ecosystem destruction, and global warming may be driving the planet toward a tipping point, after which biodiversity will be much reduced and humans may have difficulty meeting their needs for food and forest products, among other things.

Awareness of the problem has been growing. When the United States adopted the Endangered Species Act (ESA) in 1973, the goal was to protect species that were so reduced in numbers or restricted in habitat that a single untoward event could wipe them out. Both environmental groups and politicians were concerned over declining populations of some birds and plants. According to Ted Williams, "Law of Salvation," *Audubon* (November/December 2005), "Protecting the planet's genetic wealth made sense morally and economically. It was considered, rightly enough, what decent, civilized people do." The ESA, therefore, barred construction projects that would further threaten endangered species. In one famous case, construction on the Tellico Dam on the Little Tennessee River in Loudon County, Tennessee, was halted because it threatened the snail darter, a small fish. Another case involved the spotted owl, which was threatened by logging in the Northwest. Those in favor of the dam or the timber industry felt that the value of the endangered species was trivial compared to the human benefits at stake. Those in favor of the ESA argued that the loss of a single species might not matter to the world, but where one species went, others would follow. Protecting one species also protects others.

The ESA has had some notable successes despite lengthy legal battles, which have used up funds intended for protecting species, and pressures to ease restrictions on activities that might damage species or their habitat. These pressures have made it difficult to maintain the ESA. It was last authorized in 1988. Efforts to reauthorize it have repeatedly failed. According to Peter Uimonen and John Kostyack, "Unsound Economics: The Bush Administration's New Strategy for Undermining the Endangered Species Act," National Wildlife Federation (June 2004), the Bush administration reduced critical habitat protection, suppressed and distorted information on the economic benefits to local economies of habitat conservation, exaggerated costs, and reduced funding. Erik Stokstad, "What's Wrong with the Endangered Species Act?" *Science* (September 30, 2005), described the Republican-sponsored attempt to "reform" the ESA with the Threatened and Endangered Species Recovery Act as further restricting the ESA budget by requiring that landowners be compensated "for the fair market value of any development or other activity that the government vetoes because it would impact endangered species." It would also insulate landowners against the hazards of regulation as long as they have a plan in place to mitigate the effects of their actions and drastically limit the time federal agencies have to object to projects that might harm endangered species. Critics charged that the legislation was clearly more favorable to private interests than to endangered species. Proponents of reform argued that improving protection of private property rights was crucial; see Nancy Marano and Ben Lieberman, "Improving the Endangered Species Act: Balancing the Needs of Landowners and Endangered Wildlife," Heritage Foundation WebMemo #861 (September 23, 2005) (www.heritage.org/Research/EnergyandEnvironment/wm861.cfm).

After the Threatened and Endangered Species Recovery Act passed the House of Representatives in September 2005, Senator Mike Enzi (R-Wyo) introduced similar legislation in the Senate, calling it a Christmas present for "Wyoming farmers and ranchers. . . . We have a bill that will both help recover species and preserve landowner livelihood." Six U.S. senators, both Republican and Democrat, asked the Keystone Center to prepare a report on changes needed in the Endangered Species Act. The Keystone Center is a Colorado think-tank that helps "leaders from governmental, nongovernmental, industrial, and academic organizations to find productive solutions to controversial and complex public policy issues." Its report, released in April 2006 (www.keystone.org/spp/env-esa.html), concluded that the ESA's effectiveness for species recovery could be improved in several ways and that the burdens imposed by the ESA on private interests could be relieved with incentives, but reconciling the two goals is difficult and the House bill is not the way to do it. Senator Lincoln Chafee (RRI), one of the six who requested the Keystone report, said in March 2006 that Senate action was unlikely for that year. In May 2007, at a hearing of the House Natural Resources Committee, committee chairman Nick Rahall called for recommitting the ESA to being based on science, even as Republicans insisted the ESA interfered with private property rights (see Mike Deehan, "Democrats Say Science Will Guide Endangered Species Act," *CongressDaily* (May 9, 2007).

Michael J. Bean, "The Endangered Species Act under Threat," *Bioscience* (February 2006), cautioned that the House Act undermines "the government's long-standing trust responsibility to safeguard wildlife. The Senate should think long and hard before embracing the House's radical proposals." As part of that thinking process, The Ecological Society of America, together with other scientific societies, published the "Scientific Societies's Statement on the Endangered Species Act" (www.esa.org/pao/policyStatements/pdfDocuments/2-2006_finalStatement_Scientific%20Societies%20ESA.pdf) on February 27, 2006. Among other things, the statement objected to bureaucratic attempts to limit the definition of useful scientific data, called for eliminating delays in evaluating rare species for listing as threatened or endangered, improving funding for research, and restoring protections for critical habitat. It also recognized that "parties experiencing economic and social impacts from recovery activities should be included" in planning. In the end, the Senate bill never came to a vote.

Efforts to reform the ESA have not ceased, as is apparent from two 2011 congressional hearings. On October 13, 2011, the House Committee on Science, Space, and Technology, Subcommittee on Investigations and Oversight, held a hearing on the relative roles of science and policy in protecting species. Francesca T. Grifo of the Union of Concerned Scientists argued that in the past political factors have greatly interfered with implementation of the ESA. Science must play a larger role. Craig Manson, general counsel to the Westlands Water District in California's Central Valley (threatened by attempts to restrict water use in order to protect species), argued that science should play only an advisory role. "Science can tell us what is, while policy determines what ought to be done." On December 6, 2011, the House Committee on Natural Resources held a Hearing on "The Endangered Species Act: How Litigation Is Costing Jobs and Impeding True Recovery Efforts." Committee chair Doc Hastings (RWA) proposed that the Endangered Species Act be updated to make it more difficult for environmental groups to abuse the ESA with lawsuits. Hearing witnesses provide the essays for this Issue. In the YES selection, Brandon M. Middleton argues that the Endangered Species Act elevates species protection above human well-being, punishes landowners, fails to bring species back from the brink of extinction, frustrates local conservation efforts, and unfairly favors environmental groups in the courtroom. As written, the ESA encourages litigation. A more balanced approach is needed. In the NO selection James J. Tutchton of WildEarth Guardians argues that litigation is a tool to enforce the law. It is not true that litigation costs jobs or impedes true recovery efforts, and there is no need to make litigation more difficult.

Does Excessive Endangered Species Act Litigation Threaten Species Recovery, Job Creation, and Economic Growth? by Easton

79

YES ↵

Brandon M. Middleton

Testimony before the House Committee on Natural Resources Oversight Hearing on "The Endangered Species Act: How Litigation Is Costing Jobs and Impeding True Recovery Efforts"

The flaws behind the Endangered Species Act are numerous and well-known. Rather than provide incentives for conservation and environmental stewardship, the Endangered Species Act punishes those whose property contains land that might be used as habitat by endangered and threatened species. The statute's success rate is dismal, at best-few species that are classified as endangered or threatened ever return to recovered, healthy populations. Further, expansive and inflexible Endangered Species Act regulation by federal agencies often frustrates innovative local and state conservation efforts, with the result being greater conflict and less compromise.

These structural defects raise serious concerns over the Endangered Species Act's efficacy as a conservation statute and demonstrate that the statute provides little meaningful benefit to endangered and threatened species.

However, the statute's structural defects that victimize Americans in environmental litigation are particularly troubling. The Endangered Species Act elevates species protection above human well-being, benefitting extreme environmentalists and encouraging them to seek low-cost court victories at the expense of individual Americans as well as federal agencies throughout the country.

Specifically, environmental groups take full advantage of the Endangered Species Act's lenient citizen plaintiff standard. "Any person" may sue under the statute, a broad provision which has led to what the U.S. Fish and Wildlife Service has recognized as a litigation crisis.

Once environmental groups enter the courtroom, they enjoy precedent that stacks the deck in their favor. It is not difficult to win an Endangered Species Act lawsuit, but of equal concern is that courts often impose draconian and unhelpful remedies that harm businesses and property owners. The disturbing logic here is that the Endangered Species Act requires such results, no matter the costs. The fact that the Endangered Species Act gener-

ously authorizes attorneys' fees to prevailing parties further encourages environmental groups to take an overly aggressive approach to litigation without regard for the costs imposed on public and private parties.

With these structural defects in place, environmental groups would be foolish not to exploit them. Considering the state of the Nation's economy and the continuing onslaught of Endangered Species Act litigation, these defects certainly deserve the attention of the American people.

The Endangered Species Act's Lenient Standard for Becoming a Citizen Plaintiff

Numerous environmental groups thrive on bringing repeated Endangered Species Act cases to federal courtrooms. The Endangered Species Act is especially appealing to serial litigants because it provides that "any person may commence a civil suit" under the statute. Justice Scalia has criticized this expansive citizen suit provision as "an authorization of remarkable breadth when compared with the language Congress ordinarily uses," noting that in other environmental statutes, Congress has used more restrictive tests for citizen plaintiffs. Some courts have gone so far as to rule that the Endangered Species Act authorizes animals themselves to sue in their own right. . . .

To be sure, courts still demand that plaintiffs satisfy Article III of the Constitution by requiring a "case or controversy" before adjudicating a case. But the Endangered Species Act's otherwise minimal pleading requirements have resulted in what the U.S. Fish and Wildlife Service has described as a "cycle of litigation" that is "endless, and is very expensive, thus diverting resources from conservation actions that may provide relatively more benefit to imperiled species."

U.S. House of Representatives, December 6, 2011.

Indeed, in its October 2006 critical habitat designation for the Alameda whipsnake, the Service noted that such designations generally are "the subject of excessive litigation," and that "[a]s a result, critical habitat designations are driven by litigation and courts rather than biology, and made at a time and under a time frame that limits our ability to obtain and evaluate the scientific and other information required to make the designation most meaningful." The Service was clear that excessive Endangered Species Act litigation has compromised the integrity of the statute:

> We have been inundated with lawsuits for our failure to designate critical habitat, and we face a growing number of lawsuits challenging critical habitat determinations once they are made. These lawsuits have subjected the Service to an ever-increasing series of court orders and court-approved settlement agreements, compliance with which now consumes nearly the entire listing program budget. This leaves the Service with little ability to prioritize its activities to direct scarce listing resources to the listing program actions with the most biologically urgent species conservation needs.
>
> The consequence of the critical habitat litigation activity is that limited listing funds are used to defend active lawsuits, to respond to Notices of Intent (NOIs) to sue relative to critical habitat, and to comply with the growing number of adverse court orders. As a result, listing petition responses, the Service's own proposals to list critically imperiled species, and final listing determinations on existing proposals are all significantly delayed.
>
> The accelerated schedules of court-ordered designations have left the Service with limited ability to provide for public participation or to ensure a defect-free rulemaking process before making decisions on listing and critical habitat proposals, due to the risks associated with noncompliance with judicially imposed deadlines. This in turn fosters a second round of litigation in which those who fear adverse impacts from critical habitat designations challenge those designations. The cycle of litigation appears endless, and is very expensive, thus diverting resources from conservation actions that may provide relatively more benefit to imperiled species.

More recently, the Service has asked Congress to set a limit on the number of species it is authorized to consider under the Endangered Species Act petition process. Without any such limit, the tactic for environmental groups appears to be "the more, the merrier" when it comes to Endangered Species Act listing petitions. After all, given the statute's expansive citizen suit provision, multi-species petitions make sense because the Service's inability to manage an overload of documents means only that the petitions will be settled in court, with the attendant attorney's fees. As Gary Frazer, the Service's assistant director for endangered species, has noted, "[t]hese mega-petitions are putting us in a difficult spot, and they're basically going to shut down our ability to list any candidates in the foreseeable future." Mr. Frazer likewise recognized that if "all our resources are used responding to petitions, we don't have the resources to put species on the endangered species list. It's not a happy situation."

The consequences of the Endangered Species Act's friendly citizen suit provision are thus clear, albeit counterproductive. Citizen plaintiffs' easy access to courts has come at the cost of meaningful recovery and environmental progress.

Endangered Species Act Litigation Can Bring Handsome Rewards

The Endangered Species Act's attorney's fees provision defies common sense because it allows an environmental group to obtain attorney's fees even when a lawsuit is brought over a recovered and healthy species that has been recommended by the Service for delisting. In most litigation, "parties are ordinarily required to bear their own attorney's fees—the prevailing party is not entitled to collect from the loser." Federal courts "follow a general practice of not awarding fees to a prevailing party absent explicit statutory authority." . . .

The Endangered Species Act, however, provides that courts "may award costs of litigation (including reasonable attorney and expert witness fees) to any party, whenever the court determines such award is appropriate." This is an extremely charitable provision, especially considering that environmental plaintiffs need not fear an award of attorney's fees to the opposing party in the event they do not prevail. . . .

The Endangered Species Act attorney's fees provision leads to absurd results. In *Center for Biological Diversity v. Marina Point Development Co.*, a California business currently faces the prospect of paying the Center for Biological Diversity and another environmental group more than $1 million in fees and costs without proof of harm to any species. In that case, the anti-development plaintiffs sought and received an injunction to stop a commercial project based on claims the project would harm listed bald eagles. However, the U.S. Fish and Wildlife Service had already determined that bald eagles were fully recovered and should be delisted and that the challenged project would have no effect on the species. And, in fact, while the case was on appeal in the Ninth Circuit, the case became moot when the Service removed bald eagles from the list of threatened and endangered species altogether. But, while the Ninth Circuit recognized that the property owners activities did not violate the Endangered Species Act, it nonetheless ruled that the Center was entitled to fees under the statute, since the delisting of the bald eagle occurred while the Center's dubious district court victory was on appeal. . . .

Does Excessive Endangered Species Act Litigation Threaten Species Recovery, Job Creation, and Economic Growth? by Easton

81

This suit provided no benefit to any species but imposed enormous costs on a private company without any proof of violation. Common sense dictates that the property owner should not have to pay for a statutory violation that it did not commit, but the Endangered Species Act's attorney's fees provision has enabled precisely this result. Surely, this is not what Congress intended.

Did Congress Really Intend for the Endangered Species Act to Be Imposed "Whatever the Cost"?

Thanks in part to the Endangered Species Act's litigation incentives discussed above, the Natural Resources Defense Council (NRDC), Earthjustice, and other environmental groups sued in 2005 to shut down critical California water projects in order to supposedly protect an insignificant fish called the delta smelt, a species that until then had generated little interest outside the extreme environmental community. NRDC and Earthjustice won their lawsuit, leading to an unprecedented water supply crisis for the San Joaquin Valley and Southern California.

Yet, just a few years later, after the U.S. Fish and Wildlife Service capitulated to the environmental community and issued a formal delta smelt management regime that caused still more water supply uncertainty, the same federal judge who had previously ruled in favor of NRDC and Earthjustice ruled against them and the government, holding that the U.S. Fish and Wildlife Service had gone too far in its misguided effort to protect the delta smelt, and finding that federal staffers engaged in bad faith in attempting to defend delta smelt Endangered Species Act restrictions.

But what caught legal scholars' attention was Judge Wanger's remedy for the U.S. Fish and Wildlife Service's Endangered Species Act violations. Despite the protests of NRDC and Earthjustice, Judge Wanger took a common sense approach and considered the harm that would result from allowing the illegal delta smelt regulations to go forward. In his August 31, 2011, decision to enjoin delta smelt-based water restrictions, Judge Wanger ruled that where the imposition of flawed ESA regulations would "affirmatively harm human communities through the reduction of water supplies and by reducing water supply security in future years," it is appropriate for courts to balance this human hardship against the needs of protected species. As Judge Wanger wrote, "[i]f such harms cannot be considered in the balance in an ESA case, it is difficult to envision how a resource-dependent [party] would ever" prevail on an injunctive relief motion in an Endangered Species Act case.

While Judge Wanger's decision to consider human hardship in the delta smelt case deserves praise, it may seem remarkable that there was ever a question over the court's authority to consider the human costs of ill-advised Endangered Species Act regulation. Unfortunately, Judge

Wanger's decision to balance the hardships and consider the public interest in natural resources is the exception in Endangered Species Act cases, not the rule. More often than not, courts give the benefit of the doubt to environmental groups and the hundreds of species they represent, regardless of the circumstances. The deck is stacked such that environmental groups have an incentive to sue even when there would be little to no benefit to a species from litigation, and even though the harm and financial toll of such litigation may be great.

One may ask, then, how this came to be–how are environmental groups able to argue with almost universal success that courts should consider the consequences their decisions have on endangered species, but at the same time claim that courts have no authority to consider the effects their decisions will have on those who actually bear the brunt of the Endangered Species Act, *i.e.*, landowners and natural resource users?

The answer stems from the Supreme Court's notorious 1978 Supreme Court decision, *TVA v. Hill*. *TVA* concerned whether the Tennessee Valley Authority could proceed with the opening and operation of the nearly complete Tellico Dam project, notwithstanding the fact that the dam's operation would either eradicate the nearly extinct snail darter species or at the very least destroy the fish species' critical habitat. Although environmental groups contended that the Endangered Species Act required the injunction of the Tellico Dam, the district court declined to do so due to the amount of public money that had already been spent on the project, noting that "[a]t some point in time a federal project becomes so near completion and so incapable of modification that a court of equity should not apply a statute enacted long after inception of the project to produce an unreasonable result."

The Supreme Court, however, did not agree with the district court and enjoined the Tellico Dam project from going forward. Despite recognizing that "[i]t may seem curious to some that the survival of a relatively small number of three-inch fish among all the countless millions of species extant would require the permanent halting of a virtually completed dam for which Congress has expended more than $100 million," the Court concluded that "Endangered Species Act require[d] precisely that result."

TVA's long-term impact, however, is found not in the result it reached, but in the precedent it set. In his majority opinion, Chief Justice Burger purported to discern Congress's will in enacting the Endangered Species Act by suggesting a legislative intent that is found nowhere in the text of the statute: "The plain intent of Congress in enacting this statute was to halt and reverse the trend toward species extinction, whatever the cost." Similarly, "the plain language of the Act, buttressed by its legislative history, shows clearly that Congress viewed the value of endangered species as 'incalculable.'"

Even more starkly, Chief Justice Burger suggested that Congress divested federal courts of their traditional equitable discretion in Endangered Species Act cases. According to the Court, there was no "mandate from the people to strike a balance of equities on the side of the Tellico Dam. Congress has spoken in the plainest of words, making it abundantly clear that the balance has been struck in favor of affording endangered species the highest of priorities. . . ."

TVA's draconian language provided ammunition for environmental groups to use the Endangered Species Act to deprive property owners and resource users of their rights, while at the same time preventing courts from considering the hardship resulting from such an unbalanced approach. According to this view, *TVA* represents Congress's intent that the Endangered Species Act restricted federal courts' traditional equity jurisdiction. Yet in actuality, Congress did no such thing, even though it was fully capable of including an explicit provision that mandates the restriction of federal courts' traditional equity jurisdiction.

Indeed, *TVA's* precedent has led environmental groups to routinely argue that the economic impacts of an Endangered Species Act injunction are irrelevant, and that courts are forbidden from considering economic hardship when fashioning injunctive relief. The effort to exploit *TVA* has largely been successful. The Ninth Circuit, for example, holds that Congress "removed from the courts their traditional equitable discretion in injunction proceedings of balancing the parties' competing interests. The 'language, history, and structure' of the ESA demonstrates that Congress' determination that the balance of hardships and the public interest tips heavily in favor of protected species."

Similarly, in the First Circuit, courts hold that "[a]ltough it is generally true that in the preliminary injunction context that the district court is required to weigh and balance the relative harms to the non-movant if the injunction is granted and to the movant if it is not" that is not the case in Endangered Species Act litigation, as "that balancing has been answered by Congress' determination that the 'balance of hardships and the public interest tips heavily in favor of protected species.'"

Today, a primary reason for costly Endangered Species Act litigation and the injunction even of "green" energy projects can be found in *TVA's* instruction that Congress placed endangered species above all other concerns, including humans. When a federal court stopped the development of a wind energy project in West Virginia two years ago due to alleged threats to the endangered Indiana bat, it repeatedly cited *TVA* and opined that "Congress, in enacting the ESA, has unequivocally stated that endangered species must be afforded the highest priority." In California, the same attorneys who forced the injunction of the West Virginia wind project are now attempting to prevent the City of San Francisco from engaging in flood control efforts at a municipal golf course, supposedly because flood control harms the California red-legged frog. Of course, the environmental attorneys' argument is based largely on *TVA*, as they claim that *TVA* prevents the district court from balancing the hardships of increased flooding against the needs of a local amphibian.

Based on the environmentalists' "species protection whatever the costs" approach to the Endangered Species Act, it should come as no surprise that Judge Wanger's recent limitation of the *TVA* rule has found disfavor with the environmental community. While Judge Wanger allowed water users to at least have an equal voice in the delta smelt proceedings, NRDC and Earthjustice have appealed, arguing that the "district court's view of *TVA v. Hill* is wrong," and that the court "improperly balanced" the water supply impacts of Endangered Species Act regulation against delta smelt habitat concerns.

Keeping in mind Judge Wanger's admonition that, in the context of delta smelt water supply impacts, "[i]f such harms cannot be considered in the balance in an ESA case, it is difficult to envision how a resource-dependent [party] would ever" prevail on an injunctive relief motion in an Endangered Species Act case, the environmental community's protest of even the slightest limitation of *TVA* demonstrates just how much they depend on the decision's troubling precedent in cases where they seek to forestall economic development and human needs. Courts, in general, recognize the extreme viewpoint of environmentalists, but all too often they punt on engaging in a balanced approach to the Endangered Species Act. Instead, the blame for the harsh realities of Endangered Species Act litigation is placed on the legislative branch, as it was Congress who purportedly ordered that endangered species be afforded "the highest of priorities," no matter the costs.

It is misplaced, of course, for courts to blame Congress on an approach to injunctive relief never imagined or sanctioned by the legislative branch. But although the harms resulting from the "whatever the cost" approach are all too real for property owners and resource users faced with an Endangered Species Act lawsuit, addressing the problem is fortunately not difficult. As the Supreme Court itself recognized in *TVA*, "[o]nce Congress, exercising its delegated powers, has decided the order of priorities in a given area, it is for the Executive to administer the laws and for the courts to enforce them when enforcement is sought."

Thus, if Congress were to determine that the Supreme Court's interpretation of the order of priorities under the Endangered Species Act is incorrect, and that the human species is entitled to at least as much priority as allocated to any other animal species, then litigation will shift more towards a balanced approach that at least gives property owners and resources users an equal voice in the courtroom. Abandoning the "whatever the cost" mandate would deprive the environmental community of one of their greatest litigation weapons, and would result in less of a perverse incentive for regulated parties to protect endangered species. Moreover, allowing for a full balancing of harms and consideration of the public interest would not

Does Excessive Endangered Species Act Litigation Threaten Species Recovery, Job Creation, and Economic Growth? by Easton

83

preclude environmental groups from obtaining an injunction in all Endangered Species Act cases, but would instead enable a more balanced approach to the statute that better comports with traditional notions of equity and fairness.

Conclusion

Incentives matter. Unfortunately, when it comes to the Endangered Species Act, the incentives favor the environmental community without providing a meaningful benefit to the species that the statute seeks to protect.

This is especially so in the context of Endangered Species Act litigation. Numerous environmental groups enjoy successful practices that depend on Endangered Species Act restrictions of property owners, natural resource users, and government agencies alike. This is a testament to how much the statute encourages and fosters Endangered Species Act lawsuits.

Unless lawsuits become more difficult to bring and draconian injunctions more difficult to obtain, the disturbing trend of endless and ongoing Endangered Species Act litigation is likely to continue.

BRANDON M. MIDDLETON is a Staff Attorney for the Environmental Section of the Pacific Legal Foundation (www .pacificlegal.org/), a public interest legal organization that fights for limited government, property rights, individual rights, and a balanced approach to environmental protection.

James J. Tutchton

→ NO

Testimony before the House Committee on Natural Resources Oversight Hearing on "The Endangered Species Act: How Litigation Is Costing Jobs and Impeding True Recovery Efforts"

Introduction

The Endangered Species Act is our nation's primary wild-life conservation statute designed to protect biological diversity. It grew out of an emerging consensus that the protection of both charismatic animals and other lesser-known species, once deemed valueless, is necessary if we are to succeed in protecting not only the species we find charismatic, but also the ecosystems on which they, and ultimately we, depend. As human understanding has grown, we have learned that ecosystems, not unlike a woven sweater, can begin to unravel when even a single thread is pulled out. When many threads are pulled, holes develop, and what was once a warm and protective sweater no longer exists. The same is true for an ecosystem that loses its parts, even those that may at first blush seem minor. For example, scientists have recently learned that a species as imposing as the grizzly bear, monarch of the Yellowstone ecosystem, relies on a species as little noticed as the white-bark pine for its survival—and that protecting the bear alone without the pine is inadequate, for the bear would have little to eat at certain times of year. The Endangered Species Act encompasses this scientific understanding of the interconnection between species, protecting both greater species and the smaller ones that allow the great creatures to survive. In the end, by protecting the full range of the tangled, and still poorly understood, web of life the Act ultimately protects humanity itself.

Because the Act protects species, as it must, wherever they are found, regardless of land ownership, and because it protects all species great and small, regardless of their popularity or immediately perceived value to humanity, it has engendered a continuing level of controversy. However, this controversy neither indicates that the task of protecting biodiversity is unimportant or unpopular, nor that the Endangered Species Act is not working as intended.

There are two false assumptions imbedded in the title of this hearing. First, that litigation directed at enforcing the Endangered Species Act is costing jobs. Second, that litigation enforcing the Act is impeding true recovery

efforts. Both of these misguided charges obscure more meaningful inquiry into the source of the problems some members of this Committee apparently perceive.

Litigation is a tool to enforce the law. Congress writes our laws, but it generally must rely on the executive branch to enforce them. However, at times, especially when Congress is concerned about whether the executive branch is willing or able to enforce a particular law, Congress has enacted provisions encouraging private citizens to enforce, or compel the executive branch to follow, the law. These "citizen-suit" provisions, found in most environmental and civil rights statutes, represent a bedrock principle of our democracy: the idea that citizen oversight can make our government institutions better. They are most useful in situations where the volume of legal enforcement necessary to fully implement a law may outgrow the capacity of federal agencies, where the desire of private litigants to enforce the law may exceed that of federal officials, or when a law places obligations, such as deadlines for action, on federal agencies and Congress desires outside help to ensure that these federal agencies comply with the law. The citizen-suit provision in the Endangered Species Act serves all three of these functions.

Accordingly, because litigation, whether conducted solely by government prosecutors or by private citizens, is merely a tool to increase compliance with the law, a charge that litigation is costing jobs is, at base, a charge that enforcing the law is costing jobs. There is little difference between having a law that is unenforced or unenforceable, and having no law at all. Thus, to the extent some members of this Committee perceive a conflict between enforcement of the Endangered Species Act and economic activity, this Committee should not be considering whether it wants the law Congress has passed enforced via litigation, but whether it likes the law it has written or believes it should be amended. The question of whether the Endangered Species Act should be enforced is only a component of the larger issue: what does Congress think of the Act itself?

U.S. House of Representatives, December 6, 2011.

Does Excessive Endangered Species Act Litigation Threaten Species Recovery, Job Creation, and Economic Growth? by Easton

85

Similarly, the second false assumption imbedded in the title of this hearing, that litigation to enforce the Endangered Species Act is impeding the recovery of species, also serves to obscure the fundamental inquiry. The clearly stated goal of the Endangered Species Act is to recover species from the edge of extinction. Congress drafted the various provisions of the Act to achieve this end. Thus, if members of this Committee perceive a conflict between enforcing the Endangered Species Act through litigation and achieving the Act's goal of recovering species, the source of the perceived problem is not with the enforcement of the Act, but with the Act's efficacy. Enforcement is simply implementation. The Committee's concern should be with whether the law works when enforced, not with limiting enforcement. Unenforced laws are worse than meaningless because they engender disrespect for both the rule of law and the legal system.

In short, the two assumptions contained in the title of this hearing hide more fundamental questions that should be explored. The basic inquiry here is not, and should not be, whether litigation directed at enforcing the Endangered Species Act is a problem, but whether Congress wants the Endangered Species Act enforced as written and believes it is effective in meeting its goals. To focus on the litigation enforcing the law as the source of the problems some members of this Committee perceive masks the actual conflict. Simply put, if this Committee does not want the Endangered Species Act enforced—it does not want the Act. This Committee should openly acknowledge and debate the root cause of the problems some of its members perceive. Unfortunately, the title of this hearing indicates this Committee may be inappropriately focused on shooting the messenger, those who litigate to enforce the Endangered Species Act, rather than examining the questions behind the message: Are endangered species worth saving, does this nation remain committed to the saving them, and is the Endangered Species Act an effective means to achieve this end? As discussed below, the answer to these questions is clearly—yes.

The Endangered Species Act Is Needed

The Endangered Species Act Protects Valuable Natural Resources

The vast variety of species with which humans share this planet are of incalculable value to us. As stated by Representative Evans on the House floor in 1982:

> [I]t is important to understand that the contribution of wild species to the welfare of mankind in agriculture, medicine, industry, and science have been of incalculable value. These contributions will continue only if we protect our storehouse of biological diversity . . . [O]ur wild plants and animals are not only uplifting to the human spirit,

but they are absolutely essential—as a practical matter—to our continued healthy existence.

As Americans, we have celebrated the comeback of the bald eagle, the very symbol of our country, from a low of 487 nesting pairs in the continental United States to more than 9,000 nesting pairs. In large part, the Endangered Species Act is responsible for the eagle's recovery. Similarly, we now enjoy the company of approximately 3 million American alligators, a species we almost lost before it was protected under the Act and quickly recovered. The whooping crane, a symbol of wisdom, fidelity, and long life in many cultures, has also benefited from protection under the Endangered Species Act, rebounding from a low of 16 individuals to approximately 400. However, though the Act has prevented the extinction of this species, the Whooper is not yet ready to graduate from the Act's protection. Such charismatic creatures the Act has pulled back from the brink of extinction are frequently invoked in hearings on the Endangered Species Act. The law, however, does not deny its protective shield to creatures whose pictures may never grace a wildlife calendar.

While some have criticized the Endangered Species Act for protecting "bugs and weeds," these invertebrates and plants are frequently of the most utilitarian value to humans. As expressed by Harvard professor E. O. Wilson, if we do not protect the little things that run the world:

> New sources of scientific information will be lost. Vast potential biological wealth will be destroyed. Still undeveloped medicines, crops, pharmaceuticals, timber, fibers, pulp, soil-restoring vegetation, petroleum substitutes, and other products and amenities will never come to light . . . it is also easy to overlook the services that ecosystems provide humanity. They enrich the soil and create the very air we breathe. Without these amenities, the remaining tenure of the human race would be nasty and brief. The life-sustaining matrix is built of green plants with legions of microorganisms and mostly small, obscure animals—in other words, weeds and bugs. . . .

On a global scale, 25 to 40 percent of pharmaceutical products come from wild plants and animals. A full 70 percent of pharmaceutical products are modeled on a native species, despite only 0.1% of plant species having been examined for their medicinal value. Invertebrate pollinators are also of high value to humanity. A variety of pollinators, such as some butterflies and bats, are currently protected by the Endangered Species Act, although others are not. The loss of pollinators threatens ecological and economic systems across the country.

One of the Endangered Species Act's explicit purposes is "to provide a means whereby the ecosystems upon which endangered species and threatened species depend may be conserved." This vision of ecosystem protection appears frequently throughout the Act's legislative history. The economic benefits healthy ecosystems provide humanity dwarf even our national debt. Economists estimate the global value of "ecosystem services" at $33 trillion annually and in the U.S. alone at $300 billion annually. Even these dramatic estimates are conservative, as the value of ecosystems ultimately equates to the value of everything—as without ecosystems humans could not survive. Moreover, most of the services, currently provided to us for free by ecosystems, are so intricate and provided on such a massive scale that it would not be feasible to replicate them at any cost even if scientists possessed the knowledge to do so. The tremendous value of ecosystems is placed at risk by the continued erosion of the biodiversity.

Additionally, endangered species are of great aesthetic, symbolic, and recreational value. Animals and nature are ubiquitous in our children's fairly tales and stories, which inform social codes of conduct. Continued destructiveness towards nature may consequently impact human cognition and social relations. "The more we know of other forms of life, the more we enjoy and respect ourselves. Humanity is exalted not because we are so far above other living creatures, but because knowing them well elevates the very concept of life." The recreational value of wildlife is also very significant. The U.S. Fish and Wildlife Service has determined that approximately 87 million adult Americans, or 38 percent of the adult population, spend more than $120 billion in the course of wildlife-related recreation annually. These expenditures support hundreds of thousands of jobs. These jobs are every bit as valuable to those who hold them as are the jobs the Committee perceives at risk from enforcement of the Endangered Species Act. In short, the protection of biodiversity appears well worth the effort. Just as a nation should not squander its fiscal resources, it should not squander its natural ones. The Endangered Species Act is central to our national effort to conserve our irreplaceable natural resources.

The Present Rate of the Loss of Species Is Alarming

The current rate of species' extinction worldwide is estimated at 1,000 times the natural rate of extinction and is increasing. The impact of seven billion humans on species diversity is comparable to that of the asteroid that wiped out most life on Earth 65 million years ago. Like geologists do today, future intelligent beings, should there be any, will be able to mark the current human-caused extinction epoch by observing the number and diversity of fossils preserved in future rock layers. Unless these trends are reversed, by the year 2020 up to 20 percent of all extant species will no longer exist. According to the International Union for the Conservation of Nature, one in every four mammals is facing a high risk of extinction in the near future. Almost half of all tortoises and freshwater turtles are threatened. More than one-fifth of the world's birds face extinction according to Birdlife International. One third of the world's amphibians are also vanishing. At least

two out of every five species on earth will go extinct due to human-caused climate change if greenhouse gas emissions are not promptly curtailed.

Moreover, there is a trickle-down effect from species' extinction as the loss of one species leads to the loss of other dependent species. For example, researchers recently calculated that the extinction of nearly 6,300 plants listed as threatened or endangered by the International Union for the Conservation of Nature would also result in the loss of nearly 4,700 species of beetles and 136 types of butterflies.

In sum, there should be no legitimate debate over whether or not our planet's biodiversity is rapidly diminishing. There should also be little debate that this loss is attributable to human activities and dramatic human population increases:

> Human demographic success has brought the world to this crisis of biodiversity. Human beings—mammals of the 50-kilogram weight class and members of a group, the primates, otherwise noted for scarcity—have become a hundred times more numerous than any other land animal of comparable size in the history of life. By every conceivable measure, humanity is ecologically abnormal. Our species appropriates between 20 and 40 percent of the solar energy captured in organic material by land plants. There is no way that we can draw upon the resources of the planet to such a degree without drastically reducing the state of most other species.

Over 99 percent of scientists agree that a serious, world-wide loss of biodiversity is likely, very likely, or virtually certain. There is also strong scientific consensus that humans are responsible for this extinction crisis. Indeed, last year the United Nations marked the first ever International Year of Biodiversity to call attention and spur action to address this problem. The United States Endangered Species Act serves as a model for many other nations and exhibits our national commitment to the international effort to save the diversity of life on Earth.

The Endangered Species Act Enjoys Widespread Public Support

As a remedy to stem the tide of extinction and protect species for the use and enjoyment of future generations the Endangered Species Act enjoys widespread public support. Passed almost unanimously by Congress and signed into law by President Nixon in 1973, the Endangered Species Act has consistently remained popular. In 1999, university researchers concluded that 84 percent of the American public supported the current Endangered Species Act, or an even stronger version of the law. A poll commissioned by the Endangered Species Coalition and conducted by Harris Interactive between February 16–20 of this year, found that despite the ensuing decade of

Does Excessive Endangered Species Act Litigation Threaten Species Recovery, Job Creation, and Economic Growth? by Easton

87

attacks on the Act since 1999 and the controversies over its implementation and enforcement, an identical 84 percent of Americans adjusted for age, sex, race/ethnicity, education, region of the country, number of adults in the household, and number of phone lines in the household, supported or strongly supported the Endangered Species Act. While support was strongest among Democrats (93%), the majority of Republicans (74%) also supported or strongly supported the Act. The majority of Americans of both political parties (64%) also believe that the Act is a safety net providing balanced solutions to save wildlife and plants at risk of extinction. In short, the protection of endangered species is a broadly supported American value. Extinction is not.

The Endangered Species Act Is Effective

Not unlike the biblical Noah, checking off the animals boarding his Ark, two by two, the Endangered Species Act operates based on a list. Species on the list receive the Act's protections while unlisted species do not. The leading cause of species imperilment in the U.S. is habitat destruction. The protective provisions of the Endangered Species Act, particularly those that protect a listed species' designated critical habitat, are effective at stemming habitat destruction and recovering species. Listed species with a designated critical habitat are twice as likely to be recovering as those without designated critical habitat.

Additionally, research shows that as of 2006 the Endangered Species Act had prevented the extinction of at least 227 species. According to the U.S. Fish and Wildlife Service, only nine of the approximately 1,445 domestic species ever added to the Endangered Species Act list have been declared extinct. Seven of these were mostly likely extinct before they received the Act's protection. Thus, the Act has only failed two species: a success rate in preventing extinction of over 99 percent. Conversely, protection under the Act has successfully recovered at least 22 species. Accordingly, the Endangered Species Act is succeeding in recovering species at least twice as often as it is has failed. Indeed, if the seven species that were likely extinct before they were listed under the Act are discounted, the Endangered Species Act is succeeding in recovering species at a rate more than 10 times that at which it fails.

Enforcement Through Litigation Has Increased the Effectiveness of the Endangered Species Act

While the Endangered Species Act has been over 99 percent successful in preventing extinction, it is still criticized by some because 1,397 species remain on the domestic protected species list, while only 22 have been finally recovered. However, this criticism is misplaced. The task of recovering species from the edge of extinction is difficult. The Endangered Species Act has been on the job for 38 years. However, many of the species currently protected by the Act, have not been listed nearly so long, but were added more recently. Moreover, pursuant to the requirements of the Act, the U.S. Fish and Wildlife Service has estimated the costs of, and planned for, the recovery of many endangered species on long time lines often exceeding 50 years.

Perhaps more importantly for purposes of the Committee's inquiry into conflicts between the Endangered Species Act and economic activity, one must recognize that the timeline for species recovery is dependent on the resources devoted to recovery—and the strength of the protective regulations implemented to achieve recovery. Thus, increasing the rate of recovery will require additional resources and more, not less, protective regulations—the type of regulations that have the potential to affect economic activity. Any criticism of the rate of species recovery must recognize that this rate can only be increased by greater, not reduced, effort and thus calls for more effective enforcement of, or strengthening of, the Endangered Species Act.

Additionally, the rate of species recovery is also dependent on how close to the abyss of extinction species are when they are first offered the protections of the Act. For example, seven species were likely already extinct before they were first listed. Many others have been listed only when their populations have fallen to incredibly low levels. The size of a vertebrate population at the time of listing is often so low that only the establishment of captive breeding populations will avoid extinction. This occurred in the well-known cases of the Mexican wolf, the black-footed ferret, and the California condor whose protection came only after each had dwindled to fewer [than] two dozen individuals.

The majority of the cases filed by WildEarth Guardians pursuant to the citizen-suit provision of the Endangered Species Act have involved efforts to compel the federal agencies responsible for administering the Act to meet the deadlines prescribed by Congress for making listing decisions. This effort to protect all deserving species under the Act sooner rather than later increases their chances for recovery and also serves to shorten the timeline needed to recover a species. Importantly, for this Committee's inquiry into perceived conflicts between the Endangered Species Act and economic activity, adding species to the list before they are at the verge of extinction allows greater flexibility and accommodation of activities that might conflict with recovery through the Act's regulatory mechanisms.

Having an accurate and complete list of endangered species protected by the Act benefits those trying to save species, by allowing them to begin protecting and recovering deserving species sooner. It also benefits those engaged in planning economic activities that may be affected by a species listing by allowing them to modify their plans or activities to accommodate the needs of endangered species before devoting significant resources to those plans. An incomplete or inaccurate list of endangered species benefits no one. Thus, litigation directed at listing species that need the protection of the Endangered Species Act—to make the list complete and accurate—is beneficial to all parties concerned.

In short, the debate should not focus on diagnosis (listing), but on the course of treatment (protection and recovery) we apply to listed species. Diagnosis is simply information upon which future decisions can be made. We understand this when it comes to visiting the doctor's office. Accurate and timely diagnosis of disease is critical. Only once the diagnosis is made do we begin to discuss our treatment options with our doctor, with choices spanning the spectrum from intensive intervention to doing nothing. Our understanding of the Endangered Species Act, the law under which we provide emergency room care to species in need, should be no different. Accordingly, the Act provides that listing decisions must be based solely on the best available science and not account for economic impacts. The perceived conflict between economic activities and protecting endangered species should not influence listing decisions, but may be appropriately debated when we decide how to recover listed species and what level of economic dislocation we will tolerate in those efforts.

However, because this Committee appears concerned that litigation conducted by WildEarth Guardians and others is somehow interfering with species recovery, it is important to note that both Guardians and the Center for Biological Diversity have recently entered into separate, but overlapping, settlement agreements with the U.S. Fish and Wildlife Service. For those concerned that the process of listing species under the Act is overly litigious, these settlement agreements are good news. In its separate settlement, Guardians has agreed not to file litigation enforcing the Act's listing deadlines for the next five years. In return, the U.S. Fish and Wildlife Service has agreed to make final listing decisions for all the species the Agency had previously concluded warranted the protection of the Act, but for which the Service had not made final listing decisions in its 2010 Candidate Notice of Review. Thus, the Service will be making final decisions for the species which it has preliminarily concluded are most deserving of the Act's protections. Neither agreement requires the Service to list any particular species, but only to complete its analysis and make a final decision. Most of the species that will receive final listing decisions under these settlement agreements have been waiting for more than two decades for action. The agreements promise an end to this waiting and will result in a more accurate and complete endangered species list upon which future decisions can be made. Recovery efforts for the species the Service ultimately concludes deserve listing will begin sooner, and with this head start, recovery efforts should also be both more efficient and less disruptive to economic activity than if these species are allowed to continue declining without legal protection while waiting for action.

These settlement agreements would not have come to pass without litigation to enforce the Act's deadlines. In that sense, the litigation that led to the agreements benefitted both the enforcement of the Endangered Species Act and the quicker recovery of species which should in turn reduce the economic impacts of species protection. Any contrary conclusion is unwarranted.

There Are Actions That Could Increase the Rate of Species Recovery

Listing Decisions Should Be Made Promptly and in Keeping with the Endangered Species Act's Deadlines

Finally, in response to this Committee's apparent concern that species are not recovering rapidly or efficiently, there are actions Congress could take to increase the rate of recovery. As discussed above, the difficulty of recovery is proportional to the degree of imperilment a species faces when it is first added to the endangered species list. The Endangered Species Act provides a two to two-and-one-half year timeline for making a decision as to whether or not to add a species to the endangered species list once it has been petitioned for listing. The Act also provides that the responsible agencies may add a species to the list on their own initiative. In practice, Congress has failed to fund the U.S. Fish and Wildlife Service listing program at levels sufficient for it to timely address either the number of citizen petitions it receives or the number of species sliding towards extinction. Nor has the Service requested adequate funding for these tasks. Thus, the Service has been in chronic violation of the listing deadlines that Congress provided in the Act to compel agency action. These delays have caused WildEarth Guardians and others to litigate to enforce Congressional mandates and spur prompter action. Species continue to decline while the agency delays addressing their status and deciding whether or not they deserve the Act's protections, thereby rendering recovery efforts more difficult. Accordingly, if the goal of Congress is to increase the rate and potential success of recovery efforts, the first step is to fund the listing program at levels that will allow the U.S. Fish and Wildlife Service to avoid breaking the law. Identification of the problem (prompt listing action) is the first step to its resolution (quicker recovery).

Funding the Service at a rate sufficient for it to comply with the settlement agreements it recently entered with WildEarth Guardians and the Center for Biological Diversity will not only increase the recovery prospect for the species that receive final listing decisions by forcing action more promptly, but will avoid a return to litigation as the only means available to Guardians, the Center, and others to enforce the Act's deadlines.

Does Excessive Endangered Species Act Litigation Threaten Species Recovery, Job Creation, and Economic Growth? by Easton

89

Critical Habitat Designation Should Be Required for All Listed Species

As a related matter, Congress amended the Endangered Species Act in 1978 to require the Fish and Wildlife Service to designate critical habitat for a species, to the extent determinable and prudent, at the time of listing. As discussed above, listed species with a designated, and thus protected, critical habitat are twice as likely to be recovering as those without designated critical habitat. Accordingly, to increase the rate of recovery, Congress should also fund the Service at levels sufficient to allow the Agency to designate critical habitat for species at the time they are first listed. As an additional benefit, the prompt designation of critical habitat supports better planning by those entities whose economic activities might need to be modified to protect listed species. Additionally, because Congress applied the requirement to designate critical habitat only to species designated after 1978, if Congress desires to increase the rate of species' recovery it should remove the exemption for species listed prior to 1978 and require the designation of critical habitat, to the extent prudent and determinable, for all listed species, including those that have been on the list the longest.

Deadlines for the Preparation of Recovery Plans Should Be Established, Recovery Plans Should Be Made Enforceable, and Recovery Plans Should Be Fully Funded

Lastly, and again if the concern is with increasing the rate of species' recovery, Congress should focus on Section 4(f) of the Endangered Species Act, the provision that requires the preparation of recovery plans for listed species. Unlike the other provisions of Section 4, the recovery planning provision contains no deadlines. Thus, this most important task of planning for species recovery may linger incomplete for many years. The responsible agencies have developed a goal of preparing a recovery plan for each listed species within two and one-half years of listing. However, in practice this timeline is not always followed. For example, the National Marine Fisheries Service failed to prepare recovery plans for the Sperm, Fin, and Sei Whales for more than 30 years until compelled to do so by a lawsuit filed by WildEarth Guardians. Accordingly, if Congress desires recovery to occur more rapidly, it should establish deadlines requiring prompt recovery planning.

Furthermore, recovery plans are generally not enforceable by citizens. Thus, the actions the responsible agencies determine are necessary to recover species are undertaken solely at the pleasure of the agencies. Again, the agencies do not always implement the recovery plans they have prepared or delay their implementation. Accordingly, to compel agencies to carry out the tasks they have determined are necessary to recover listed species, Congress should consider making the development and implementation of recovery plans more enforceable by citizens.

An unenforceable or unimplemented plan that simply gathers dust in an agency's file cabinet is of little utility. Thus, conversely, the problem this Committee perceives with delayed recovery efforts is not caused by too much litigation, but by the inability of citizens to force federal agencies to do what they said they should and would do—through litigation forcing the implementation of recovery plans.

Section 4(f) does require the responsible agencies to prepare timelines and estimate the costs of recovery actions. The success of these plans and the adherence to their timelines for action thus hinge on the amount of funding available. Accordingly, if this Committee desires to increase the rate of species recovery, Congress can drive that effort through funding, and it should take steps to insure both that the agencies request sufficient funding to meet their recovery plans and that Congress provides it.

Conclusion

The Endangered Species Act is this nation's commitment that the tragic and irreparable extinctions of species that occurred prior to the Act's passage will not be repeated. In passing the Act, Congress not only recognized that sharing this world with the vast variety of species on it increases human joy and well-being, but is, in the end, essential to human life. Existence without our fellow companions on this planet would not only be lonely, it would be impossible. The protection of fragile and unique species is not without cost. Frequently, these species have been driven to the edge of the abyss by untempered human expansion and monopolization of resources. Allowing for their survival requires a measure of restraint on our part. However, the perception that saving species from extinction costs jobs is shortsighted. Saving species is not only of substantial economic benefit, it allows for sustainable economic development by preserving resources so that they may be enjoyed and used by future generations. Our children will not forgive us if they are able only to learn of the wolf's howl, the prairie chicken's dance, or the bear's roar in museums. More importantly, our descendents will not survive, or will survive only in a more hostile and unforgiving world, without all the little things, the bugs and weeds, that drive our ecosystems and allow the larger forms of life to thrive. Humans cannot pollinate their crops without the assistance of beetles, bees, butterflies, and bats. And humans will suffer if the mysterious storehouse of adaptations and unique properties found in plants and animals [is] thrown away without understanding. Driving species to extinction is not unlike burning a library. Driving species to extinction before we even begin to understand them is like burning the library without once reading the books.

Fortunately, extinction is not an American value. Since its passage, and despite numerous controversies, the American people have consistently and overwhelming supported the Endangered Species Act. This support cuts across all lines that might otherwise divide us. The Act is working, and will work even better with increased enforcement and renewed effort. Litigation, the focus of this hearing, is nothing more than a means to enforce the Act. More importantly, litigation has shown success in ensuring the Act is implemented as Congress intended. Though litigation is adversarial, such disagreements in a civil society are necessary to promote change, force action, and reach resolution. Congress recognized as much when it provided mechanisms for, and requested citizens to help, the government implement the Act and meet the obligations it placed on itself. It is inappropriate to denigrate successful litigation, brought by citizens, that has forced the government and others to follow the law. To do so, is to attack the law itself. If Congress does not want a law enforced it should not have such a law. WildEarth Guardians does not believe that this nation wants to abandon the Endangered Species Act, and it is proud of its efforts to enforce our bedrock national commitment to never again drive a species to extinction. Rather, Guardians believes this nation is committed to ensuring our rich flora and fauna, and the ecosystems on which they depend, survive and flourish for future generations.

James J. Tutchton is General Counsel for WildEarth Guardians (www.wildearthguardians.org), an organization that works to protect and restore wildlife, wild rivers, and wild places in the American West.

Does Excessive Endangered Species Act Litigation Threaten Species Recovery, Job Creation, and Economic Growth? by Easton

91

EXPLORING THE ISSUE

Does Excessive Endangered Species Act Litigation Threaten Species Recovery, Job Creation, and Economic Growth?

Critical Thinking and Reflection

1. Why do business interests object to lawsuits aimed at enforcing the Endangered Species Act?
2. In what way does litigation help or hurt the functioning of the Endangered Species Act?
3. What are "trade-offs"? In what way may they play a part in protecting endangered species?

Is There Common Ground?

Neither side in this debate says it wants the Endangered Species Act (ESA) repealed. Middleton wants its priorities adjusted to give more weight to human concerns and to make it more difficult to litigate. Tutchton says the ESA is effective as it stands. Tutchton also says that despite the value of litigation as a means of pressuring enforcement of the ESA, his organization, WildEarth Guardians, recognizes the concerns that there is too much litigation and has agreed not to file lawsuits seeking enforcement of listing deadlines for five years.

1. Do you think this litigation moratorium is likely to satisfy those who think there is too much litigation? Why or why not?
2. Do you think this litigation moratorium will make much difference to the effectiveness of the ESA? Why or why not?
3. The litigation moratorium limits only lawsuits dealing with one third of the ESA process

(listing, protection, and recovery). One might expect that the focus of litigation will shift to recovery plans. Why is this unlikely to happen?

Create Central

www.mhhe.com/createcentral

Additional Resources

A. D. Barnosky, et al., "Has the Earth's Sixth Mass Extinction Already Arrived?" *Nature* (March 3, 2011)

A. D. Barnosky, et al., "Approaching a State Shift in Earth's Biosphere," *Nature* (June 7, 2012)

Michael J. Bean, "The Endangered Species Act under Threat," *Bioscience* (February 2006)

Brian Gratwicke, Thomas E. Lovejoy, and David E. Wildt, "Will Amphibians Croak under the Endangered Species Act?" *Bioscience* (February 2012)

Internet References . . .

Environmental Defense Fund

www.edf.org

U.S. Fish and Wildlife Service—Endangered Species

www.fws.gov/endangered/

World Wildlife Fund

worldwildlife.org

Selected, Edited, and with Issue Framing Material by:
Thomas Easton, *Thomas College*

ISSUE

Can "Green" Marketing Claims Be Believed?

YES: **Jessica Tsai**, from "Marketing the New Green," *Customer Relationship Management* (April 2010)

NO: **Richard Dahl**, from "Green Washing: Do You Know What You're Buying?" *Environmental Health Perspectives* (June 2010)

<table>
<tr><td>

Learning Outcomes

After reading this issue, you will be able to:

- Explain what greenwashing is and how it misleads consumers.
- Explain how to make "green" marketing claims without being guilty of greenwashing.
- Describe why corporations have in the past made excessive "green" advertising claims.
- Explain why better definitions of "green" advertising terms are needed.

</td></tr>
</table>

ISSUE SUMMARY

YES: Jessica Tsai argues that even though marketing is all about gaining attention in a very noisy environment, it is possible to improve brand image by being environmentally responsible without being guilty of greenwashing.

NO: Richard Dahl argues that consumers are reluctant to believe corporate claims of environmental responsibility because in the past such claims have been so overblown as to amount to "greenwashing."

With the birth of the environmental movement, it quickly became apparent that people were interested in the environmentally related behavior of corporations. Initially, this interest took the form of criticism, some of which involved public protest, invasions of corporate offices, and even dumping barrels of toxic goop, collected from the pipes delivering a corporation's wastes to a local river or lake, on the carpet of the CEO's office. The point was to shame corporations into cleaning up their act. Sometimes it worked. More often, it took a while—even years—for corporations to realize that it was good public relations to look environmentally responsible. The result was more emphasis on landscaping around factories and office buildings, even to the extent of providing public walking trails, tree-planting projects, and more. As clean air and water laws were passed, corporations made much of the way they were cleaning up air and water pollution, usually without mentioning that they were being forced to do so by law. Product ingredients that had always been derived from tree barks and other natural sources were now touted as "natural." When sustainability became an environmental buzzword, it also became a PR buzzword whose definition was remarkably elastic. Is using coal more sustainable than using oil? Well, the supply of coal will certainly last longer, but the effect on the environment hardly fits the idea of meeting present needs without impairing the ability of future generations to meet theirs. Is "clean coal" green coal? Many think the term an oxymoron, coined mostly to clean up perceptions of an inescapably dirty energy technology.

The term that has come to be used for such behavior is "greenwashing," by analogy with "whitewashing," which means painting over dirt and blemishes in a fence or a reputation with whitewash (an early form of white paint). See Wendy Priesnitz, "Greenwash: When the Green Is Just Veneer," *Natural Life* (May/June 2008). Perhaps unfortunately, greenwashing works. Satyendra Singh, Demetris Vrontis, and Alkis Thrassou, "Green Marketing and Consumer Behavior: The Case of Gasoline Products," *Journal of Transnational Management* (2011), found that consumers preferentially seek out gasoline brands perceived as greener.

In 2010, the Business Reference and Services Section (BRASS) Program of the American Library Association's annual meeting discussed "Clean, Green, and Not So

Mean: Can Business Save the World?" *Reference & User Services Quarterly* (Winter 2010). The theme was corporate social responsibility, of which environmental responsibility is one component, expressed as sustainable marketing, defined as 'meeting the . . . definition of marketing in a way that meets both the organizational goals and customers' needs while preserving, benefiting, and replenishing both society and the environment." How real is the environmental benefit? Jonathan Latham, "Way Beyond Greenwashing," *Dollars & Sense* (March/April 2012), is skeptical, noting that "many of the biggest conservation nonprofits including Conservation International and the Nature Conservancy have already agreed to a series of global bargains with international agribusiness. In exchange for vague promises of habitat protection, sustainability, and social justice, these conservation groups are offering to greenwash industrial commodity agriculture."

Jason Daley, "Green Fallout," *Entrepreneur* (August 2010), notes the essential hypocrisy of many businesses when he describes the way the Deepwater Horizon oil spill in May 2010 tarnished the image of BP (or British Petroleum), the "company that had cultivated the greenest image in the oil industry":

> The BP blowout was the swan song of an old style of green marketing, one in which companies could make green claims and hope that no one would look over their shoulders. In the last five years, a new type of green marketing has taken hold, and it has high standards.
>
> It's no longer enough to say you're green in your advertising. It's not even enough to have one or two flagship green products in your line or to screw in a few compact fluorescents and send out a press release.

(See also Miriam A. Cherry and Judd F. Sneirson, "Beyond Profit: Rethinking Corporate Social Responsibility and Greenwashing After the BP Oil Disaster," *Tulane Law Review*, 2011.) Not that BP was getting away with much before the oil spill, for *European Environment & Packaging Law Weekly* (May 15, 2009), could title an article "Shell and BP in Poll Position on 'Greenwash' Blacklist."

The basic problem has not yet vanished. According to Greenpeace:

> Corporations are falling all over themselves to demonstrate to current and potential customers that they are not only ecologically conscious, but also environmentally correct.
>
> Some businesses are genuinely committed to making the world a better, greener place. But for far too many others, environmentalism is little more than a convenient slogan. Buy our products, they say, and you will end global warming, improve air quality, and save the oceans. At best, such statements stretch the truth; at worst, they help conceal corporate behavior that is environmentally harmful by any standard.

The average citizen is finding it more and more difficult to tell the difference between those companies genuinely dedicated to making a difference and those that are using a green curtain to conceal dark motives. Consumers are constantly bombarded by corporate campaigns touting green goals, programs, and accomplishments. Even when corporations voluntarily strengthen their record on the environment, they often use multi-million dollar advertising campaigns to exaggerate these minor improvements as major achievements.

Sometimes, not even the intentions are genuine. Some companies, when forced by legislation or a court decision to improve their environmental track record, promote the resulting changes as if they had taken the step voluntarily. And at the same time that many corporations are touting their new green image (and their CEOs are giving lectures on corporate ecological ethics), their lobbyists are working night and day in Washington to gut environmental protections.

As Greenpeace notes, not every corporation that makes "green" claims is trying to fool the public. But some are, so that the question for the consumer becomes how to tell the difference. Awareness of the problem has grown, and the Federal Trade Commission (FTC) published the first of its "Guides for the Use of Environmental Marketing Claims" (or "Green Guides") in 1992. In October 2010, the FTC announced proposed revisions to the Green Guides; see Lynn L. Bergeson, "Selling Green: US FTC Releases Proposed Revisions to the 'Green Guides,'" *Environmental Quality Management* (Spring 2011). The proposed revisions address the use of product certifications and seals of approval and claims relating to "renewable energy," "renewable materials," and "carbon offsets." Before proposing the revisions, the FTC studied consumer perceptions of the terms often used in such claims (such as "biodegradable") to help draw the line between valid and misleading claims. Since perception plays an important role in this issue, some—including some in the public relations and advertising industry—have objected that the revisions do not go far enough. The Cone company filed comments saying that it "had hoped to see the Commission take a more definitive stance on general environmental benefit claims, perhaps even prohibiting the use of words such as 'sustainable' or 'earth friendly.'" It also wished for attention to the imagery used in ads, which can sometimes convey very strong messages that might not be defensible if put into words. See "Cone Files Comment to FTC in Response to Proposed Green Guides Revisions" at www.coneinc.com/ftcgreenguides. Jessica E. Fliegelman, "The Next Generation of Greenwash: Diminishing Consumer Confusion Through a National Eco-Labeling Program," *Fordham Urban Law Journal* (October 2010), calls for "a nationwide eco-labeling program [to] reduce consumer confusion." Such a program would be enforced by the FTC and the states. Robert B. White, "Preemption in

Green Marketing: The Case for Uniform Federal Marketing Definitions," *Indiana Law Journal* (Winter 2010), argues that "federal definitions for green-marketing terms that have the force of law and expressly preempt all state definitions should be promulgated, thereby increasing green-marketing regulations' certainty and uniformity." The problem is similar in Britain; Josh Naish, "Lies . . . Damned Lies . . . and Green Lies," *Ecologist* (June 2008), notes that there the number of relevant regulations and definitions lead to consumer confusion. Simplification and even more government regulations are required. Susie Staerk Ekstrand and Lilholt Kristine Nilsson, "Greenwashing?" *European Food & Feed Law Review* (2011), discuss Danish guidelines intended to limit "how environmental and ethical claims can be used in marketing." According to *Ecos* (February 2012), "Tougher laws are needed to combat greenwashing."

In the YES selection, Jessica Tsai argues that even though marketing is all about gaining attention in a very noisy environment, it is possible to improve brand image by being environmentally responsible without being guilty of greenwashing. In the NO selection, Richard Dahl argues that consumers are reluctant to believe corporate claims of environmental responsibility because in the past such claims have been so overblown as to amount to "greenwashing."

YES ↵

Jessica Tsai

Marketing the New Green

Companies should be rightfully proud of their environmental improvements. So why can't they market those achievements without seeming mercenary?

Unless your company happens to be in an industry where the environment is literally part of the business model—solar-cell manufacturer, forestry-supply firm, waste-management consultancy—then it's unlikely to have had an enterprisewide sustainability process woven into its corporate DNA. In fact, most companies have never even used the word *sustainability,* and the others probably reverse-engineered some half-hearted "green" initiative as a mere afterthought. The reality is that, among existing companies, green processes are usually just a byproduct of cost-saving or efficiency-improving projects. According to Mark Smith, executive vice president at customer engagement specialist Portrait Software, the rare businesses that have truly tackled sustainability "got there as a side effect of their primary goal—making money or saving money. In terms of 'green marketing'? No one started there." Glitz and glamour, bigger and better: From a traditional marketer's perspective, Smith says, "it's all about gaining attention in a very noisy marketplace." Whether that involves toxic paint to ensure colors "pop" or nonbiodegradable packaging materials designed to endure a rough-and-tumble delivery, marketers have long embraced any and all environmental catastrophes that help get the message across. The recession, however, made marketing budgets themselves the scene of a catastophe—drastic cuts were the order of the day. In early 2009, 71 percent of respondents in Forrester Research's Global CMO Recession Online Survey reported reduced budgets compared to the year prior, with more than half of those bloodied reporting cuts of

20 percent or more. "The interesting side effect [of this] is that those big, grand [campaigns] have been cut back and reined in," Smith says. The added bonus? It's had a great effect on the environment as well. Still, sustainability has never been a top consideration for marketing, at least not on the departmental or corporate level. It's usually a single person or small group of people who have a personal stake in environmental activism.

That pretty much describes Seema Haji, the senior product marketing manager at Actuate, a provider of business intelligence solutions. Actuate's companywide green initiatives—and the eventual launch of its sustainability management product—can be traced back to the passion of Haji, who calls herself a "pseudo-environmentalist in a corporate world," having started her own blog, *World-Saving Tips for the Lazy (and Busy)*, a few years before joining Actuate. (You can find her efforts at bleedinggrass.blogspot .com.)

Founded in 1993, Actuate now has more than 4,400 customers worldwide, but it wasn't until 2008 that the company began thinking seriously about going green—and helping its customers do the same. "We were all brainstorming," Haji recalls. "My bosses turned to me and said, 'Oh, yeah—she's a hippie. Let her do this, she'll have fun with it.' And I really did."

Word had been trickling up from users that going green was not only emerging as a huge competitive advantage but as a way to cut energy costs and improve brand image. Despite the enthusiasm, though, users didn't know what steps to take.

SIX WAYS TO PROMOTE GREEN MARKETING

- **Integrating green initiatives into every aspect of the organization:** Companies are trying to link the corporate brand to efforts in social responsibility, Edwards says—and environmental stewardship can affect the bottom line as it improves customer relationships. At United Parcel Service (UPS), for example, new mapping systems enabled a "No Left Turn" rule to eliminate costly left turns from drivers' routes. According to *The New York Times,* UPS spokeswoman Heather Robinson reported that the company shortened delivery routes by 28.5 million miles, saving 3 million gallons of gas and reducing carbon emissions by 31,000 metric tons. "They're tying their brand image to efficiency and environmental savings," Edwards says.
- **Using ecolabels and ecologos on products or marketing materials:** Perhaps the most well-known ecolabel is the recycling symbol composed of chasing arrows, created in 1970 by Gary Anderson, who won a graphics and

design competition hosted by the Container Corporation of America. Since then, a significant number of labels have popped up, some of which have contributed to an industry malfeasance known as "greenwashing.". . . Other widely recognized symbols include the USDA Organic, which signifies the use of organic ingredients in food; Forest Stewardship Council (FSC) indicates wood and paper products produced in methods that advocate responsible forestry; and Energy Star identifies home, building and construction, and electronics that are energy efficient.

When adopting these labels, Edwards warns marketers to be careful—while it can help inform consumers, over-saturation of labels in the market has resulted in label blindness. For the most part, consumers today only recognize a handful of labels. Therefore, marketers must identify whether the logo: a) is credible; b) is meaningful and recognizable by the intended audience; and c) fits with the organization's message.

- **Engaging customers in green marketing:** Companies are looking to motivate consumers by encouraging them to participate and engage in the campaign or directly with the product. Marketers that send out direct mail pieces can put links directing marketers to participate in green programs online, or do something as simple as ask customers to recycle the mail after reading. Edwards has seen largely positive feedback from marketers who've attempted to bring customers into the mix; the number of those doing so is growing but still pretty small, Edwards admits. Only about 100 marketers have enlisted in the DMA's "Recycle Please" program—a nationwide public education campaign where DMA members are asked to display a "Recycle Please" logo in catalogues and direct mail pieces.
- **Asking and respecting customer choices and preferences:** Segmentation is a practice that goes back to Marketing 101. Companies that are leveraging customer data and respecting their preferences will inevitably have fewer unnecessary mailings. . . . In October 2007, the DMA launched its Commitment to Consumer Choice policy, which among other stipulations, requires all DMA members to provide existing and prospective customers and donors with notice of an opportunity to modify or opt out of commercial communications. By giving consumers this choice, companies are not only acting environmentally responsible, but also reinforcing their corporate responsibilities.
- **Adopting a lifecycle approach:** Companies are selecting green materials and products for their marketing materials and adopting a lifecycle approach that looks at the whole of the campaign, thereby foreseeing areas of potential waste. Edwards sees more marketers adopting recycled and FSC-approved papers and printing, vegetable and soybased inks, smaller formats and trim sizes, and a reduction in paper use overall. Aromatherapy and skincare treatment provider Decleor now only uses Programme for the Endorsement of Forest Certification and FSC-certified paper, despite the fact that it's 3 percent to 7 percent more costly. This year, the company stopped printing its logo on gold foil and changed it to a deep, eggplant color, in order to ensure that its paper products are 100 percent recyclable. Moreover, the company only maintains relationships with FSC-certified printers and has actually stopped doing business with a printer that wasn't—until that printer came back six months later newly certified.
- **Shifting to the online space:** Digital marketing was projected to reach $25.6 billion in 2009, and reach $55 billion, 21 percent of all marketing spend, by 2014, according to Forrester Research's United States Interactive Marketing Spend report. Channels included in this report were mobile marketing, social media, email marketing, display advertising, and search marketing. More and more companies are requiring that employees remind email recipients to think about the environment before printing.

After discussions with customers and partners, Actuate developed a set of more than 100 metrics intended to help people understand what they should be tracking: carbon-credit usage, average water consumption by facility, average electrical consumption per employee, to name just a few. At the same time, Actuate wanted users to have the flexibility to define custom metrics according to the needs of their respective businesses and industries. The end result—Actuate for Sustainability Management—deploys interactive dashboards, sustainability scorecards, and strategy maps to help companies measure overall employee satisfaction, environmental impact, access to training and education, and community engagement.

At Portrait, going green wasn't a corporate mandate, but support spread like wildfire. "A year or so ago, we had a huge green initiative where we plastered stickers all over the place saying we were a green company," Smith says.

The company installed recycling bins around the office; its United Kingdom branch composted food and waste; energy use for computers and lights was more closely monitored and regulated; new window shades were installed to help reduce the need for air conditioning.

"It all started as a grassroots thing," Smith says, "but then our CFO came around and said it was fantastic. Our electricity and power bills were lower—I don't think he was expecting it at all." Cost-cutting certainly wasn't the main objective, but the savings have earned the attention—and support—of seniorlevel executives. Even initiatives that have cost more—it's cheaper and easier to discard than to recycle—are still being encouraged as the company strives to be even more ecoconscious.

Still, other than looking at comparisons between one electric bill and the next, Portrait couldn't speak to a particular technology that measures a business's environmental impact—let alone any attendant cost savings. At press time, even Actuate was only just a few months into implementing its own sustainability application. "The stuff they ask us to measure is all about dollars, response rates, mailing volumes," Smith says. "Not many [companies] have particularly brought it to that next level."

For some, however, becoming a sustainable enterprise is simply motivated by the desire to do the right thing. Portrait hasn't yet utilized its green initiatives to enhance its brand image, although Smith admits it's something he'll definitely be considering in the future.

Meta Brophy, director of publishing operations for Consumers Union (CU), the parent of *Consumer Reports*, says that her company doesn't herald its green initiatives in marketing solicitations to its consumers, focusing instead on its industry peers. "CU has worked to lead by example . . . in hopes that other direct marketers will follow suit and help to build a more sustainable direct marketing community," Brophy says.

CU abides by the Direct Marketing Association's Green 15 Standards & Environmental Action Program, which has led to changes in the six principal areas of list hygiene and data management, design, paper procurement and usage, printing, recycling, and pollution reduction. For five years, CU has been a green advocate, sharing its progress and best practices at industry events such as National Postal Forums, New England Mail Expo, New York Nonprofit Conference, and the World Environment Center Roundtable. "We incorporate such initiatives because it's the right thing to do," she says, "and generally speaking we save money."

Gregory Unruh, author of *Earth, Inc.*, has been involved in bridging the gap between the business sector and the sustainability movement. As a professor of global business and director of the Lincoln Center for Ethics in Global Management at Thunderbird School of Global Management, Unruh has found that the only way to get companies to stick to sustainability is by proving that it's profitable. "The ultimate goal here is to embed sustainability and make it a standard business practice," Unruh says. "You start with one [practice], set up in a way that—after you complete that one—you create cost-reduction and profitability opportunities that then provide the momentum you need," he says. "If it makes a profit, companies will do it.". . .

Building-products provider BlueLinx has always had a focus on sustainability—by necessity, considering its business is heavily dependent on wood. BlueLinx long ago made the connection between practicing sustainable forestry and maintaining a sustainable—and profitable—business. Nevertheless, Shiloh Kelly, the firm's communications and national sustainability lead, contends that being green "wasn't something to try and sell an extra piece of wood over." With the launch of its virtual trade show, however, BlueLinx was able to promote a new sustainability program and brand of ecoproducts. . . .

"The whole aspect of sustainability is to get every last drop from everything you can," Kelly says. "Not only from

a resource standpoint, but from a financial standpoint and a viral standpoint." To that end, Kelly admits she was especially blown away by the viral impact of the BlueLinx virtual event, and in particular its effectiveness around the company brand.

"Prior to . . . [our] rollout, our customers didn't know the depth of our resources or expertise," Kelly explains. "The calls concerning these matters were few and far between." After the virtual event, however, top vendors, customers, and media publications were pouring in, soliciting BlueLinx as a primary source of information around sustainable-building products and industry programs. . . .

Decleor, a French provider of aromatherapy and skincare treatments, has used only essential oils and plant-based ingredients since its founding 35 years ago—factors that Cindy Willette, the company's director of marketing, admits were taken for granted during most of that time. The company has had certain advantages simply by virtue of its products. For instance, essential oils corrode plastic, so glass bottles were the most natural packaging. However, it wasn't until the company saw its competitors taking away market share that Decleor realized the importance of strengthening its green marketing.

Similar to BlueLinx, Decleor realized that it, too, needed to be responsible for replenishing the resources its products consumed. In 2008, the company launched a three-year, responsible-development project in Madagascar, whereby Decleor would partner with Association Madagascar to contribute to the region's reforestation, install solar electrification in schools, and provide medical equipment to hospitals. (As of press time, 40,000 trees had been planted and a school with 600 students had been equipped with solar electrification.) As part of the promotion, the company sold 100-percent-natural shopping bags at spas and retail stores and promised to plant five trees for every bag purchased. These initiatives have captured the attention of both the consumer and trade audiences, Willette says, citing a 20 percent increase in inquiries coming through the company Web site last year.

And yet, despite the company's ecofriendly roots, some traditional practices remain difficult to weed out. "There are ways of doing things that get ingrained in corporate culture," Willette says. "We need to have this material, this piece, to support this launch in this way. As we started looking at the brand from a bigger picture . . . [we said], 'Let's take a harder look at what we're doing and decide if we need to do that anymore. Just because it's always been done, doesn't mean it needs to be done.'"

In scrutinizing its marketing and overall business processes, Decleor has not only changed its printing and paper policies . . ., but overall energy consumption fell by 10 percent in the last two years; industrial waste declined by 6.6 percent; paper usage dropped by 14 percent; gas usage was down 23 percent; water usage was down 21 percent; all while production increased 4.7 percent during the same time frame. Seeing the benefits of going green, Decleor is unlikely to revert back to the traditional ways.

THE 7 SINS OF GREENWASHING

In partnership with EcoLogo, a program that provides a certification mark approved by the Global Ecolabelling Network, environmental marketing firm TerraChoice released a report last April describing the seven sins of "greenwashing." TerraChoice defines this concept as "the act of misleading consumers regarding the environmental practices of a company or the environmental benefits of a product or service."

In order to avoid committing greenwashing—which can lead to consumer mistrust, delay true innovation, stir market skepticism, and, thus, damage credibility—the firm warns companies against the following "sins":

- **The Hidden Trade-Off:** Fully understand your product's environmental impact across its entire lifecycle and continue to improve it. Don't overemphasize one facet to hide the drawbacks of another.
- **Lack of Proof:** Back up your claims with scientific evidence and make that information readily accessible to the public.
- **Vagueness:** Clearly articulate how your product or service is environmentally beneficial in a language your customer can understand.
- **Worship of False Labels:** Select eco-labels that are from accredited programs, preferably ones that address a product's entire lifecycle.
- **Irrelevance:** Don't claim to be something you're not or to be something that all or most of your competition shares.
- **Preference for the Lesser of Two Evils:** Connect consumers with the right product rather than pitching a product that may still be harmful and unnecessary but may be more "green" than the next.
- **Dishonesty:** Don't lie—ever. Don't even exaggerate.

Sources: TerraChoice, EcoLogo

"Our mantra here is 'less is more,'" Willette says, "so rather than create more, we try to create better."

As a certain amphibian once said, it's not easy being green. A culture that's been trained to be high in both consumption and waste production took centuries to build—and will be difficult to tear down.

"Start small," Haji advises. "Pick the most important things that will make the most impact and start with that. Start in one department [and] spread this through the rest of the organization."

Reducing paper consumption may be the first and easiest step to take. Encourage your marketing and sales teams to send materials digitally through email or USB flash drives. (If you *really* want to go green, there are bamboo drives, wood drives, and biodegradable, leadfree drives available.)

Sure, you could say that even walking around isn't 100 percent green—the energy to fuel that movement came from food, which was prepared at a restaurant after being processed at (and shipped from) a factory, and so on. But Kelly says that perspective picks at lesser evils and distracts from the bigger issue, one that companies can finally tackle—and marketers can sincerely boast about.

"With the way technology and innovation has changed, there are huge differences between the evils now," Kelly says. "Anything you do out there has a sustainable effect. It's just whether you're going to ignore it, or do something about it."

JESSICA TSAI is an Assistant Editor at *CRM* magazine.

Richard Dahl   ➔ **NO**

Green Washing: Do You Know What You're Buying?

In a United States where climate change legislation, concerns about foreign oil dependence, and mandatory curbside recycling are becoming the "new normal," companies across a variety of sectors are seeing the benefit of promoting their "greenness" in advertisements. Many lay vague and dubious claims to environmental stewardship. Others are more specific but still raise questions about what their claims really mean. The term for ads and labels that promise more environmental benefit than they deliver is "greenwashing." Today, some critics are asking whether the impact of greenwashing can go beyond a breach of marketing ethics—can greenwashing actually harm health?

Greenwash: Growing (Almost) Unchecked

Greenwashing is not a recent phenomenon; since the mid-1980s the term has gained broad recognition and acceptance to describe the practice of making unwarranted or overblown claims of sustainability or environmental friendliness in an attempt to gain market share.

Although greenwashing has been around for many years, its use has escalated sharply in recent years as companies have strived to meet escalating consumer demand for greener products and services, according to advertising consultancy TerraChoice Environmental Marketing. Last year TerraChoice issued its second report on the subject, identifying 2,219 products making green claims—an increase of 79% over the company's first report two years earlier. TerraChoice also concluded that 98% of those products were guilty of greenwashing. Furthermore, according to TerraChoice vice president Scot Case, the problem is escalating.

TerraChoice also measured green advertising in major magazines and found that between 2006 and 2009, the number mushroomed from about 3.5% of all ads to just over 10%; today, Case says, the number is probably higher still. Case says researchers are currently working on another update that will be released later this year, and he predicts the number of products making dubious green claims will double.

Compounding the problem is the fact that environmental advertising—in the United States, at least—is not tightly regulated. The Federal Trade Commission (FTC), the agency responsible for protecting the public from unsubstantiated or unscrupulous advertising, does have a set of environmental marketing guidelines known as the Green Guides. Published under Title 16 of the *Code of Federal Regulations*, the Green Guides were created in 1992 and most recently updated in 1998. According to Laura DeMartino, assistant director of the FTC Division of Enforcement, the proliferation of green claims in the marketplace includes claims that are not currently addressed in the Green Guides, and updated guidance currently is being developed.

The FTC originally planned to begin a review of the Green Guides in 2009, but the commission moved the schedule up, according to DeMartino, in response to a changing landscape in environmental marketing. "The reason, at least anecdotally, was an increase in environmental marketing claims in many different sectors of the economy and newer claims that were not common, and therefore not addressed, in the existing Guides," she says. "These are things like carbon offsets or carbon-neutrality claims, terms like 'sustainable' or 'made with renewable materials.'"

The FTC held a series of workshops in 2008, holding separate events for each of three areas: carbon-offset and renewable-energy claims, green packaging, and buildings and textiles. In association with each workshop, the FTC asked for comments to help shed light on consumer perception of green advertising, but DeMartino says the commission received very few. The FTC responded to this gap by commissioning a research firm, Harris Interactive, to provide that information. DeMartino says that research has been completed, and a report on it will accompany the revision announcement, which is expected soon.

How Updated Guidance Might Look

In aspiring to revise its environmental marketing guidelines, the FTC is following a trend that has been evident in other nations. In 2008, the Canadian Competition Bureau (a government agency similar in function to the FTC) updated its environmental marketing guidelines to reduce green misinformation, and the Australian Competition and Consumer Commission took a similar step. In March 2010, the U.K. Committee of Advertising Practice and Broadcast Committee of Advertising Practice announced an update to their codes of practice designed to curtail greenwashing.

Dahl, Richard. From *Environmental Health Perspectives*, June 2010, pp. A247–A252. Published in 2010 by National Institute of Environmental Health Sciences. http://ehp.niehs.nih.gov/

All three updates are "remarkably similar," Case says, but he suggests the Canadian revisions might provide the best "sneak peak" at what the FTC might do because the two agencies have a long history of working together on cross-border consumer matters. Attorney Randi W. Singer, a litigation partner at New York's Weil Gotschal who has defended companies accused of false advertising, agrees the moves made in Canada but also in Australia and the United Kingdom may provide a good look at what is to come in the United States. Those changes, coupled with her own analysis of the FTC workshop discussions, provide the basis for her to make several predictions about what the new U.S. regulatory scheme might look like.

Singer predicts the revisions will probably contain new definitional language for terms such as "carbon neutral" and "sustainable." She also expects the FTC will address the issue of third-party certifications—that is, the plethora of green labels consumers see on their products. She says the workshop discussions included "a lot of talk about the need for standardization of certifications, a need to have a process for certifications so it's not just people registering themselves, a need to standardize the iconography and the testing."

According to Case, there are now more than 500 green labels in the United States, and some are "significantly more meaningful" than others. "I testified before Congress last summer and I pointed out a certain lawyer in Florida who set up a website and is 'certifying' products. He doesn't need to see the product, he doesn't need test results. He just needs to see your credit card number," Case says.

Meanwhile, the FTC has begun to step up enforcement regarding claims that it considers clear violations of the existing Green Guides, last year charging three companies with false and unsupportable claims that a variety of paper plates, wipes, and towels were biodegradable. "When consumers see a 'biodegradable' claim they think that product will degrade completely in a reasonably short period of time after it has been customarily disposed," DeMartino says. But for about 91% of the waste in the United States, the FTC wrote in its 2009 decisions, customary disposal means disposal in a landfill, where conditions prevent even a theoretically biodegradable item from degrading quickly.

In another instance, the FTC charged four sellers of clothing and other textiles with deceptively advertising and labeling various textile items as biodegradable bamboo that had been grown in a more sustainable fashion than conventional cotton, when, in fact, the items were rayon, a heavily processed fiber. In January 2010, the FTC sent letters to 78 additional sellers of clothing and textiles warning them they may be breaking the law by advertising and labeling textile products as bamboo.

The Health Impact of Greenwash

One major result of greenwashing, say Case and others, is public confusion. But can greenwashing also pose a threat to the environment and even to public health? Critics say greenwashing is indeed harmful, and they cite examples.

THE SEVEN SINS OF GREENWASHING

In the course of assessing thousands of products in the United States and Canada, TerraChoice Environmental Marketing categorized marketing claims into the following "seven sins of greenwashing":

1. Sin of the hidden trade-off: committed by suggesting a product is "green" based on an unreasonably narrow set of attributes without attention to other important environmental issues (e.g., paper produced from a sustainably harvested forest may still yield significant energy and pollution costs).
2. Sin of no proof: committed by an environmental claim that cannot be substantiated by easily accessible supporting information of by a reliable third-party certification (e.g., paper products that claim various percentages of postconsumer recycled content without providing any evidence).
3. Sin of vagueness: committed by every claim that is so poorly defined or broad that its real meaning is likely to be misunderstood by the consumer (e.g., "all-natural").
4. Sin of irrelevance: committed by making an environmental claim that may be truthful but is unimportant or unhelpful for consumers seeking environmentally preferable products (e.g., "CFC-free" is meaningless given that chlorofluorocarbons are already banned by law).
5. Sin of lesser of two evils: committed by claims that may be true within the product category, but that risk distracting the consumer from the greater health or environmental impacts of the category as a whole (e.g., organic cigarettes).
6. Sin of fibbing: committed by making environmental claims that are simply false (e.g., products falsely claiming to be Energy Star certified).
7. Sin of false labels: committed by exploiting consumers' demand for third-party certification with fake labels or claims of third-party endorsement (e.g., certification-like images with green jargon such as "eco-preferred").

Adapted from: The Seven Sins of Greenwashing: Environmental Claims in Consumer Markets.

In 2008, the Malaysia Palm Oil Council produced a TV commercial touting itself in very general terms as eco-friendly; a voiceover stated "Malaysia Palm Oil. Its trees give life and help our planet breathe, and give home to hundreds of species of flora and fauna. Malaysia Palm Oil. A gift from narure, a gift for life." But according to Friends of the Earth and other critics of the ad, palm oil plantations are linked to rainforest species extinction, habitat loss, pollution from burning to clear the land, destruction of flood buffer zones along rivers, and other adverse effects. The U.K. Advertising Standards Authority agreed, declaring the ad in violation of its advertising standards; contrary to the message of the ad, the authority ruled, "there was not a consensus that there was a net benefit to the environment from Malaysia's palm oil plantations."

In 2008, the authority rebuked Dutch energy giant Shell for misleading the public about the environmental effects of its oil sands development project in Canada in the course of advertising its efforts to "secure a profitable and sustainable future." While acknowledging the term "sustainable" is "used and understood in a variety of ways by governmental and non-governmental organisations, researchers, public and corporate bodies and members of the public," the authority also noted that Shell provided no evidence backing up the "sustainability" of the oil sands project, which has been criticized widely for its environmental impact.

Case contends that makers of indoor cleaning products are among the worst greenwash offenders. "People are attempting to buy cleaning chemicals that have reduced environmental and health impacts, but [manufacturers] are using greenwashing to either confuse or mislead them," he says. "People aren't really well-equipped to navigate the eco-babble, and so they end up buying products that don't have the environmental or human-health performances that they expect."

TerraChoice's 2009 report concluded that of 397 cleaners and paper cleaning products assessed, only 3 made no unsubstantiated or unverifiable green claims. The report noted that cleaners, along with cosmetics and children's products, are particularly prone to greenwashing—a worrisome state, given that these items are "among the most common of products in most households."

While companies see consumers' growing demands for green products as an opportunity to increase sales by making perhaps dubious environmental claims, they may also be doing so in an attempt to avoid regulation, says Bruno. In addition to the FTC's promises to tightening up its rules on environmental advertising, broader governmental pressures increasingly place greater burdens on producers to ensure their products are environmentally sound.

"A single ad or ad campaign may be an attempt to sway a customer. But the preponderance of green image ads, many of which are not even attempting to sell a product, combined with lobbying efforts to avoid regulation, add up to a political project that I call 'deep greenwash,'"

Bruno says. "Deep greenwash is the campaign to assuage the concerns of the public, deflect blame away from polluting corporations, and promote voluntary measures over bona fide regulation."

However, several corporate and marketing professionals warn that growing consumer cynicism about these kinds of general campaigns make them risky ventures for companies who engage in them. Keith Miller, manager of environmental initiatives and sustainability at 3M, last year addressed a seminar of The Conference Board, a business-management organization, about what his company does to avoid greenwashing allegations. Summarizing his presentation for the business blog CSR Perspective, Miller said that, based on 3M's experiences, he encouraged companies to avoid making "broad environmental claims" and that any claims made should be specific to products and backed up by "compelling" data.

Ogilvy & Mather advertising agency recently released a handbook designed to guide managers in how to avoid greenwashing charges and called upon them to adopt a policy of "radical transparency" in green advertising campaigns. Business for Social Responsibility, a consulting and research organization, has also published a handbook, *Understanding and Preventing Greenwash: A Business Guide*, which also emphasized the need for transparency as well as for bolstering any environmental claims with independent verification.

Reining in Greenwash

In the absence of a strong regulatory scheme, consumer and environmental groups have stepped into the vacuum to keep an eye on corporate use of greenwashing. Greenpeace was one of the first groups to do so, creating a separate anti-greenwash group, stopgreenwash.org, which monitors alleged greenwash ads and provides other information on identifying and combating greenwash. The University of Oregon School of Journalism and Communication and EnviroMedia Social Marketing operate greenwashingindex.com, where people may post suspected greenwash print or electronic ads and rank them on a scale of 1 to 5 (1 is "authentic," 5 is "bogus").

Claudette Juska, a research specialist at Greenpeace, also points to numerous antigreenwash blogs that have emerged. The result, she says, is that "there's been a lot of analysis of greenwashing, and the public has caught on to it. I think in general people have become skeptical of any environmental claims. They don't know what's valid and what isn't, so they disregard most of them."

Thomas P. Lyon, a business professor at the University of Michigan who has written and spoken extensively about greenwashing, agrees. He says companies are aware they may be criticized or mocked for making even valid claims, so they're starting to grow skittish about making green claims of any kind. "That's why companies, I think, want to see the FTC act—to give them some certainty," he says.

David Mallen, associate director of the National Advertising Division of the Council of Better Business Bureaus, the advertising industry's self-regulatory body, says companies are growing increasingly aware of the dangers of greenwashing. Although some of the matters his office handles are initiated by consumers, the large majority are prompted by companies disputing competitors' claims.

"We're definitely seeing a rise in challenges about the truth and accuracy of green marketing and environmental marketing," he says. "It's certainly taking up a greater percentage of the kinds of advertising cases that we look at. Because green advertising is so ubiquitous now, there's so much greater potential for confusion, misunderstanding, and uncertainty about what messages mean and how to substantiate them."

Typically, he says, a company will be attacked for making a broad or general claim about a product being environmentally friendly "based only on a single attribute, which might not even be a meaningful one." But he says many other cases focus on a competitor's use of a word such as "biodegradable" or "renewable." He adds, "We're also seeing these aggressive, competitive green advertisements where a company will say 'Not only are we green, not only are we making significant efforts toward sustainability, but our competitors aren't.'"

Lyon says he's found the companies that are most likely to engage in greenwashing are the dirtiest ones, because dirty companies know they have a bad reputation, so little is lost in making a green claim if the opportunity arises. At the same time, he and coauthor John W. Maxwell wrote in 2006, "[P]ublic outrage over corporate greenwash is more likely to induce a firm to become more open and transparent if the firm operates in an industry that is likely to have socially or environmentally damaging impacts, and if the firm is relatively well informed about its environmental social impacts."

"It's somewhat counterintuitive, but the clean guy is likely to shut up altogether," Lyon says. "The rationale is: if you're clean and people already think you're a green company, you don't need to bother touting it so much—and if touting it puts you at risk of being attacked, just shut up and let people think you're clean."

Making Green Claims Work

But when a clean company pulls in its horns over the risks of backlash from a cynical public, Lyon believes an opportunity has been lost. He suggests clean companies can be effective green marketers if they take certain steps. First, he says, they might incorporate a full-blown environmental management system (EMS), which would detail its full environmental program in a comprehensive manner. "When a company has an EMS in place, you have a greater expectation that they actually do know what their environmental results are," Lyon explains. EMSs themselves are supposed to meet an international standard called ISO 14001 developed by the nongovernmental International Organization for Standardization in Geneva, which sets out a variety of voluntary environmental standards.

Another step is to take part in the Global Reporting Initiative (GRI), an international organization that has pioneered the world's most widely used corporate sustainability reporting framework. The GRI was launched in 1997 by a nonprofit U.S. group called Ceres—a network of investors, environmental organizations and other public interest groups—in partnership with the United Nations Environment Programme. Lyon says the GRI can provide good green credibility at the company level.

The FTC's attention, however, is directed at products—not companies. And Lyon is one of several experts questioning how effective the looming changes to the Green Guides will be in modifying greenwashing. "Honestly, I don't think the FTC Green Guides are going to block much activity," he says. "All the FTC can do is force companies not to provide materially false information. They could potentially go into the domain of what's misleading as well, but that's very tricky. But they could . . . require companies to give you a more complete story."

To Lyon, the ideal system for regulating green marketing claims would entail comprehensive labeling and certification requirements. "You could picture a system that would be a little like the nutrition labeling that we get for food," he explains. "But whether or not that would be helpful is really unclear to me. From what I understand, there's not a lot of evidence that those nutrition labels have changed America's eating habits."

Among hundreds of green labels available today, a few are broadly recognized as highly reliable. One of them is Green Seal, which awards its seal to companies that meet standards that examine a product's environmental impact along every step of the production process, including its supply of raw materials. "It's a differentiator," says Linda Chipperfield, vice president of marketing and outreach at Green Seal. "If you're really walking the walk, you should be able to tell your customers about it."

Other labels are attained via self-certification—that is, if a company wants the label, they can buy it—and aren't so reliable. The Government Accountability Office (GAO) recently proved that in an investigation of Energy Star, a joint program of the U.S. Environmental Protection Agency (EPA) and Department of Energy.

Energy Star provides labels to companies who submit data about products and seek the stamp of approval to place on their packages. "Currently, in a majority of categories [Energy Star] is a self-certification by the manufacturer, which leaves it vulnerable to fraud and abuse by unscrupulous companies," says Jonathan Meyer, an assistant director in the GAO's Dallas office. Indeed, over a nine-month period, GAO investigators gained Energy Star labels for 15 bogus products, including a gas-powered alarm clock the size of a portable generator. In addition, two of the bogus firms that GAO created as "manufacturers" of the products received phone calls from real companies that wanted to purchase products because the fake companies were listed as Energy Star partners.

The EPA and DOE subsequently issued a joint statement pledging to strengthen the program. The GAO report has also prompted responses from consumers and industry alike that a strong and reliable federal certification program is needed. In a story on the investigation *The New York Times* quoted the director of customer energy efficiency at Southern California Edison as saying industries affected by Energy Star hope the report will be "a wake-up call to whip [the program] into shape."

Case believes an improved regulatory scheme does require some kind of certification and labeling. "I think there is room for some kind of unifying green label," he says. "But I'm not sure if the government wants to get into the business of putting 'approved' stickers on good products." He proposes that the function of providing environmental labels be handled by a new office of the EPA. Under this plan, the EPA would combine several existing environmental labels (such as Energy Star and Green Seal) under a single brand to make it easier for consumers to identify more environmentally preferable goods and services. He points to the U.S. Department of Agriculture's affirming label on organic foods as a model.

Toward a Unified Approach

The growing demands of society for greener products and corporate America's desires to meet it and make a profit make for "a fascinating interaction with cultural change," says Lyon. "The norms have really started to shift. I think that's our hope for information and labeling—that it will create a new floor that keeps rising. I don't think we're anywhere close to that yet, but I think it's starting to happen."

Case says he is "somewhat hopeful" that all involved are moving toward a unified approach to solving the challenges posed by greenwashing. "The huge danger of greenwashing is if consumers get so skeptical that they don't believe any green claims," he says. "Then we've lost an incredibly powerful tool for generating environmental improvements. So we don't want consumers to get too skeptical."

Richard Dahl is a Boston freelance writer who writes periodically for MIT.

EXPLORING THE ISSUE

Can "Green" Marketing Claims Be Believed?

Critical Thinking and Reflection

1. Should corporations be allowed to claim they or their products are kind to the environment even when they are not?
2. Why are cleaning supplies particularly prone to greenwashing?
3. How can a consumer reliably choose environmentally benign products?

Is There Common Ground?

No one seems willing to defend the practice of greenwashing. The debate is between those who say green marketing cannot be trusted and those who say, "It can too! At least if it's done right." Yet it seems clear that greenwashing happens and one can easily suspect that there must be internal corporate documents that say something like "Go ahead and lie."

1. How might such internal documents be revealed?
2. Are there similar cases of corporate mendacity on record? See Rebecca Leung, "Battling Big Tobacco," *60 Minutes* (February 11, 2009) (www.cbsnews.com/stories/2005/01/13/60ll/main666867.shtml). For a list of the top ten whistle-blowers, see http://www.toptenz.net/top-10-whistle-blowers.php.

Create Central

www.mhhe.com/createcentral

Additional Resources

Miriam A. Cherry and Judd F. Sneirson, "Beyond Profit: Rethinking Corporate Social Responsibility and Greenwashing After the BP Oil Disaster," *Tulane Law Review* (2011)

Wendy Priesnitz, "Greenwash: When the Green Is Just Veneer," Natural Life (May/June 2008)

Satyendra Singh, Demetris Vrontis, and Alkis Thrassou, "Green Marketing and Consumer Behavior: The Case of Gasoline Products," *Journal of Transnational Management* (2011)

Internet References . . .

Federal Trade Commission

www.ftc.gov/bcp/edu/pubs/consumer/general/gen02.shtm

Greenpeace—Greenwashing

www.stopgreenwash.org/

SourceWatch

www.sourcewatch.org/index.php/SourceWatch

Unit 3

Energy Issues

*H*umans cannot live and society cannot exist without producing environmental impacts. The reason is very simple: Humans cannot live and society cannot exist without using resources (e.g., soil, water, ore, wood, space, plants, animals, oil, sunlight), and these resources come from the environment. Many of these resources (e.g., wood, oil, coal, water, wind, sunlight, uranium) have to do with energy. The environmental impacts come from what must be done to obtain these resources and what must be done to dispose of the wastes generated in the process of obtaining and using them. The issues that arise are whether and how we should obtain these resources, whether and how we should deal with the wastes, and whether alternative answers to these questions may be preferable to the answers that experts think they already have.

In 2007, the Intergovernment Panel on Climate Change released its fourth assessment report, summarizing the scientific consensus as the climate is warming, human activities are responsible for it, and the impact on human well-being and ecosystems will be severe. (The fifth report will be released in stages through late 2013 and 2014.) This brought energy issues to the fore with unprecedented urgency. The issues presented here are by no means the only issues related to energy, but they will serve to demonstrate the vigor and variety of current energy debates.

Selected, Edited, and with Issue Framing Material by:
Thomas Easton, *Thomas College*

ISSUE

Do We Need Research Guidelines for Geoengineering?

YES: M. Granger Morgan, Robert R. Nordhaus, and Paul Gottlieb, from "Needed: Research Guidelines for Solar Radiation Management," *Issues in Science and Technology* (Spring 2013)

NO: Jane C. S. Long and Dane Scott, from "Vested Interests and Geoengineering Research," *Issues in Science and Technology* (Spring 2013)

Learning Outcomes

After reading this issue, you will be able to:

- Describe some of the ways climate change may harm people around the world.
- Describe some of the methods that have been proposed for stopping or limiting climate change.
- Describe what is meant by "solar radiation management" or "geoengineering" and why it may be a useful approach to combating climate change.
- Describe how various vested interests prevent society from effectively addressing the risks of climate change.

ISSUE SUMMARY

YES: M. Granger Morgan, Robert R. Nordhaus, and Paul Gottlieb argue that before we can embark on geoengineering a great deal of research will be needed. First, however, we need a plan to guide research by developing standards and ensuring open access to research results.

NO: Jane C. S. Long and Dane Scott argue that though we need to do much research into geoengineering, not all issues are technical. Vested interests (whose fortunes may be threatened by change, who may fear consequences, who may be driven by the craving for fame, or whose thinking may be dominated by ideology rather than facts), mismanagement, and human weakness must be addressed before engaging in geoengineering.

Exactly what will global warming do to the world and its people? Projections for the future have grown steadily worse. Effects include rising sea level, more extreme weather events (Richard A. Kerr, "Humans Are Driving Extreme Weather; Time to Prepare," *Science*, November 25, 2011), reduced global harvests (Constance Holden, "Higher Temperatures Seen Reducing Global Harvests," *Science*, January 9, 2009), and threats to the economies and security of nations (Michael T. Klare, "Global Warming Battlefields: How Climate Change Threatens Security," *Current History*, November 2007; and Scott G. Bergerson, "Arctic Meltdown: The Economic and Security Implications of Global Warming," *Foreign Affairs*, March/April 2008). As rainfall patterns change and the seas rise, millions of people will flee their homelands; see Alex de Sherbinin, Koko Warner, and Charles Erhart, "Casualties of Climate Change," *Scientific American* (January 2011). Perhaps worst of all, even

if we somehow stopped emitting greenhouse gases today, the effects would continue for 1,000 years or more, during which sea level rise may exceed "several meters"; see Susan Solomon et al., "Irreversible Climate Change Due to Carbon Dioxide Emissions," *Proceedings of the National Academy of Sciences* (February 10, 2009). And more and more climate scientists are now saying that all projections to date have been far too conservative; see John Carey, "Global Warming: Faster than Expected?" *Scientific American* (November 2012).

It seems clear that something must be done, but what? How urgently? And with what aim? Should we be trying to reduce or prevent human suffering, or to avoid political conflicts, or to protect the global economy—meaning standards of living, jobs, and businesses? The humanitarian and economic approaches are obviously connected, for protecting jobs certainly has much to do with easing or preventing suffering. However, these approaches can also conflict. In October 2009, the Government Accountability Office (GAO) released "Climate Change Adaptation: Strategic Federal Planning Could Help Government Officials Make More Informed Decisions" (GAO-10-113; http://www.gao.gov/products/GAO-10-113), which noted the need for multi-agency coordination and strategic (long-term) planning, both of which are often resisted by bureaucrats and politicians. Robert Engelman, *Population, Climate Change, and Women's Lives* (Worldwatch Institute, 2010), notes that addressing population size and growth would help but "Despite its key contribution to climate change, population plays little role in current discussions on how to address this serious challenge."

The U.S. Climate Change Science Program's "Scientific Assessment of the Effects of Global Change on the United States, a Report of the Committee on Environment and Natural Resources, National Science and Technology Council" (May 29, 2008; available at http://www.whitehouse.gov/files/documents/ostp/NSTC%20Reports/Scientific%20Assessment%20FULL%20Report.pdf) describes the current and potential impacts of climate change. In sum, it says that the evidence is clear and getting clearer that global warming is "very likely" due to greenhouse gases largely released by human activity and there will be consequent changes in precipitation, storms, droughts, sea level, food production, fisheries, and more. Dealing with these effects may require changes in many areas, particularly relating to energy use.

President Barack Obama has indicated that his administration will take global warming more seriously. In June 2009, the U.S. House of Representatives passed an Energy and Climate bill that promised to cap carbon emissions and stimulate use of renewable energy. The Senate version of the bill failed to pass; see Daniel Stone, "Who Killed the Climate and Energy Bill?" *Newsweek* (September 15, 2010). There are few signs that the world is ready to take the extensive actions deemed necessary by many; see, e.g., Janet L. Sawin and William R. Moomaw, "Renewing the Future and Protecting the Climate," *World Watch* (July/August 2010). The December 2011 United Nations Framework Convention on Climate Change, held in Durban, South Africa, came to no conclusion more substantive than an agreement to come up by 2015 with a firm international agreement to deal with the problem. "Denialists," largely funded by the fossil fuel industry, insist that the specter of global warming is a political bogeyman; for an example, see Steve Goreham's *The Mad, Mad, Mad World of Climatism* (New Lenox Books, 2013), a book that has been promoted intensively by the Heartland Institute.

What can be done—if we ever decide to do anything? It seems possible to make enormous headway; according to Jane C. S. Long and Jeffery Greenblatt, "The 80% Solution: Radical Carbon Emission Cuts for California," *Issues in Science and Technology* (Spring 2012), the 2005 call by California's governor to reduce carbon dioxide emissions to 80 percent below the 1990 level by 2050 is achievable; much of the necessary technology already exists, making the problem one of practical implementation. But there remain immense political obstacles. Many are discussed by Bill McKibben, "Global Warming's Terrifying New Math," *Rolling Stone* (August 2, 2012), who finds that we are poised to burn vastly more fossil fuels and add much more carbon to the atmosphere when we need to be burning and adding less. Reinforcing his point are new projections that say we have more reserves of oil and gas, largely due to fracking of oil shale, and can continue burning fossil fuels even longer before we run out; see Richard A. Kerr, "Are World Oil's Prospects Not Declining All That Fast?" *Science* (August 10, 2012), and "Supply Shock from North American Oil Rippling through Global Markets," International Energy Agency press release (May 14, 2013) (http://iea.org/newsroomandevents/pressreleases/2013/may/name,38080,en.html).

The same may be said of technological solutions such as geoengineering. The basic idea behind geoengineering springs from the realization that natural events such as volcanic eruptions can cool climate, sometimes dramatically, by injecting large quantities of dust and sulfates into the stratosphere, where they serve as a "sunshade" that reflects a portion of solar heat back into space before it can warm the Earth (see, e.g., Clive Oppenheimer, "Climatic, Environmental and Human Consequences of the Largest Known Historic Eruption: Tambora Volcano (Indonesia) 1815," *Progress in Physical Geography* (June 2003).

Such effects have prompted many researchers to think that global warming is not just a matter of increased atmospheric content of greenhouse gases such as carbon dioxide (which slow the loss of heat to space and thus warm the planet) but also of the amount of sunlight that reaches Earth from the Sun. So far most attempts to find a solution to global warming have focused on reducing human emissions of greenhouse gases. But it does not seem unreasonable to consider the other side of the problem, the energy that reaches Earth from the Sun. After all, if you are too warm in bed at night, you can remove the blanket *or* turn down the furnace. Suggestions that something similar might be done on a global scale go back more than 40 years; see Robert Kunzig, "A Sunshade for Planet Earth," *Scientific American* (November 2008).

Paul Crutzen suggested in "Albedo Enhancement by Stratospheric Sulfur Injections: A Contribution to Resolve a Policy Dilemma?" *Climate Change* (August 2006) that adding sulfur compounds to the stratosphere (as volcanoes have done) could reflect some solar energy and help relieve the problem. Such measures would not be cheap, and at present there is no way to tell whether they would have undesirable side-effects, although G. Bala, P. B. Duffy, and K. E. Taylor, "Impact of Goengineering Schemes on the Global Hydrological Cycle," *Proceedings of the National Academy of Sciences of the United States of America* (June 3, 2008), suggest it is likely that precipitation would be significantly reduced.

Roger Angel, "Feasibility of Cooling the Earth with a Cloud of Small Spacecraft Near the Inner Lagrange

Point (L1)," *Proceedings of the National Academy of Sciences of the United States of America* (November 14, 2006), argues that if dangerous changes in global climate become inevitable, despite greenhouse gas controls, it may be possible to solve the problem by reducing the amount of solar energy that hits the Earth by using reflective spacecraft. He does not suggest that climate engineering solutions such as this or injecting sulfur compounds into the stratosphere should be tried *instead of* reducing greenhouse gas emissions. Rather, he suggests that such solutions should be evaluated for use *in extremis*, if greenhouse gas reductions are not sufficient or if global warming runs out of control. This position may make a good deal of sense. In January 2009, a survey of climate scientists found broad support for exploring the geoengineering approach and even developing techniques; see "What Can We Do to Save Our Planet?" *The Independent* (January 2, 2009) (http://www.independent.co.uk/environment/climate-change/what-can-we-do-to-save-our-planet-1221097.html). However, any such program will require much more international cooperation than is usually available; see Jason J. Blackstock and Jane C. S. Long, "The Politics of Geoengineering," *Science* (January 29, 2010). Jamais Cascio, "The Potential and Risks of Geoengineering," *Futurist* (May/June 2010), argues that there will be a place in international diplomacy for efforts "to control climate engineering technologies and deal with their consequences." Among those consequences may be major changes in the amount and distribution of rain and snow; see Gabriele C. Hegerl and Susan Solomon, "Risks of Climate Engineering," *Science* (August 21, 2009).

James R. Fleming, "The Climate Engineers," *Wilson Quarterly* (Spring 2007), argues that climate engineers fail to consider both the risks of unintended consequences to human life and political relationships and the ethics of the human relationship to nature. They also, he says, display signs of overconfidence in technology as a solution of first resort. A similar criticism has been levied against S. Matthew Liao, Anders Sandberg, and Rebecca Roache, "Human Engineering and Climate Change," *Ethics, Policy and the Environment* (Summer 2012), who argue that because behavioral, market-based, and technological solutions to human-caused climate change do not appear able or likely to solve the problem, it is appropriate to consider alternatives such as human engineering (meaning using embryo selection to reduce the average size of humans and using drugs to create an aversion to meat or increased altruism). Thomas Sterner et al., "Quick Fixes for the Environment: Part of the Solution or Part of the Problem?" *Environment* (December 2006), say that "Quick fixes are sometimes appropriate because they work sufficiently well and/or buy time to design longer term solutions . . . [but] When quick fixes are deployed, it is useful to tie them to long-run abatement measures." Fundamental solutions (such as reducing emissions of greenhouse gases to solve the global warming problem) are to be preferred, but they may be opposed because of "lack of understanding of ecological mechanisms, failure to recognize the gravity of the problem, vested interests, and absence of institutions to address public goods and intergenerational choices effectively."

It is worth noting that much of the discussion of climate engineering presumes that we will try as a global society to reduce carbon emissions but that our efforts will be insufficient. However, the United Nations Global Climate Change Conference, held in Copenhagen in December 2009, failed to produce a binding agreement to start reducing emissions. Instead, it achieved only an agreement to meet again in 2010 and "start tackling climate change and step up work toward a legally binding treaty." Many people feel that if we do not move faster, geoengineering may be our only option. Robert B. Jackson and James Salzman, "Pursuing Geoengineering for Atmospheric Restoration," *Issues in Science and Technology* (Summer 2010), argue that we need to move toward increased energy efficiency and greater use of renewable energy and explore ways to remove carbon from the atmosphere. If we are successful, we will be less likely to need "sunshades" approaches to geoengineering; however, research into such approaches is essential. Ken Caldeira and David W. Keith, "The Need for Climate Engineering Research," *Issues in Science and Technology* (Fall 2010), say that climate engineering may prove to be "the only affordable and fast-acting option to avoid a global catastrophe." Research into the subject is essential because "the stakes are too high for us to think that ignorance is a good policy."

In the YES selection, M. Granger Morgan, Robert R. Nordhaus, and Paul Gottlieb argue that before we can embark on geoengineering a great deal of research will be needed. First, however, we need a plan to guide research by developing standards and ensuring open access to research results. In the NO selection, Jane C. S. Long and Dane Scott argue that though we need to do much research into geoengineering, not all issues are technical. Vested interests (whose fortunes may be threatened by change, who may fear consequences, who may be driven by the craving for fame, or whose thinking may be dominated by ideology rather than facts), mismanagement, and human weakness must be addressed before engaging in geoengineering.

YES

M. Granger Morgan, Robert R. Nordhaus, and Paul Gottlieb

Needed: Research Guidelines for Solar Radiation Management

As this approach to geoengineering gains attention, a coordinated plan for research will make it possible to understand how it might work and what dangers it could present.

Emissions of carbon dioxide (CO_2) and other greenhouse gases (GHGs) continue to rise. The effects of climate change are becoming ever more apparent. Yet prospects for reducing global emissions of CO_2 by an order of magnitude, as would be needed to reduce threats of climate change, seem more remote than ever.

When emissions of air pollutants, such as sulfur dioxide and oxides of nitrogen, are reduced, improvements occur in a matter of days or weeks, because the gases quickly disappear from the atmosphere. This is not true for GHGs. Once emitted, they remain in the atmosphere for many decades or centuries. As a result, to stabilize atmospheric concentrations, emissions must be dramatically reduced. Further, there is inertia in the earth-ocean system, so the full effects of the emissions that have already occurred have yet to be felt. If the planet is to avoid serious climate change and its largely adverse consequences, global emissions of GHSs will have to fall by 80 to 90% over the next few decades.

Because the world has already lost so much time, and because it does not appear that serious efforts will be made to reduce emissions in the major economies any time soon, interest has been growing in the possibility that warming might be offset by engineering the planet: a concept called geoengineering. The term solar radiation management (SRM) is used to refer to a number of strategies that might be used to increase the fraction of sunlight reflected back into space by just a couple of percentage points in order to offset the temperature increase caused by rising atmospheric concentrations of CO_2 and other GHGs. Of these strategies, the one that appears to be most affordable and most capable of being quickly implemented involves injecting small reflective particles into the stratosphere.

There is nothing theoretical about whether SRM could cool the planet. Every time a large volcano explodes and injects tons of material into the stratosphere, Earth's average temperature drops. When Mount Pinatubo exploded in 1991, the result was a global-scale cooling that averaged about half a degree centigrade for more than a year.

So SRM could work. As undesirable impacts from climate changes mount up, the temptation to engage in SRM will grow. But what if someone tries to do it before we knew if it will work, or what dangers might come with it? The time has come for serious research that can get the world answers before it is too late. To that end, we offer a plan.

Variable Effects—and Benefits

SRM could be designed to bring average temperatures around the world back to something close to their present levels. But because particles injected into the stratosphere distribute themselves around the planet, it is doubtful whether strategies can be found to cool just some vulnerable region, such as the Arctic. Even with a uniform distribution of particles, the spatial distribution of the temperature reductions will not be uniform. For example, work by Katharine Ricke, then at Carnegie Mellon University and now at the Carnegie Institution for Science, has shown that over many decades the level of SRM that might be optimal for China will move further away from the level that might be optimal for India, although in both cases the regional climates would be closer to today's climate than they would have been without SRM.

Change in precipitation patterns induced by climate change might present a particularly strong inducement to undertake SRM. But here again, there are some variables and some unknowns. Although the best current estimates suggest that SRM, on average, could probably restore precipitation patterns to approximately those of today, the ability of climate models to predict the details of precipitation is still not very good. Also, some parts of the world are likely to find at least a little bit of warming or other climate change to be beneficial, and so later in this century countries in those regions might not want to return to the climate of the past few centuries, even if they could. In the short term, modest warming and elevated CO_2 will probably enhance some agricultural production, although with further warming most agriculture will suffer.

Although SRM could offset future warming, it does nothing to slow the steadily rising atmospheric concentrations of CO_2. The higher concentration of CO_2 in the atmosphere is already having notable effects on terrestrial and oceanic ecosystems. Some plant species are able to metabolize CO_2 much more efficiently than others, giving them a comparative advantage in a high-CO_2 world. This is beginning to disrupt and shift the makeup of terrestrial ecosystems.

Over a third of the CO_2 that human activities are adding to the atmosphere is being absorbed by the world's oceans. Today the oceans are roughly 30% more acidic than they were in preindustrial times. Sarah Cooley and colleagues at the Woods Hole Oceanographic Institution have estimated that by late in this century, there will be a dramatic drop in harvest yields of molluscs, resulting in a serious decline in the protein available to low-income coastal populations. Also, acidification is already affecting the ability of many coral species to make reef structures. Many marine experts believe that if emissions and ocean acidification continue to increase, most coral reefs will be gone by the end of this century. In addition to being aesthetically and economically important, reefs (along with coastal mangroves) provide the breeding grounds for many oceanic species and form the base of many oceanic food chains.

Political Landscape

Today in the United States, there are many people who doubt that climate change is occurring, or if it is, that those changes result from human action. Congress is no longer pursuing legislation to mandate reduced emissions of GHGs, and many political leaders have been avoiding the issue.

Federal regulatory actions to advance energy efficiency and reduce emissions from coal-fired power plants are making modest contributions to reducing emissions of GHGs, as is the tightening of the Corporate Average Fuel Economy, or CAFE, standards covering vehicles. Indeed, as Dallas Burtraw and Matthew Woerman of Resources for the Future recently observed, these regulations, together with the dramatic growth in the use of natural gas, have placed the United States on a path to achieve President Obama's goal for reducing U.S. emissions of GHGs. The goal calls for cutting emissions by 2020 to a level that is 17% below levels emitted in 2005. Of course, a 17% reduction does not come close to the U.S. share of reductions needed to stabilize the climate. A few states, most notably California and some in the northeast, are taking direct steps to reduce emissions. Overall, however, the United States shows no signs of being ready to adopt policies to implement the large economy-wide emission reductions necessary to deal with climate change.

Explicit climate policy has progressed further in Europe, where there is a widely shared understanding of the reality of climate change and the risks that it holds. But even as Europe has taken steps to begin reducing emissions of GHGs, these efforts also remain modest when compared with what will be needed to stabilize climate. In the 27 nations that comprise the European Union, per capita CO_2 emissions are roughly half those of the United States. However, Europe's present economic difficulties, together with Germany's growing dependence on coal as it moves to abandon nuclear power, have resulted in a rate of emissions reduction that now lags that of the United States.

Across the major developing nations—China, India, and Brazil—the primary focus is, of course, on economic growth. China is actively developing wind and solar power, as well as technologies for carbon capture and sequestration. China is doing this because it faces local and regional air pollution that is prematurely killing millions of people, because the government realizes that the country will need to wean itself from coal, and because the government assumes that sooner or later the rest of the world will get serious about reducing emissions and, when that happens, China wants to be a strong player in the international markets.

A Tempting Quick Fix

Although subtle impacts from climate change have been apparent for decades, it is only recently that changes have become more obvious and widespread. Over the next few decades, such changes will become ever more apparent. Because reducing atmospheric concentrations of GHGs is inherently slow and expensive, as more and more people and nations grow concerned, SRM could become a tempting quick fix.

SRM is a technology that has enormous leverage. Recent analysis by a university/industry team of researchers, led by Justin McClellan of the Aurora Flight Science Corporation, suggests that a small fleet of specially designed aircraft could deliver enough mass to the stratosphere in the form of small reflecting particles to offset all of the warming anticipated by the end of this century for a cost of less than $10 billion per year, or roughly one ten-thousandth of today's global gross domestic product of $70 trillion (in U.S. dollars). In contrast, estimates by the Intergovernmental Panel on Climate Change in its fourth assessment report suggest that the annual cost of controlling emissions of GHGs to a level sufficient to limit warming will be between half a percent and a few percent of global gross domestic product.

Clearly, given this enormous cost difference, as the impacts of warming and other climate change become more apparent, SRM is going to look increasingly tempting to countries and policymakers who face serious adverse impacts from climate change. Adding to this temptation is the fact that implementing SRM could be done unilaterally by any major nation, which is far from the case with reducing global emissions of GHGs, which would require cooperation among a number of sovereign nations around the world.

Planning a Research Agenda

Although it is well established scientifically that adding fine particles to the stratosphere would, on average, cool Earth, science cannot be at all sure about what else might happen. For example, science cannot be confident about the fate and transport of particles (or precursor materials) once they are injected. It is unknown whether and how the distribution of particles could best be maintained. The surfaces of some types of particles could provide catalytic reaction sites for ozone depletion, but again details are uncertain. Researchers have documented the transient effects of large volcanic injections, but it is not known whether a planned continuous injection of particles might produce large and unanticipated dynamic effects. In short, if the United States or some other actor were to undertake SRM today, it would be "jumping off a cliff" without knowing much about where it, and the planet, would land. Humans have a long tradition of overconfidence and hubris in considering such matters. In our view, anyone who undertook SRM based on what is known today would be imposing an unacceptably large risk on the entire planet.

The climate science community has been aware of the possibility of performing SRM for decades. However, most researchers have shied away from working in this area, in part because of a concern that the more that is known, the greater the chance that someone will try to do it. Although such concerns may have been valid in the past, we believe that the world has now passed a tipping point. In our view, the risks today of not knowing whether and how SRM might work are greater than any risks associated with performing such research.

We reach this conclusion for two reasons. First, the chances are growing that some major state might choose to embark on such a program. If science has not studied SRM and its consequences before that happens, the rest of the world will not be in a position to engage in informed discourse, or mount vigorous scientifically informed opposition if the risks are seen as too great. Second, given the slow pace at which efforts to abate global emissions of GHGs have been proceeding, the chances are growing that when the world does finally get serious about abatement, the United States and other nations may in fact need to collectively engage in a bit of SRM, if it can be done safely, in order to limit damages, while simultaneously scrambling to reduce emissions rapidly and perhaps also scrub CO_2 from the atmosphere.

There have been several calls for a significantly expanded research program on SRM. For example, the House Science Committee and an analogous committee in the UK's Parliament have explored the issue. The United Kingdom has also undertaken a modest program of research support. A task force of the Bipartisan Policy Center, an independent think tank based in Washington, DC, recently developed recommendations for a program of research by the U.S. government. However, most of the limited research now under way in the United States is occurring as part of existing programs that focus on climate and atmospheric science more generally.

Because SRM could rapidly modify the climate of the entire planet at a very modest cost, and because it holds the potential to have profound impacts on all living things, we believe that there is an urgent need for research to clarify its potential impacts and consequences and to provide sufficient reliable information to enable the establishment of appropriate regulatory controls. Building on the work of the Bipartisan Policy Center, the scientific community needs to develop a robust SRM research agenda and obtain the public and private funding necessary to carry it out. In parallel, the community needs to develop guidelines that ensure that such research is responsibly carried out. Finally, as we discuss in detail below, SRM research should be conducted in an open and transparent manner by providing public notification of proposed field experiments and providing decisionmakers and the public with full access to the results of the research.

Except for limited U.S. authority under the National Weather Modification Reporting Act to require notification and reporting of "weather modification" activities, neither U.S. nor international law provides readily useable authority to prohibit, regulate, or report on the conduct of SRM research or field experiments. Our recommendation is to develop and implement a voluntary research code before attempting to impose any regulatory mandates with respect to SRM research, for two reasons. First, a voluntary code can address and work out the various definitional and policy questions we discuss below. Second, a clumsy U.S. attempt to require notice and reporting of SRM research may simply delay or drive that research abroad, frustrating the ultimate objective of open access to responsibly conducted SRM research. A voluntary code, in our view, is the most sensible first step. The United States should take the lead by developing and implementing a code of best SRM research practices and a set of rules governing federally funded SRM research. After doing that, it should then undertake formal governmental steps and informal steps through scientific channels to urge other international players to promptly do the same.

A key component of a significant SRM research program is to develop a fully articulated research agenda. This might be done under the auspices of the U.S. National Academies, drawing on researchers from major universities, national laboratories, and federal agencies, with input from the international research community.

Code of Best Practices

In parallel with, or even before, developing a full research agenda, there is a pressing need to develop what we will call a code of best SRM research practices. This code will need three components. The first would comprise guidelines to provide open access to SRM knowledge by making research results available to decisionmakers and the public. The second would be the delineation of categories of field experiments that are unlikely to have adverse impacts on health, safety, or the environment (that is, experiments conducted within an agreed-upon "allowed zone" of experimental parameters and expected effects on the stratosphere.) The third component would be agreement that any field research to be conducted outside the allowed zone will not be undertaken before a clear national and international governance framework has been developed.

The development of this code will require a convening entity and sufficient resources to support activities. Federal funding through an Executive Branch agency might be secured for such an undertaking. For example, Congress could fund a National Research Council study to develop a set of clear definitions and research norms. Perhaps a faster way to get this done would be to persuade a well-respected private foundation to provide the necessary resources. The National Academies or the National Research Council would be appropriate organizations to convene the effort. Alternatively, the American Geophysical Union might do this as part of its recently expanded set of activities in public policy.

Formulating guidelines for SRM research and a policy to advance open access to SRM research must address a set of key issues of definition and scope:

- First we need to define what counts as SRM. Is the technology to be deployed only for SRM, such as a specific type of specially engineered reflective particle, or does it also include multiuse technology, such as high-altitude aircraft designed to deliver mass to the stratosphere but also capable of performing a variety of other missions that are completely unrelated to SRM? To the extent that SRM overlaps with fields of use that do not raise concerns, non-SRM commercial activity might be affected by efforts to single out SRM activity for special attention. What about research on "incidental" SRM? Such current or proposed research might include, for example, geophysical studies of future volcanic eruptions; studies of the atmospheric effects of "black carbon," the strongly light-absorbing particulate matter emitted by the incomplete combustion of fossil fuels; and studies of the behavior of sulfur dioxide emitted from stationary industrial sources. Any SRM open-access program must define SRM in order, among other things, to minimize its impact on related but uncontroversial commercial activity.
- Next we need to agree on what constitutes SRM research. Does it include theoretical research, literature searches, term papers, and legal memoranda, or should it be limited to experimental research, and if so, does it extend to laboratory research or should it be limited to only field experiments? If the focus is limited to field experiments, how should (and could) basic studies in atmospheric science be differentiated from studies that are more specifically focused on improved understanding of SRM? Trying to make such a demarcation on the basis of experimenters' intent strikes us as deeply problematic; objective criteria will be needed.
- Activities that should be subject to a requirement of prior notification of SRM research need to be

defined. At what stage of a project (planning, approval, or funding) should public notification occur, and in how much detail? Also, what medium or media (for example, a dedicated public Internet site, a Federal Register notice, or a proposal submitted to a designated governmental entity) should be used?
- Any policy respecting public access to SRM research needs to spell out the type of research it covers. Does it cover only completed peer-reviewed research? What about studies whose results are not published or that are in progress but have not reached the stage of publication? What about industry research and abandoned or unsuccessful projects? How can public access be bounded in such a way as to preserve valid commercial interests while providing the appropriate level of public disclosure?
- The allowed zone stipulated for experiments will have to be defined, based on results of existing scientific knowledge. This will include careful delineation of areas of permissible field studies and of a protocol for determining that a proposed field study lies within the allowed zone.
- Finally, there are a series of important policy questions that must be addressed. Should an open-access policy be a voluntary undertaking by researchers? Should the policy be incorporated into federal grants and contracts? Is it feasible to prescribe regulations that require open access to SRM research? Do any of these policies interfere with academic freedom or intellectual property rights?

Importance of Open Access

As part of the effort to develop a code of best SRM research practices, the United States should develop strategies that ensure that the knowledge developed through SRM research is available to the general public and to national and international policymakers to support informed policy discourse and decisionmaking. The creditability and usefulness of a research program can best be advanced by providing the public with advance notice of SRM field experiments and public access to research results.

The SRM research code of best practices should include a commitment to make public the existence of all SRM research activities, perhaps through a mechanism as simple as posting to a common Web site. It should include an agreement that results from prescribed types of research will be made public (preferably through publication in refereed journals). It should provide guidance on the types of field studies that can be undertaken without any special oversight or approval. And it should express an understanding with respect to privately held intellectual property, as discussed below.

Because most federally funded research would probably already have been described in publicly assessable proposals, posting announcements with an abstract of plans to conduct specific field studies on a common public Web site is not likely to present a significant problem for most investigators. However, asking investigators to post preliminary findings on such a site could be more problematic. This is because some leading journals adopt a strict interpretation with respect to the definition of "prior publication." We believe that in the interests of promoting open access to SRM knowledge, an effort should be made to induce several top journals to adopt a more lenient policy in the case of work related to SRM.

In developing a voluntary code for research conduct, comment and advice should be sought from federal agencies, universities and other research institutions, and nongovernmental organizations and companies likely to conduct SRM research. To maximize its acceptance, the code should probably draw a line between research results that are to be publicly disclosed and those that do not need to be publicly disclosed so as to protect the commercial interests of technologies with multiple non-SRM uses. The expectation is that once the code is finalized, its recommendation could be incorporated into approval requirements in government and private nonprofit funding arrangements for SRM research and promoted as a model for industrial researchers and non-U.S. researchers.

U.S. Government Support

Although there has already been some modest support of SRM research from private sources, if a concerted SRM research program is undertaken in the United States, it most likely will involve funding by the federal government as well as some use of the unique capabilities of federal equipment and laboratories. Federal research activities that meet the definition of SRM research should include provisions requiring that an abstract describing the research to be performed be made publicly available upon execution of the underlying agreement. In the case of research involving field experiments, the National Environmental Policy Act may require an Environmental Impact Assessment, unless the proposed project fits into a category excused from such assessment. If an assessment is required and prepared, the public will have ample notice and opportunity for comment.

Federal research agreements should include provisions requiring delivery to the government of publicly releasable research results, commensurate with the SRM research code of best practice. Federal agencies have experience in negotiating lists in each of their research agreements of specifically identified publicly releasable data that would meet the standards set by the SRM code while at the same time, in appropriate agreements, excluding data whose restriction on public release would not be inconsistent with the SRM research code.

Federal research agreements also typically include a patent rights clause that usually provides that the agreement awardee has the option to elect to retain title to its new inventions made under the agreement. In order to lessen the incentive for private commercial interests to influence the direction of the pursuit of SRM, it would be desirable to restrict the assertion of such private intellectual property rights to technical fields other than SRM. Federal agencies already have statutory authority to take prescribed action to restrict or partially restrict the patent rights of awardees. For example, in order to control commercialization, the Department of Energy has provided for federal government ownership of inventions made by its research contractors in the field of uranium enrichment.

A uniform standard can be applied across transactions involving multiple agencies, through mechanisms such as Federal Acquisition Regulations, Office of Management and Budget circulars, and presidential executive orders. Because the promulgation of government-wide guidance may take some time, individual agencies can act on their own initiative if they feel that their mission justifies such action. Individual agency action may lay the groundwork for broader action across the government. If a lead agency is identified to conduct SRM research, that agency should take such an initiative, in the same way that the National Institutes of Health required that investigators who received its support to conduct analysis of genetic variation in a study population submit descriptive information about their studies to a publicly accessible database. The United States could also use international cooperative research agreements as a means to encourage other countries to follow the code of best SRM research practices.

Action by the U.S. government would set a powerful precedent by a major player in the world economy and world research community, giving the nation better standing to advocate for international action in SRM research. Specific U.S. action, developed with input from stakeholders including public interest groups, would establish a model that ensures appropriate public availability of information without unnecessarily affecting commercial interests.

Understand Before Regulating

The approach we have advocated would have the United States take the lead in developing a set of norms for good research practice for SRM. We have proposed that once developed, these norms should be adopted by federal research programs and urged upon all privately funded research. Once the norms are developed and implemented, it should be possible to persuade others across the international research community to adopt similar norms. Organizations such as the International Council of Scientific Unions and the national academies of science in various countries are well positioned to promote such adoption.

As we noted above, the U.S. National Weather Modification Reporting Act provides a statutory framework for making an SRM open-access research policy mandatory in the United States, at least insofar as the research entails field experiments that are conducted domestically and are of such a scale that they could actually affect climate or weather. Our recommendation, however, is to develop and implement a voluntary research code before attempting to use this authority to implement federal rules governing SRM research.

There is also the question of whether considerations should attempt to go beyond open-access policies for SRM research (that is, notice and reporting) and impose substantive regulation, such as permit requirements or performance or work practice standards. We believe that it is premature today to embark on the development and implementation of substantive regulatory requirements. But as the prospect of large-scale field studies—or actual implementation—of SRM becomes more real, the need for and pressure to develop such regulation will grow. Because future regulations should be based on solid well-developed science, the creation of a serious program of SRM research, combined with procedures to ensure open access to SRM knowledge, is now urgent.

M. Granger Morgan is Lord Chair Professor and Head of the Department of Engineering and Public Policy at Carnegie Mellon University.

Robert R. Nordhaus is a partner in the Van Ness Feldman law firm. He was formerly general counsel of the U.S. Department of Energy and of the Federal Energy Regulatory Commission.

Paul Gottlieb is a business consultant. He was formerly the assistant general counsel for technology transfer and intellectual property for the U.S. Department of Energy.

Jane C. S. Long and Dane Scott

Vested Interests and Geoengineering Research

Much remains uncertain about geoengineering, which may offer important benefits—or risks. In moving ahead, there is a set of guidelines that should prove valuable.

On March 11, 2011, Japan suffered one of the most devastating earthquakes in its history, followed by a massive tsunami that engulfed reactors at the Fukushima Daiichi nuclear power plant located near the coast. In Japan, the government body that regulates nuclear power is not highly independent of the utilities it oversees, and regulators had failed to address known safety issues with the reactors. After the crisis, Japan lurched toward a nuclear-free ideology. How can you blame them?

But the catastrophe was not fundamentally caused by a lack of technical information or know-how. Reviewers of the incident found that during the crisis, regulators and company officials made some highly questionable management decisions that were influenced by fears of financial loss and of losing face. Indeed, the catastrophe could have been avoided if good decisions had been made based on available data without the influence of vested interests. As a result, Japan's energy future has been delimited by human foibles and the resulting breakdown of trust in nuclear energy.

If the tsunami in Japan flooded one coastline and several nuclear reactors, climate change may flood all coastlines and cause worldwide dislocations of people, failures of agriculture, and destruction of industries. The likelihood of these impacts has lent legitimacy to the investigation of intentional climate management, or "geoengineering." Society may, at some future time, attempt geoengineering in order to stave off the worst, most unbearable effects of climate change. The technical challenge alone is enormous, but Fukushima provides a cautionary tale for managing the endeavor. Is it possible to develop the trustworthy capacity to manage the climate of Earth?

Incentives for Manipulation

The potential opportunities, benefits, harms, and risks of geoengineering the climate will almost certainly create incentives to manipulate geoengineering choices, and the stakes will be enormous. Societies globally would be wise to face these potential vested interests as they begin to consider researching geoengineering.

Vested interests, in this realm, relate to fortune, fear, fame, and fanaticism, and what to do about them. In moderation, seeking fortune or fame, exercising caution, or being guided by philosophy are appropriate and can lead to innovation and good decisions. However, these attributes may become liabilities when nations, institutions, or individuals seek to manipulate the decisionmaking process to make money, enhance stature, save face, or influence decisions based on fanatical ideology. Society can and should expect people to act with honesty and integrity, but should also plan for dealing with vested interests.

Before moving to planning, it is first worthwhile to examine the forces at work.

Fortune. Parties who stand to gain or lose fortunes by promoting or opposing a geoengineering decision have a vested interest in manipulating that decisionmaking process. Researchers or companies with a financial stake in experiments or possible deployments may seek to push research or deployment in a direction that is ill-advised for society as a whole. Recently, for example, a company desiring to sell carbon credits for sequestering carbon in the ocean conducted a rogue experiment on iron fertilization (the Haida experiment) off the west coast of Canada without obtaining permission or giving due consideration to potential environmental impacts. At this time, there is no legal framework in place to protect society's interests from a financially motivated company attempting such a geoengineering experiment. In the history of environmental remediation, companies that made money from remediation activities have at times fought changes in regulation that would obviate the need for remediation. For example, California used to require the excavation of soil that had been contaminated by leaking gasoline tanks, until researchers documented that naturally occurring soil bacteria would eventually consume the leaked gasoline, thereby obviating expensive excavation. Companies that stood to make a profit from excavation fought this change in regulation. Similarly, a company with contracts to perform geoengineering would have a vested interest in continued deployment.

Countries that produce fossil fuels and companies comprising the fossil fuel industry may view geoengineering as a way to delay or distract attention from mitigation efforts and thus promote the technology to protect their interests. The chief executive officer of Exxon Corporation, Rex Tillerson, articulated his opinion about climate change, glibly commenting: ". . . we'll adapt to that. It's an engineering problem and it has engineering solutions." The opinion espoused by Tillerson reflects his company's vested interests. Investigators have documented cases where companies with vested financial interests have bought studies to suppress or manipulate data related to climate change, smoking, and pharmaceuticals in order to obtain favorable opinions, decrease funding, or delay the publication of research. In *Merchants of Doubt,* Naomi Oreskis, a professor of history and science studies at the University of California, San Diego, described what she saw as the fossil fuel industry's efforts to manipulate the scientific process and conduct extensive misinformation campaigns related to climate science. These lessons reinforce the idea that the design of a geoengineering enterprise should limit the influence of financial incentives.

Fear. The idea that humans can control the climate is fundamentally hubristic. Individuals involved in geoengineering should be appropriately fearful of this technology and should have great humility and healthy self-doubt that they can control the consequences of intervention.

But there are inappropriate fears that should be avoided. Those involved in geoengineering should not fear losing face when they point out problems or discover negative results. Scientific journals should publish negative results, which for geoengineering are equally as important as positive results. Society surely needs to know if a proposed technology is ineffective or inadvisable.

An institution charged solely with managing geoengineering research would have a vested interest in having geoengineering accepted and deployed, because its continued existence would depend on the approach under consideration being a viable course of action. The institution might be tempted to overstate the benefits of the technologies if it fears losing funding. An institution whose focus is on geoengineering might not want to listen to minority opinions that could slow the momentum of research funding.

As a case in point drawn from recent events, during the economic collapse, there were minority positions within the Bush administration that could have saved the national economy from disaster. For example, the chief of the Commodity Futures Trading Commission, Brooksley Born, repeatedly warned of the dangers of the unregulated derivative market. Her prescient minority voice was suppressed by powerful groupthink within the administration that was vested in economic growth, and she eventually resigned. One cannot help but wonder how many minority voices were suppressed out of fear in light of Born's experiences. People who are in the minority and sense problems or dangers must not be afraid to speak against the majority or powerful figures who might become invested in the success of geoengineering research.

Fear is a powerful human motivator and often drives institutional culture. It would be a grave mistake to create institutions and power structures in which people are motivated to become overconfident about their ability to control the climate or fear speaking out when they represent minority opinions or are bearers of bad news.

Fame. Perhaps universally, humans have a desire for recognition. Scientists and engineers and other advocates are not immune from wanting to become a Nobel Prize winner, or be called on by the media, or even just have an enviable publication record. The desire for recognition can become a vested interest that leads to a loss of perspective.

Individuals developing geoengineering concepts are likely to know more about the subject than anyone else, and their expertise has tremendous value for society. However, it is always better to have a fresh pair of eyes on a difficult and consequential subject. Society should not depend solely on the developers of technology to assess the effectiveness and advisability of their proposals.

Fanaticism. Unlike climate change, geoengineering is not yet an ideologically polarized partisan issue, but it could become so. Society would clearly benefit from a debate over geoengineering that is grounded in quality information and reasonable dialogue. Fanaticism would polarize and distort the debate and the sound decision-making that society requires.

A reasonable ideological position drifts into fanaticism when it hardens into a rigid devotion. Most people have ideological positions on matters of importance, but human philosophies are incomplete and imperfect. For example, in a moment of surprising candor in the aftermath of the 2008 financial crisis, the former chairman of the Federal Reserve Board, Alan Greenspan, famously testified that there was a flaw in his free-market ideology, and that the flaw helped cause the crisis. The tendency to adhere too rigidly to one's worldview can put one in danger of sliding into fanaticism. Fanatics often use unreasonable and unscrupulous means to promote their causes. To state the tragically obvious: Fanatics can sincerely do much harm.

Many groups and people will oppose the very idea of geoengineering for legitimate philosophical reasons and use honest means to argue against such research. They will raise important issues that need to be debated. However, motivated segments with a vested interest in their ideology or world-view can behave like fanatics, ignoring or misrepresenting factual information and using questionable techniques to create distrust, a situation that could in turn lead to an inability to act strategically in the face of climate catastrophes.

On the right side of the political spectrum, for example, an individualistic free-market ideology might lead to fanatical positions that see geoengineering as an alternative to "heavy-handed" government regulations to mitigate greenhouse gases. For example, Larry Bell, an endowed professor at the University of Houston and frequent commenter on energy-related matters, remarked in his latest book, *Climate Corruption*, that for many on the right, climate change "has little to do with the state of the environment and much to do with shackling capitalism and transforming the American way of life in the interests of global wealth redistribution." Their vested position, aligned with Exxon's, could be "we will just engineer our way out of this problem."

On the left, rigid environmental or antitechnology ideologies might lead some groups to oppose any discussion of geoengineering. Geoengineering is born out of a fundamental concern for global environmental health, but as with the climate problem in general, it has conflicts with an environmental ideology that narrowly focuses on species preservation, regional conservation, and what is called the precautionary principle. (One version of the precautionary principle states: "When an activity raises threats of harm to human health or the environment, precautionary measures should be taken even if some cause and effect relationships are not fully established scientifically.") Based on an ideology that centers on species preservation, some environmental groups oppose the development of renewable power plants that are designed to help provide energy without emissions but that could cause climate change that could wipe out many species around the globe. In the case of geoengineering, there are environmental groups, such as ETC, that cite the precautionary principle as grounds for banning all geoengineering research, which it sees as a threat to biodiversity. In an ironic twist, rigid antitechnology ideology might become a wedge used by some environmental groups to reject any consideration of geoengineering, even when research is motivated by a desire to preserve biodiversity.

Addressing the Four F's

Moderating the corrupting effects of fortune, fear, fame, and fanaticism should be integral to the development of future geoengineering choices. Society can pay attention to institutional and policy issues that would prevent vested interests from doing harm and provide a counterbalance to human foibles. In this spirit, we offer some guidance for transparency, institutional design, research management, public deliberation, and independent advisory functions. Our suggestions reflect and expand on ideas presented in the Oxford Principles, issued in 2011 by a team of scholars in the United Kingdom as a first effort at producing a code of ethics to guide geoengineering research, and in a report on geoengineering and climate remediation research published in 2011 by the Bipartisan Policy Center, an independent think tank based in Washington, DC. These overlapping strategies each deal with more than one vested interest and could help build genuine trust among scientists, policymakers, and the public.

Transparency. U.S. Supreme Court Justice Louis Brandeis famously said, "Sunlight is the best disinfectant." When all parties present accurate information clearly and forthrightly, vested interests become less influential. To enable effective public accountability and deliberation, information must be transmitted in a way that is comprehensive, but useful to the lay public. Information users need an accurate understanding of such things as funding priorities, research results, limitations, predictions, plans, and errors. This can be referred to as "functional" transparency, to emphasize that the meaning and significance of the information made transparent should also be transparent, not obfuscated in a blizzard of data.

Functional transparency presents challenges. Scientists use many specialized caveats to express what is known about the climate that are understood by and important for scientists, but can obscure the significance of information and be misleading to nonspecialists. For example, climate scientists have had a difficult time articulating the connections between extreme weather events, such as storms Irene and Sandy, and global climate change. The relationship between weather and climate is complex, and scientists know that such extreme storms have a finite likelihood of occurring with or without climate change. Yet understanding the connections between extreme weather events and climate change is important for effective public deliberations on climate mitigation and adaptation. Scientists' need for cautious, complex caveats often clouds the issue in public deliberations.

Also, the public is more likely than the scientific community to focus on the context for research. For example, public discourse on geoengineering research, especially outdoor research, is likely to focus on the purpose of the experiment (in fact, some public deliberation never gets beyond this issue), on alternatives to the experiment, on the potential benefits of the methods being researched, and on the potential risks of the experiments being used. This may be especially true whenever there is the possibility that vested interests may be involved, in which case people are wisely concerned about the motives and goals of research. As a case in point drawn from agricultural biotechnology, the debate has largely centered on the motives and goals of research. Proponents of biotechnology often claim that their goals are to address the problems of world hunger and agricultural sustainability. Opponents question these motives, charging that the real goal is a singular focus on increasing the wealth of researchers and their corporate sponsors.

Scientists, however, do not always make the purpose of research completely transparent. For example, some highly legitimate and important climate science research—say, on cloud behavior—simultaneously informs geoengineering concepts. This research spawns the publication of geoengineering analyses, even though geoengineering is not explicitly named as a purpose of the research. Investigators can and do purposely downplay or obfuscate geoengineering as a purpose of research, because this topic is controversial. Just as in the debate about biotechnology, the lack of transparency about the purpose of the research may eventually erode trust and undermine public deliberation about geoengineering. Norms for research transparency should include forthright statements about the purpose of research. There should be clear and understandable assessments of the scope and state of knowledge and expected gains in understanding that could come from research. The transparent release of research information should be designed to inform public deliberation.

As a good example of bridging the divide between scientific discourses and public deliberations, Sweden's nuclear waste program conducted a study, called a "safety case," of a proposed repository for nuclear wastes. The study proved a primary tool in developing a public dialogue on the topic, and the process resulted in a publicly approved, licensed facility. The safety case communicated in lay language the technical arguments about why the proposed repository was thought to be safe. It also described the quality of the information used in the argument; that is, how well the factors contributing to safety were understood. The document laid out future plans about what would be done to improve understanding, the expected outcome of these efforts, and how previous efforts to improve understanding performed as expected or not. At a follow-up iteration of the safety case, the results of recent experiments were compared with previously predicted results. Over time, the transparency of this process enabled everyone, including the public, to see that the scientists investigating the future behavior of the proposed repository had an increasingly accurate understanding of its performance.

What lesson does this experience hold for geoengineering? Whereas the goal of this nuclear waste research was to build a successful repository, the goal of geoengineering research is not to successfully deploy geoengineering but rather to provide the best information possible to a decision process about whether to deploy. Nevertheless, the safety case provides a useful example for satisfying the norm of transparency required for effective public deliberations on scientific issues.

There is reason to hope that the propensity for ideological decisionmaking can be limited by transparency in geoengineering research. The experience at Fukushima, however, suggests that the opposite is true: Vested interests can drive nontransparent and poor management decisions that destroy public trust and encourage more extreme, fanatical responses. To engender trust, the people or groups conducting or managing research should explain clearly what they are trying to accomplish, what they know and do not know, and the quality of the information they have. They should reveal intentions, point out vested interests, and admit mistakes, and do all of this in a way that is frank and understandable—all examples of actions that enhance trust. Any subsequent modifications of plans and processes should be transparent and informed by independent assessments of purpose, data, processes, analyses, results, and conclusions.

Institutional design. Institutional design can foster standards of practice and appropriate regulations that will counteract many vested interests. Public funding of research is the first act of research governance, as it implies a public decision to do research in the first place. If research is publicly funded, then democratically elected officials can be held accountable for it. Although public funding would not by itself prevent privately funded research, it would fill a vacuum that private money has so far filled. Furthermore, publicly funded research should not lead to patenting that would produce financial vested interests. Geoengineering should be managed as a public good in the public interest.

Governments should charter the institutions charged with developing geoengineering research to be rewarded for exposing methods that are bad ideas as well as good. One way to obviate institutional vested interest in the success of a method would be to create institutions responsible for a wide spectrum of climate strategies. If an institution investigates an array of alternatives, it would have great freedom to reject inferior choices. In an ideal world, institutions would be created to develop technical strategies for dealing with climate change in general, the defining problem of our time, and these institutions would be given broad purview over mitigation efforts, adaption requirements, and the evaluation of geoengineering.

Just as institutions should not be punished for admitting to failed concepts, individual scientists involved in geoengineering research should not have their careers depend on positive versus negative results. If they discover adverse information, it should be valued appropriately as adding to overall understanding. Organizations that fund research and universities and laboratories that conduct research should publicize and reward research results demonstrating the ineffectiveness, inadvisability, or implausibility of a geoengineering idea. Just as NASA scientists in the early years applauded when a rocket (unmanned, of course) blew up, institutions should reward curiosity and courage in the face of failures.

Research management. Most research in the United States today is "investigator-driven," in which funding agencies, such as the National Science Foundation, may design a general call for proposals, but the investigators generate the research topics. Funding agencies may convene workshops to explore strategic research needs that subsequently become part of a programmatic call for research proposals. Workshops help to illuminate research that will contribute to an overall goal, but this process does not organize research to achieve a mission per se. There are important previous instances when research with a large-scale public goal was conducted in a collaborative "mission-driven" manner. Now the nation rarely uses this model, and investigator-driven research is the norm.

Geoengineering research (and climate research in general) might benefit from rediscovering, and perhaps reinventing, collaborative mission-driven research modes that focus on a structured investigation of all interconnected parts of the Earth-human-biosphere systems of interest. Interconnections, key failure modes, and critical information needs would be among critical factors to be systematically identified and addressed. The complex, potentially powerful, and intricate problem of intentional management of the climate requires a systems approach. As well, collaborative mission-driven research management would serve to balance the motivation of individuals by rewarding success in meeting the overall goals of research and would compensate for the somewhat random focus of investigator-driven research.

Initial reactions to this suggestion may tend toward the negative, given how mission-driven research was conducted during the Cold War. In the United States and the former Soviet Union, this style of research resulted in massive radioactive releases into the environment, causing extensive contamination of soil, sediments, and surface- and groundwater throughout weapons complexes. Learning from this experience, and using the suggestions offered here, could lead to a reinvention of mission research for geoengineering that would be open, transparent, publically accountable, and environmentally motivated.

There also may be a need to revise the current method, peer review, for assessing the outcomes of research. Peer review will remain necessary—in part, to help balance the exuberance of individual scientists—but by itself will probably be inadequate. Peer review of journal articles would cover geoengineering projects in pieces, without taking into consideration their tight connection to the context and the entirety of the system problem. There is a potentially useful alternative method that was developed for assessing the results of research on nuclear weapons systems that could not be published for security reasons. With severe limitations on access to peer review, laboratories conducting the research pitted two teams against each other to provide checks and scrutiny on research results. In this "red team/blue team" model, one team develops the research, and the other tries to ferret out all the problems. This approach balances a team that might represent institutional and personal vested interests in promoting a technology with a team whose vested interest is in finding out what is wrong with the idea. For evaluating geoengineering research and results, the red team/blue team approach could be considered a more systematic form of peer review.

Public deliberation. Effective public deliberation of the issues, benefits, risks, liabilities, ethics, costs, and other relevant issues will expose and help to neutralize any vested interests that might be in play. Public deliberation can highlight inappropriate profit-making concerns, point out unbalanced scientific positions, call attention to hubris and institutional bias, and counter the influence of partisan positioning on decisionmaking. Public deliberation will be enhanced and facilitated by research that is conducted transparently in trusted institutions that are managed to produce outcomes in the public interest; that is, through all of the suggestions described here. Public deliberation is perhaps one of the few approaches that can help expose ideologies for what they are, whether they come from the political right, which often obfuscates and denies climate science, or from extreme environmentalism, which often uses scare tactics to stop any technological choice anywhere, anytime. No group or individual should get a pass for mendacity in the face of the choices the nation will have to make regarding climate. Public discourse and deliberation will help prevent manipulative dialogue sponsored by ideologues from becoming decisive.

Deliberative dialogue facilitates real-life decisions about setting and prioritizing research goals and selecting the most appropriate means to achieve those goals. For geoengineering, society needs discussions that characterize ethical and social goals; examine competing alternatives; discuss practical obstacles; consider unwanted side effects; assess the technology, including its effectiveness and advisability; and ultimately produce policy recommendations. The deliberative process requires placing scientific research and technological developments in a larger social and ethical context, using this analysis to select intelligent and ethical goals, and identifying appropriate and effective means to achieve those goals.

A recent project in the United Kingdom is a successful example of effective public deliberation on geoengineering. In 2011, a team of researchers planned an experiment to investigate the feasibility of using tethered balloons to release small aerosol particles into the atmosphere that might reflect a few percent of incoming solar radiation and thereby cool things down a bit. The field test would be part of a larger research project, called Stratospheric Particle Injection for Climate Engineering, or SPICE, that involved laboratory and computer analysis of several geoengineering techniques. Although the proposed experiment was nearly risk-free—the plan was to spray a small amount of water into the air—public deliberation about the plans revealed that this work looked like a "dash to deployment" for an immature solar radiation management technology. Research on deployment was not deemed to be necessary or important at this stage. Deliberation also exposed the fact that one of the investigators had intellectual property interests in the balloon technology, and this seemed to violate the principle that geoengineering should be conducted as a public good. Consequently, the investigators themselves stopped the experiment. This honest response helped the investigators accrue credibility and build trust, because their decision responded appropriately to public deliberation.

Independent advisory functions. An independent, broadly based advisory group would facilitate all of these suggested strategies for addressing vested interests. Such a group could help develop standards and norms for transparency, assess and evaluate institutional design, help to develop norms for research management, and lead the way in developing modes and norms of public deliberation. Because the issues raised by geoengineering go far beyond science, an advisory body should also be able to address a variety of broader issues, such as ethics, public deliberation, and international implications. This expanded charge implies that a board's membership should also go beyond scientific expertise.

Forming independent advisory boards will face a number of barriers. Indeed, the need for an advisory function highlights the inherent controversial nature of geoengineering research, a fact that can make the political choice to start research even more difficult. In the United States, a public board of this type would probably have to meet the standards of the Federal Advisory Committee Act, which is intended to ensure that advisory committees are objective and transparent to the public. Ironically, the act's requirements effectively inhibit the formation of advisory committees, because they require funding, which is now scarce. There are other potential problems as well. Much current research on geoengineering is very preliminary, and perhaps all of the techniques identified so far will be ineffective, will have unacceptable side effects, or will be impossible to deploy under real-world conditions. Some people in the geoengineering field are concerned that having an advisory board to oversee research might interfere with the research before it has demonstrated that there is anything—positive or negative—that is worthy of oversight. Also, it is not clear what agency or person in the government should form such a board and to whom it should report.

As a practical example of how an advisory body might prove useful, consider again the SPICE balloon/ aerosol experiment in the United Kingdom. If there had been an advisory board in place, it might well have recommended that the government cancel the experiment. Such a recommendation would have facilitated government action to stop the experiment, rather than leaving the decision up to the scientists involved, and this step would have given a rather different message to the public about managing controversial research.

The potential value of advisory boards also has been backed up by the research community itself. At a 2011 workshop on geoengineering governance sponsored by the Solar Radiation Management Initiative (an international project supported by the Royal Society in the United Kingdom, the Environmental Defense Fund, and the Third World Academies of Science), participants were asked to consider various forms of organizing geoengineering research. All of them favored a requirement for an independent advisory group, perhaps the only conclusion of this meeting that had unanimous agreement.

In practical terms, consideration of such advisory boards will give rise to many questions about their membership, scope, and authority. To whom should a board report, and how should a national advisory board relate to the international community? How should an advisory board relate to the many governmental and intergovernmental agencies that would almost surely be involved in geoengineering research of one kind or another? Review boards that deal with research involving human subjects cannot actually authorize such research, but they do have the authority to stop research deemed unethical. Should a similar authority be developed for advisory boards on geoengineering research? Answers to these questions should evolve over time, perhaps starting with informal, nonbinding discussions among the various agencies involved.

Just as scientists do not yet know very much about the effectiveness, advisability, and practicality of possible geoengineering technologies, society also does not know very much about how to manage knowledge as it emerges from geoengineering research. If society is to govern this effort without the ill effects of vested interests, it will be necessary to learn how to govern at the same time as researchers are gathering information about the science and engineering of the various concepts. So the early formation of advisory boards or commissions to guide the development of governance is perhaps the first and most important action in countering the potential adverse effects of vested interests and in ensuring that any decisions to pursue or not pursue geoengineering remain legitimate societal choices.

Preparing in Advance

Although the future of geoengineering remains uncertain although tantalizing, one thing is clear. It is not too early to begin the conversation about the human weaknesses, vested interests, and frightening possibilities of mismanaging geoengineering. The Fukushima disaster is just one in a long list of reminders of the consequences of not anticipating and moderating the effects of such all too human foibles as fortune, fear, fame, and fanaticism. It is unthinkable that geoengineering should be added to this list of human-caused technological tragedies.

And though much remains to be learned, it is also clear that a number of approaches are already available to moderate the corrupting effects of vested interests: norms for transparency, institutions designed for honest evaluation, management of research in the public interest, public deliberation to expose vested interests and counter fanaticism, and independent advisory boards to highlight and recommend specifics in all of these areas.

The challenge, then, is to get started. Earth is facing ever greater climate threats. Solutions need to be identified and implemented, with all appropriate speed. For many people, geoengineering may offer important help—if the nation, and the world, proceed in a deliberate, thoughtful manner in conducting research and applying the lessons learned.

JANE C. S. LONG is the principal Associate Director at large and Fellow in the LLNL Center for Global Strategic Research for Lawrence Livermore National Laboratory.

DANE SCOTT is the Director of the Mansfield Ethics and Public Affairs Program at the University of Montana.

EXPLORING THE ISSUE

Do We Need Research Guidelines for Geoengineering?

Critical Thinking and Reflection

1. Which "vested interests" seem most likely or least likely to interfere with progress in geoengineering or solar radiation management?
2. In what sense is dealing with climate change a matter of ethics?
3. What measures should be taken in the near future to best prepare for the long-term impacts of global warming?
4. Should geoengineering or solar radiation management be reserved for use as a last resort?

Is There Common Ground?

Both sides in this issue agree that geoengineering holds promise and that research efforts are appropriate in advance of actual use of the technique. They differ in their views of what form the first of those efforts should take.

1. Can technological tragedies be prevented by efforts to anticipate human weaknesses, vested interests, and mismanagement?
2. Are there other areas of society where similar efforts would be appropriate before action is taken?
3. Morgan, Nordhaus, and Gottlieb mention that possible regulation of geoengineering or solar radiation management must be based on solid science. In what sense is regulation an effort to deal with human weaknesses, vested interests, and mismanagement?

Create Central

www.mhhe.com/createcentral

Additional Resources

Scott G. Bergerson, "Arctic Meltdown: The Economic and Security Implications of Global Warming," *Foreign Affairs* (March/April 2008)

K. A. Brent and J. McGee, "The Regulation of Geoengineering: A Gathering Storm for International Climate Change Policy?" *Journal of Air Quality and Climate Change* (November 2012)

Constance Holden, "Higher Temperatures Seen Reducing Global Harvests," *Science* (January 9, 2009)

Robert B. Jackson and James Salzman, "Pursuing Geoengineering for Atmospheric Restoration," *Issues in Science and Technology* (Summer 2010)

Richard A. Kerr, "Humans Are Driving Extreme Weather; Time to Prepare," *Science* (November 25, 2011)

Michael T. Klare, "Global Warming Battlefields: How Climate Change Threatens Security," *Current History* (November 2007)

Alex de Sherbinin, Koko Warner, and Charles Erhart, "Casualties of Climate Change," *Scientific American* (January 2011)

Internet References . . .

350.org

    www.350.org/

Climate Change

    www.unep.org/climatechange/

Global Warming

    www.epa.gov/climatechange/index.html

Intergovernmental Panel on Climate Change

    www.ipcc.ch

University Corporation for Atmospheric Research

    www.ucar.edu/

Selected, Edited, and with Issue Framing Material by:
Thomas Easton, *Thomas College*

ISSUE

Should We Continue to Rely on Fossil Fuels?

YES: Mark J. Perry, from Testimony at Committee on House Oversight and Government Reform Hearing on "Administration Energy Policies" (May 31, 2012)

NO: Daniel J. Weiss, from Testimony at Committee on House Oversight and Government Reform Hearing on "Administration Energy Policies" (May 31, 2012)

Learning Outcomes
After reading this issue, you will be able to: • Explain what an "all-of-the-above" energy plan entails.   • Explain why the United States should invest in renewable energy technologies.   • Explain what should be done to deal with emissions of mercury, lead, and arsenic by coal-burning power plants.

ISSUE SUMMARY

YES: Mark J. Perry argues that the Obama administration's "All-of-the-Above" energy policy shows unwarranted favoritism toward alternative energy sources. Fossil fuel production is up and supplies are ample. Oil is not at all the "energy of the past," for fossil fuels will continue to power the American economy for generations to come.

NO: Daniel J. Weiss argues that though we do need to develop current energy resources, we must also use less of them, develop cleaner energy technologies, and reduce fossil fuel pollution. The Obama administration's "All-of-the-Above" energy policy is what is needed, not the "oil-above-all" policy favored by the U.S. House of Representatives, as shown in repeated votes cutting investments in alternative energy sources and pollution control and extending fossil-fuel subsidies.

Petroleum was once known as "black gold" for the wealth it delivered to those who found rich deposits. Initially those deposits were located on land, in places such as Pennsylvania, Texas, Oklahoma, California, and Saudi Arabia. As demand for oil rose, so did the search for more deposits, and it was not long before they were being found under the waters of the North Sea and the Gulf of Mexico, and even off the California beaches. For the "History of Offshore Oil," see the *Congressional Digest* (June 2010).

Unfortunately, we eventually became aware that oil—and other fossil fuels—could cause serious problems such as air pollution, acid rain, climate change, and of course oil spills. In 1969, a drilling rig off Santa Barbara, California, suffered a blowout, releasing more than three million gallons of oil and fouling 35 miles of the coast with tarry goo. John Bratland, "Externalities, Conflict, and Offshore Lands," *Independent Review* (Spring 2004), calls this incident the origin of the modern conflict over offshore drilling for oil. The *Exxon Valdez* spill was another striking

example; see John Terry, "Oil on the Rocks—the 1989 Alaskan Oil Spill," *Journal of Biological Education* (Winter 1991). Underwater oil releases have largely been prevented by the development of blowout-prevention technology. But Santa Barbara has not forgotten, and residents do not trust blowout prevention technology. See William M. Welch, "Calif.'s Memories of 1969 Oil Disaster Far from Faded," *USA Today* (July 14, 2008). The most recent spill occurred when the BP Deepwater Horizon drilling rig's Macondo well in the Gulf of Mexico exploded on April 20, 2010, resulting in the release of 200 million gallons of oil into the Gulf of Mexico with environmental impacts that are still being assessed. This record-setting disaster is reviewed in Joel K. Bourne, Jr., "The Gulf of Oil: The Deep Dilemma," *National Geographic* (October 2010). See also Mac Margolis, "Drilling Deep," *Discover* (September 2010).

As a result of the BP disaster, many people and organizations are saying that the government's approach to regulating offshore drilling needs to be changed; see "National Environmental Health Association (NEHA)

Position on Offshore Oil Drilling," *Journal of Environmental Health* (September 2010). Some changes have already been announced. The U.S. government banned offshore oil drilling at the time and extended the ban in 2012 (see Thomas J. Pyle, "Energy Department Sneaks Offshore Moratorium Past Public," *Washington Times*, July 9, 2012). In September 2010, the Department of Interior announced that "new rules and the aggressive reform agenda we have undertaken are raising the bar for the oil and gas industry's safety and environmental practices on the Outer Continental Shelf. Under these new rules, operators will need to comply with tougher requirements for everything from well design and cementing practices to blowout preventers and employee training. They will also need to develop comprehensive plans to manage risks and hazards at every step of the drilling process, so as to reduce the risk of human error." See "Salazar Announces Regulations to Strengthen Drilling Safety, Reduce Risk of Human Error on Offshore Oil and Gas Operations" (http://www.doi.gov/news/pressreleases/Salazar-Announces-Regulations-to-Strengthen-Drilling-Safety-Reduce-Risk-of-Human-Error-on-Offshore-Oil-and-Gas-Operations.cfm). On May 17, 2011, Salazar testified before the U.S. Senate Committee on Energy and Natural Resources that progress is being made and Congress must "provide the tools for the federal government to effectively oversee offshore oil and gas development, including codifying [new] safety and environmental standards" (http://www.doi.gov/news/pressreleases/Salazar-Hayes-Bromwich-Testify-on-Safe-Responsible-Domestic-Oil-and-Gas-Production.cfm).

Some people do not seem disturbed by the prospect of oil blowouts or spills. Ted Falgout, Director of the port at Port Fourchon, Louisiana, looks at the forest of oil rigs in the Gulf of Mexico and "sees green: the color of money that comes from the nation's busiest haven of offshore drilling. 'It's OK to have an ugly spot in your backyard,' Falgout says, 'if that spot has oil coming out of it.'" See Rick Jervis, William M. Welch, and Richard Wolf, "Worth the Risk? Debate on Offshore Drilling Heats Up," *USA Today* (July 13, 2008).

In 2008, oil and gasoline prices reached record highs. Many people were concerned that prices would continue to rise, with the result being rapid investment in alternative energy sources such as wind. At the same time, those who favored increased drilling, both onshore and offshore, began to call for the government to open up more land for exploration. In its last few months in office, the Bush administration issued leases for lands near national parks and monuments in Utah and lifted an executive order banning offshore drilling. Both measures were considered justified because they would reduce dependence on foreign sources of oil, ease a growing balance of payments problem, and ensure a continuing supply of oil. Critics pointed out that any oil from new wells, on land or at sea, would not reach the market for a decade or more. They also stressed the risks to the environment and called for more attention to alternative energy technologies.

The debate over whether to expand drilling for offshore oil is by no means over. In the wake of President Bush's lifting of the executive order banning offshore drilling, the federal Minerals Management Service (MMS) proposed 31 oil and gas lease sales in areas of the nation's Outer Continental Shelf. These areas are estimated to contain at least 86 billion barrels of oil and 420 trillion cubic feet of natural gas, although specific deposits had not yet been discovered. According to the MMS press release, MMS Director Randall Luthi said, "We're basically giving the next Administration a two-year head start. This is a multi-step, multi-year process with a full environmental review and several opportunities for input from the states, other government agencies and interested parties, and the general public." Reactions from states off whose shores these areas lie have been largely negative. Shortly after President Obama took office, the new Interior Department Secretary, Ken Salazar, canceled the Bush Administration's Utah oil and gas leases and announced a new strategy for offshore energy development that would include oil, gas, and renewable resources such as wind power. Unfortunately, the MMS subsequently issued many permits without adequate review. One of those permits covered the site of the BP disaster.

One factor feeding into the debate is a growing awareness that oil and other fossil fuels will not last forever. New technology is helping to maintain the flow, but not on a worldwide basis; see Richard A. Kerr, "Technology Is Turning U.S. Oil Around but not the World's," *Science* (February 3, 2012). What is approaching is known as "peak oil," the point where the amount of oil that can be pumped in a year reaches a maximum and must decline thereafter. Jim Motavalli, "The Outlook on Oil," *E Magazine* (January/February 2006), says that "one conclusion is irrefutable: The age of cheap oil is definitely over, and even as our appetite for it seems insatiable (with world demand likely to grow 50 percent by 2025), petroleum itself will end up downsizing." Peak oil may in fact have already happened, in 2005; if it hasn't, it will soon; see Robert L. Hirsch, Roger H. Bezdek, and Robert M. Wendling, "Peaking Oil Production: Sooner Rather than Later?" *Issues in Science and Technology* (Spring 2005). On the other hand, new fracking technology is expanding supplies; see Richard A. Kerr, "Are World Oil's Prospects Not Declining All That Fast?" *Science* (August 10, 2012), and "Supply Shock from North American Oil Rippling through Global Markets," International Energy Agency press release (May 14, 2013) (http://iea.org/newsroomandevents/pressreleases/2013/may/name,38080,en.html).

Another factor in the debate is the undeniable need to reduce carbon emissions to limit the damage due to global warming. It is therefore clear that we are going to need alternatives to oil and other fossil fuels. According to Anthony Lopez, Billy Roberts, Donna Heimiller, Nate Blair, and Gian Porro, "U.S. Renewable Energy Technical Potentials: A GIS-Based Analysis," National Renewable Energy Laboratory report NREL/TP-6A20-51946 (July 2012), renewable energy could supply as much as 128 times the amount of electricity consumed in the United States. If we are wise, we will develop those alternatives before the need becomes urgent. Toward that end, President Obama announced early in 2012 an "All-of-the-Above" energy plan that boosted support for alternative energy technologies such as wind, promoted efficiency, and continued the moratorium on most offshore oil drilling. Not surprisingly, the fossil fuel industry and conservatives in general oppose the plan.

In the YES selection, Mark J. Perry argues that the Obama administration's "All-of-the-Above" energy policy shows unwarranted favoritism toward alternative energy sources. Fossil fuel production is up and supplies are ample. Oil is not at all the "energy of the past," for fossil fuels will continue to power the American economy for generations to come. In the NO selection, Daniel J. Weiss argues that though we do need to develop current energy resources, we must also use less of them, develop cleaner energy technologies, and reduce fossil fuel pollution. The Obama administration's "All-of-the-Above" energy policy is what is needed, not the "oil-above-all" policy favored by the U.S. House of Representatives, as shown in repeated votes cutting investments in alternative energy sources and pollution control and extending fossil-fuel subsidies.

YES ↵

Mark J. Perry

Testimony at Committee on House Oversight and Government Reform Hearing on "Administration Energy Policies"

One of the topics I have been writing about frequently over the last several years is the U.S. energy revolution, including tracking domestic energy statistics on production and prices, fracking technology and horizontal drilling, the shale revolution, energy-related job creation, etc., and it's because of my interest and frequent writing on energy issues that I have been invited to provide testimony to your committee today on the topic of whether President Obama really supports an "all-of-the-above" energy strategy. To summarize my conclusion, I would say it would be more accurate to describe the president's strategy as "some of the above" rather than "all-of-the-above" with favoritism being directed toward alternative energy over traditional energy sources.

Introduction: The Factual Record on Domestic Oil Production

In his January State of the Union address, and in several subsequent speeches, President Obama said that the country needs an "all-out, all-of-the-above strategy that develops every available source of American energy a strategy that's cleaner, cheaper, and full of new jobs." Further, the president boasted that "under my administration, America is producing more oil today than at any time in the last eight years." I'd like to start by helping to clarify the factual record on domestic oil production.

First, the president failed to mention that the increases in oil drilling on federal lands in 2009 and 2010 reflected leases and permits that were approved before his administration took office, and that oil production on federal lands fell by 14% in 2011.

Further, it's true that total domestic oil production was higher in 2011 than in any year since 2002, but that's because oil production has increased most significantly on state and private lands, not federal lands. And those increases in U.S. crude oil production have continued this year, and in February reached their highest monthly level (more than six million barrels per day) since 1998, but those increases have taken place in locations like the Bakken region of North Dakota and the Eagle Ford Share formation in Texas, and mostly on private lands. Those ongoing increases in domestic oil production are largely because of technology advances in 3D seismic imaging, hydraulic fracturing and horizontal drilling, and not because of any intentional energy policy.

If we focus on the production of all fossil fuels (coal, oil and natural gas) on federal and Indian lands, fossil fuel production fell to a nine-year low in 2011 according to the Department of Energy. In fiscal year 2011, crude oil production on federal lands actually fell by 14%, the largest annual decrease in at least a decade, natural gas production on federal lands fell by more than 9%, and coal production fell by 1%. So in the most recent year available, the "all of the above strategy" has actually resulted in declines in fossil fuel production on federal lands. In other words, the increases in oil production in recent years referenced by the president were largely from drilling on state and private lands, and happened in spite of Obama's restrictive energy policies, not because of them.

Preferences for Alternative Energies in FY 2013 Budget

Shortly after he called for an "all-of-the-above" energy policy, the president then dismissed oil in a speech as an "energy of the past," and instead urged Americans to embrace alternative energies as "energy sources of the future." Those statements suggest that there is a clear preference in the White House for "some of the above" energy sources over others. The president's proposed budget for fiscal year 2013 reflects those preferences for some energy sources—the politically favored "green" energy sector gets preferential treatment over fossil fuel energy, in the form of numerous tax subsidies, tax credits, public expenditures, procurement preferences and grants for alternative energy. Below are the administration's top nine budget provisions for green energy in the proposed fiscal year 2013 budget:

> Extending the production tax credit for wind energy through calendar year 2013.
>
> Extending the Treasury Cash Grant Program (Section 1603 of the American Recovery and Reinvestment Act) to assist small renewable energy companies through 2012, extending tax credits (for renewable

U.S. House of Representatives, May 31, 2012.

companies able to use the credits) for one year, and converting the program into a refundable tax credit through 2016.

Increasing research and development funding to $350 million for advanced energy technologies (up from $40 million disbursed by the U.S. Department of Energy over the last two years).

Expenditures for clean domestic manufacturing, with $290 million for improving industrial processes and materials, and $5 billion for the "48C" clean energy tax credit available to manufacturers of "cleantech" products.

Expenditures for solar and wind energy, providing $310 million for the SunShot Initiative, a program designed to make solar energy cost competitive with fossil fuel energy without government subsidies by 2020, and $95 million for wind energy, including expansion in offshore wind technologies.

Expenditures for energy efficiency, including an 80 percent increase in funding to promote energy efficiency in commercial buildings and industries.

A 10 percent increase in funding for the U.S. Environmental Protection Agency's FY 2013 budget for implementation and enforcement of federal environmental safeguards, and $222 million for the U.S. Department of the Interior's newly formed Bureau of Safety and Environmental Protection.

Expanding Department of Defense clean energy initiatives, including doubling (to $1 billion more than the FY 2012 budget) expenditures for efficiency retrofitting of buildings and meeting efficiency standards for new facilities.

Maintaining funding (at the FY 2012 budget level) for international climate financing, with at least $833 million to support sustainable landscapes, clean energy, and adaptation to climate change in developing countries.

Targeting Fossil Fuels in the FY 2013 Budget with Higher Taxes

In stark contrast, the administration's fiscal year 2013 budget targets oil and natural gas companies with eight proposals for higher taxes, including plans to repeal a) the expensing of intangible drilling costs, b) "last-in, first-out" (LIFO) accounting in favor of the higher-taxed "first-in, first-out" accounting methodology, c) the deduction for tertiary injectants (fluids, gases, and chemicals) that are used in unconventional drilling, and d) the percentage depletion allowance to recover costs for capital investments. Additional tax increases on the oil and natural gas industry would come from proposed modifications of the dual capacity rule (a U.S. tax policy that prevents the double taxation of foreign earnings), increasing the

amortization period for exploration costs, and reinstating Superfund taxes.

Taken together, it is estimated by the American Petroleum Institute that all eight targeted proposals of the administration's FY 2013 budget would burden the oil and gas industry with almost $86 billion in higher taxes over the next ten years.

Drilling Restrictions

In addition to the tax proposals favoring alternative energies over fossil fuel energy sources, the administration's preferences for alternative energy sources are also reflected in drilling restrictions or limited permitting for oil and natural gas that continue in places like:

off the Mid-Atlantic coast

much of the eastern Gulf of Mexico

in the broader Gulf of Mexico (where drilling in 2012 is expected to drop 30% below premoratorium forecasts)

in the Arctic National Wildlife Refuge

on federal lands in the Rockies (where leases are down 70 percent since 2009).

Other actions taken by the administration, including rejecting the Keystone XL pipeline, cancelling millions of acres in offshore lease sales, and closing the majority of the Outer Continental Shelf to new energy production for the next five years demonstrate an administration that does not support an "all-of-the-above energy strategy" that includes increasing domestic production of fossil fuels that will remain critical to America's energy and economic future for many decades.

Meeting Future Energy Demands

Based on the Obama administration's ongoing focus on developing alternative energy sources, energy sources as the future of an America no longer dependent on traditional hydrocarbon energy, the average American would believe that the nation's need for substantial production levels of oil, natural gas, and coal will soon be a distant memory. The reality, however, is much different.

In its most recent forecast in January 2012, the U.S. Department of Energy estimated that the importance of fossil fuels (oil, gas and coal) for meeting the energy demands of the U.S. economy will decline only modestly over the next several decades, from 83% of total U.S. energy consumption in 2010, to 77% in 2035. In contrast, despite all of the attention, preferences from the Obama administration, loan guarantees, and taxpayer subsidies for renewable energy, their contribution to U.S. energy consumption of 7.3% in 2012 was barely higher than the 7.1% share back in both 1996 and 1997, and even less than the 8.9% share in 1983. Current estimates from the Department of Energy suggest that even by 2035, the renewable share of U.S. energy consumption will be slightly less than 11%.

And the most recent Department of Energy estimates may not even yet include new oil and natural gas reserves that have just recently increased significantly in importance in places like Eagle Ford Shale in South Texas, the Green River Formation in Wyoming, Utah, and Colorado, and the Mississippian Lime formation in south Kansas. And it also may not yet account for new technological advances under development by oil companies known as "superfracking," which will move drilling technology from fracking to super-fracking. This new wave of innovation has the potential to significantly raise the efficiency of domestic drilling, and will extend the current wave of fracking technology, leading to potentially huge increases in domestic oil and gas production in the near future.

The key point here is that even the government's own forecasts predict that renewable energy will continue to play a relatively minor role as an energy source over the next several decades out to the year 2035. And traditional energy sources like oil, gas, and coal will continue to provide the overwhelming share (more than three-quarters) of the fuel required to meet U.S. energy demand through the next three decades at least.

Natural Gas

Turning from oil to natural gas, we see a similar story of proven success and future promise. Domestic natural gas production has soared by more than 21% since 2005, but has fallen on federal lands by 24% over that period, and by almost 17% since Obama took office. Like oil, the increases in natural gas production have taken place on state and private lands and have happened because of the significant technological advances in drilling (3D seismic imaging, hydraulic fracturing and horizontal drilling), and not because of any energy policies of the Obama administration.

Further, the significant increases in domestic natural gas production in the last six years have brought inflation-adjusted natural gas prices to their lowest levels in several decades. Although there are some differences between crude oil and natural gas markets, the dramatic price declines in response to increased drilling for natural gas suggest that we should be skeptical of President Obama's claim that "We can't just drill our way to lower gas prices." In the case of natural gas, it was clearly the case that we did exactly that—drill our way to lower gas prices.

It's important to emphasize several key economic factors relating to shale gas production in the U.S. over the last five years.

Huge cost savings. There has been a powerful $250 billion economic stimulus to the economy from lower prices over the last three years for natural gas customers (residential, commercial, industrial and electric utilities), according to the American Gas Association.

Significant job creation from increased natural gas production has provided another energy-related economic stimulus to the U.S. economy.

Lower natural gas prices are sparking an American manufacturing renaissance that promises to create up to one million new U.S. jobs by 2025 in energy-intensive manufacturing sectors like chemicals, fertilizers, ethylene, iron and steel.

Clean natural gas has contributed to significant reduction in CO_2 emissions in the U.S.

Concluding Remarks

At a critical time for America's energy future, Obama's proposed energy platform that so heavily favors high-cost, subsidy-dependent alternative energies is likely to damage the economy, drive energy prices higher, and move us further away from energy independence and economic security. Behind Obama's claim that he supports a "sustained, all-of-the-above strategy that develops every available source of American energy," lays a war against traditional fossil-based energy sources like oil, which he has publicly dismissed to be a "fuel of the past." When it comes to evaluating different energy sources, it should be recognized that fossil fuels have delivered a significant "energy stimulus" to the U.S. economy over the last four years at a critical time for America. Even today, while we struggle through another jobless, sub-par recovery, America's energy sector has been one of the strongest sectors, delivering thousands of shovel-ready, energy-related jobs in places like North Dakota, Texas, and Pennsylvania.

While the U.S. economy is still more than four million payroll jobs below the prerecession 2007 levels, oil and gas extraction employment has increased by more than 37% during the same period. North Dakota has been labeled as the "Economic Miracle State" for its economic success over the last four years, and boasts jobless rates below 1% in cities and counties located in the heart of the Bakken oil region. That oil prosperity is now spreading to places like Eagle Ford Shale in Texas and south Kansas bringing thousands of new jobs, rising incomes, and growing wealth. Likewise, the shale gas revolution has brought energy-related prosperity to the Marcellus region of Pennsylvania and West Virginia, and in the process brought such an abundance of natural gas to the market that prices have fallen to historic lows, saving Americans billions of dollars in energy costs. Now the low energy costs are also sparking an industrial revolution in energy-intensive industries like chemical, fertilizers, iron, and steel, in addition to lowering carbon emissions in the process.

Importantly, this powerful energy-related stimulus has happened as a result of technological advances and entrepreneurship, not as part of any intentional energy policy in Washington, and has not even required any direct taxpayer subsidies in the process. Therefore, when it comes to creating "shovel-ready" jobs at no direct cost to the taxpayer, hydrocarbon energy like oil and gas have a proven track record of delivering significant benefits to the U.S. economy in the form of jobs, stable energy prices, and the shale revolution is moving us closer to energy independence every year now. When President Obama called for an energy strategy in his State of the Union address that's "cleaner, cheaper and full of new jobs," he could easily have been describing the shale revolution that has clearly already delivered on all three points.

In conclusion, the reality is that fossil fuel energy sources will continue to play a dominant role in providing stable supplies of affordable energy to America for decades to come, despite Obama's embrace of alternative energies as the "energies of the future" and his claim that oil is the "fuel of the past." Hydrocarbon energy is America's future, and it's the energy treasures beneath our feet that will continue to power the U.S. economy for many generations. By favoring new, costly, subsidy-dependent alternative energy sources over traditional sources, and by not fully supporting the proven, job-creating, low-cost fossil fuels, it would be more accurate to describe President Obama's costly energy strategy as "some of the above" instead of "all-of-the-above." Obama might wish for an energy future of alternative energy, but the scientific and economic realities suggest that the "fuels of the future" will mostly be the same as the "fuels of the past" dependable and low-cost oil, natural gas, coal, and nuclear.

MARK J. PERRY is a Professor of Economics at the University of Michigan, Flint, and a scholar at the American Enterprise Institute.

Daniel J. Weiss   ➔ **NO**

Testimony at Committee on House Oversight and Government Reform Hearing on "Administration Energy Policies"

What is an "all of the above" energy strategy? To most Americans, it means we must do three things:

- Develop the energy resources of today while using less of them.
- Invest in the new, cleaner technologies of tomorrow.
- Reduce the public health threat from pollution generated by producing and burning coal, oil, and natural gas.

President Obama, employing the tools provided to him by the 110th, 111th and previous Congresses, has accomplished all of these goals. The United States is producing more oil and gas from private and federal lands. We are importing and using less oil. We are investing in efficiency, wind, solar, and other new technologies of the future. And the administration's reductions in smog, acid rain, and toxic air pollutions will prevent up to 45,000 premature deaths annually.

Let's review the record that demonstrates that President Obama is successfully pursuing an "all of the above" energy strategy.

Develop the Energy Resources of Today

Oil and Gas Production Is Up

There has been a lot of rhetoric about this topic that has crowded out the record. The truth, however, is that the United States is producing more oil while using and importing less. Here are some facts on oil and gas production:

- U.S. oil production is at its highest rate since 1998. The Energy Information Administration predicts that it will reach 6.2 million barrels/day by the end of this year.
- Oil production from federal lands and waters is higher. The Energy Information Administration, or EIA, determined that in 2011 the United States generated 3.7 quadrillion BTUs of energy from crude oil produced from federal lands and waters

compared to 3.3 quadrillion BTUs in 2008, a 12 percent increase in production.
- The EIA determined that natural gas production in the United States increased by 15 percent between 2008 and 2011, with a record 24.2 trillion cubic feet of natural gas production last year.
- According to Bureau of Labor Statistics data, there were 75,000 more oil and gas jobs in 2011 compared to 2009.

Additionally a *National Journal* poll of 1,004 adults found significant bipartisan support for banning or regulating hydraulic fracking that produces shale gas. A majority (53 percent) supported an "increase [in] regulation of fracking to protect the environment, but NOT ban it," while 15 percent wanted to "ban fracking altogether because it's not safe for the environment."

Only one-quarter of poll subjects wanted to "reduce regulation of fracking to encourage more natural gas production." A clear majority—55 percent—of Republicans wanted either a fracking ban or more regulation; only 41 percent of Republicans wanted to reduce regulation on fracking.

Oil Use and Imports Are Down

As stated above, the United States is using and importing less oil. This has reduced the transfer of income to other countries too. U.S. oil consumption is down by 1 percent between 2008 and 2011, according to EIA data. Expenditures on foreign oil were $4.5 billion lower in 2011 than in 2008, even though oil prices were higher.

In 2011 the United States imported only 45 percent of oil, the lowest rate since 1997. In 2008 we imported 57 percent of our oil, according to the EIA. President Obama also modernized fuel economy standards for the first time since 1987. After the implementation of the second round of improvements in 2025, the United States will use 2.2 million fewer barrels of oil per day, and drivers will save $8,000 per car in lower gasoline purchases.

Because of the fuel economy standards that will take effect from 2011 to 2016, the EIA predicts that passenger (light duty) vehicle miles traveled will increase by 16 percent from 2009 to 2019, while oil use will increase by only

U.S. House of Representatives, May 31, 2012.

3 percent. This does not include the proposed standards that will further modernize fuel economy between 2017 and 2025. In addition to saving oil, domestic biofuels will provide nearly 1 million barrels of fuel per day in 2012, according to the EIA.

Investments in buses, subways, and trains can also reduce our dependence on oil and create jobs. Public transportation saves 4.2 billion gallons of gasoline annually. Every $1 billion of investment in public transportation supports 36,000 jobs.

Big Oil Companies Make Record Profits due to High Prices

High oil and gasoline prices increase oil company profits, and oil prices averaged a near-record $103 per barrel in 2011. It's little surprise, then, that the big five oil companies BP, Chevron, ConocoPhillips, ExxonMobil, and Royal Dutch Shell made a combined record profit of $137 billion last year. And from 2001 to 2011, these companies made more than $1 trillion in profits (2011 dollars). These same five companies made $33.5 billion or $368 million per day in the first quarter of 2012. Although these companies made hundreds of billions of dollars in profits, four of the five are producing less oil. Between 2006 and 2011 these five companies produced 12 percent fewer barrels of oil.

High oil and gasoline prices help offset these five companies' decline in production. CAP conducted an analysis of gasoline prices and big five oil company profit data and found that from 2008 to 2011, every one-cent increase in the price of gasoline translated into $200 million in profits for the big five companies, which explains why high prices increased their profits even as their oil production fell.

Also, despite their demand to open fragile, previously protected places for oil and gas production, oil and gas companies are not developing many of the leases that they already hold. The Department of the Interior recently determined that: There are approximately 26 million leased acres offshore and over 20 million leased acres onshore that are currently idle—that is, not undergoing exploration, development, or production.

Leased areas in the Gulf of Mexico—that are not producing or not subject to pending or approved exploration and development plans—are estimated to contain 17.9 billion barrels of UTRR [Undiscovered Technically Recoverable Resources] oil and 49.7 trillion cubic feet of UTRR natural gas. According to a March 2012 report from the Department of Interior, "more than 70 percent of the tens of millions of offshore acres under lease are inactive." This includes almost 24 million acres that do not have "approved exploration or development plans" in the Gulf of Mexico. This area has an estimated 11.6 billion barrels of oil and 50 trillion cubic feet of natural gas.

The Department of Interior held "three of the top five largest [lease] sales in the agency's history" last year, while 56 percent of the public lands leased to the oil and gas industry in the lower 48 states were not producing any fossil fuels or being explored.

Big Oil Companies Receive Billions of Dollars of Tax Breaks

Despite their trillion-plus dollars of profits earned over the past decade due to high oil and gasoline prices, Big Oil companies still receive $40 billion per decade in federal tax breaks. One of these provisions "expensing of intangible drilling costs" originated in 1916 and costs taxpayers $12.5 billion over 10 years.

President George W. Bush, a former oil man, actually supported the elimination of Big Oil tax provisions in 2005 because they were unnecessary. He said: I will tell you with $55 oil, we don't need incentives to the oil and gas companies to explore. There are plenty of incentives. What we need is to put a strategy in place that will help this country over time become less dependent.

Big Oil's Tax Break Defense Is Full of Holes

Big Oil companies and the American Petroleum Institute, or API, their lobbying arm, have misleading or wrong defenses for these tax breaks.

Rhetoric: "The industry receives not ONE subsidy, and it is one of the largest contributors of revenue to our government of any industry in America." Jack Gerard, API president and CEO, February 23, 2012.

Record: Numerous Republican leaders have noted that a tax break is the same as a direct government subsidy, in a different form. This includes former President Ronald Reagan's chief economic advisor, Martin Feldstein; former Senate Budget Committee Chair Pete Domenici (R-NM); House Ways and Means Committee Chair Dave Camp (R-MI); and Speaker of the House John Boehner (R-OH).

- Feldstein: "These tax rules because they result in the loss of revenue that would otherwise be collected by the government are equivalent to direct government expenditures."
- Domenici: "Many tax expenditures substitute for programs that easily could be structured as direct spending. When structured as tax credits, they appear as reductions of taxes, even though they provide the same type of subsidy that a direct spending program would."
- Rep. Camp: "'Tax expenditures' [are] provisions that technically reduce someone's tax liability, but that in reality amount to spending through the tax code."
- Rep. Boehner: "What Washington sometimes calls tax cuts are really just poorly disguised spending programs."

Rhetoric: "Raising taxes will not lower energy prices for American families and businesses in fact, the Congressional Research Service says this plan could cause gasoline prices to go higher."

Record: A May 2011 Congressional Research Service memo, "Tax Policy and Gasoline Prices," to Senate Majority Leader Harry Reid (D-NV) determined that eliminating tax breaks for Big Oil companies would have little impact on the price of gasoline. Here is a summary of CRS's conclusion of the impact of eliminating specific tax breaks for Big Oil: Section 199: With current prices at, or near, $100 per barrel in the United States, it is unlikely that firms will slow production, or close wells with the loss of the Section 199 deduction.

Intangible drilling costs: The Wood MacKenzie study did not conclude that U.S. gasoline prices would be affected by the tax changes.

Dual Capacity Rules: [Elimination of] this provision . . . should have no effect on the firms output or pricing decisions, and therefore no effect on the price of gasoline. General Considerations: The total expected tax revenues are only 5% of the earnings of the five largest firms in the industry and a smaller percentage of the total industry.

Rhetoric: Reducing or eliminating these tax breaks will reduce oil production or cost jobs.

Record: Even with the tax breaks, oil production and employment by the big five companies is lower. As previously noted, the big five companies produced 12 percent less oil in 2011 compared to 2006. And despite earning more than $1 trillion in profits between 2001 and 2011, the big five oil companies have shed more than 11,000 U.S. jobs over the past few years, according to "Profits and Pink Slips: How Big Oil and Gas Companies Are Not Creating U.S. Jobs or Paying Their Fair Share" by the House Natural Resource Committee Democrats.

Rhetoric: Big Oil already pays its fair share of taxes.

Record: The biggest oil companies claim that they pay a large amount of taxes. Reuters found that they support this claim by lumping various fees, payments, and taxes together: The industry lumps together U.S. and foreign taxes. It includes taxes that are deferred and thus not paid yet. U.S. companies must pay taxes on profits earned abroad, but they can defer these taxes until they bring the cash into the country. Reuters also determined that in 2011, "Exxon Mobil paid 13 percent of its U.S. income in taxes after deductions and benefits in 2011, according to a Reuters calculation of securities filings. Chevron paid about 19 percent."

And Reuters reports that ConocoPhillips paid an effective federal tax rate of 18 percent last year. These tax rates, Reuters concludes, are "a far cry from the 35 percent top corporate tax rate." To further put this into perspective, the average American household paid an effective federal tax rate of 20 percent in 2007, the last year for which data are available.

Big Oil Receives Far More Subsidies Than Renewables

Despite Big Oil's trillions of dollars of earnings, and billions of dollars of tax breaks dating back 100 years, some

Big Oil allies claim that these companies need these tax breaks. Meanwhile, important incentives to invest in clean, emerging renewable technologies are under attack. For example, the production tax credit for wind energy will expire at the end of this year. Its demise threatens 37,000 jobs. In addition, it would surrender the growing market for clean tech to our economic competitors.

It is important to note that Big Oil and nuclear energy have received vastly more federal assistance than wind, solar, and other renewable energy sources. According to a DBL Investors analysis from 2011:

> In inflation adjusted dollars, nuclear spending averaged $3.3 billion over the first 15 years of subsidy life, and O&G subsidies averages $1.8 billion, while renewables averaged less than $0.4 billion. . . . federal incentives for early fossil fuel production and the nuclear industry were much more robust than the support provided to renewables today.

First New Nuclear Reactors Approved in 30 Years

The first two new nuclear reactors in a generation were approved in February 2012 at Plant Vogtle in Waynesboro, Georgia. Two more reactors in South Carolina were approved in March. The Georgia reactors are in the process of receiving a federal loan guarantee from the Department of Energy.

Coal Mining Jobs Are Up

Coal companies, some utilities, and the coal industry's lobbying arm claim that there is a so-called "War on Coal" because the Environmental Protection Agency is requiring power plants to reduce their pollution (see below for more details). Despite their high profits, these companies want to avoid reducing their smog, acid rain, toxic, and carbon pollution.

This alleged war is little more than a myth. Coal employment has been growing. The U.S. Mine Safety and Health Administration reports that there were more coal miners employed in the United States in 2011 since 1997, and nearly 3 percent more compared to 2008. This includes more miners in 2010 in Pennsylvania and Virginia, according to the Energy Information Administration. There are also 1,500 more coal miners in West Virginia since President Obama took office, according to the West Virginia Center on Budget & Policy.

Coal production in Colorado and Utah rose 25 percent in the third quarter of 2011 compared with the same period in 2010. Craig, Colorado, "a northwest Colorado town based on an economy powered largely by the surrounding county's coal mines, is doing relatively well, according to the mayor," reported Politico. . . .

There has been a reduction in coal production over the last several years, but protecting children's health isn't the reason. The *West Virginia Gazette*, however, reports that coal companies "have most frequently cited competition from low natural gas prices, a warm winter and the sluggish economy—not tougher environmental rules—as the central reasons for production cutbacks."

Invest in the Cleaner Technologies of Tomorrow

Investments in Renewables Are Vital to U.S. Economic Competitiveness

The United States is competing with China, Germany, and other nations to produce the clean energy technologies of the future that the world will demand to reduce the carbon pollution responsible for climate change. By 2020 clean energy will be one of the world's biggest industries, totaling as much as $2.3 trillion. Of the seven strategic emerging industries identified by China's State Council as focal points for government investment in economic growth, five are related to the clean energy economy.

The growing clean energy industry is very attractive to investors. Reuters just reported that "Goldman Sachs Group Inc. plans to channel investments totaling $40 billion over the next decade into renewable energy projects, an area the investment bank called one of the biggest profit opportunities." The question is whether there is a friendly or hostile economic climate in the United States that encourages Goldman Sachs and others to invest in renewable energy here at home. Opposition to incentives and other forms of government support could drive these companies to invest in other nations instead.

Renewable Electricity Has Nearly Doubled under Obama

Under President Obama, the United States made investments in renewable energy and they are paying off. In 2011, "U.S. clean energy investment moved back ahead of China for the first time since 2008," according to Bloomberg New Energy Finance. And federal loans or guarantees are a good deal for taxpayers. For every $100 the government lends or guarantees, the program only costs taxpayers 94 cents.

Thanks to such investments, the generation of non-hydro renewable electricity will nearly double from 108 gigawatts in 2008 to 196 gigawatts in 2012, based on EIA data. This includes nearly tripling wind-generated electricity and more than doubling solar electricity.

Wind Energy Is a Growing Source of Electricity

One of the fastest growing electricity sources of any kind is wind generation. According to the American Wind Energy Association: The U.S. wind industry now totals 48,611 MW of cumulative wind capacity through the end of the first quarter of 2012. The U.S. wind industry has added over 35% of all new generating capacity over the past 5 years, second only to natural gas, and more than nuclear and coal combined. Currently, total wind generation is enough to power more than 12 million homes.

The production tax credit for wind energy, which became law in 1992, "has generated $15 billion to $20 billion a year in private investment over the past five years, in the process becoming one of the fastest growing U.S. manufacturing industries," according to the American Wind Energy Association, or AWEA.

Clean Energy Investments Create Jobs

Federal investments in clean energy technologies beginning in 2009 "created or save[d] nearly 1 million jobs [through 2010], according to a report from the Economic Policy Institute and the BlueGreen Alliance." The Bureau of Labor Statistics recently determined that, "In 2010, 3.1 million jobs in the United States were associated with the production of green goods and services."

The wind industry employs 75,000 people, according to AWEA. Jobs in the solar industry will grow by one-third to 124,000 between 2010 and 2012, according to the National Solar Jobs Census 2011. This includes an 11 percent increase in manufacturing jobs. Investments in home energy efficiency save families money The Department of Energy's Weatherization Assistance Program has supported the weatherization of more than 750,000 low-income homes over the past three years.

The program provides:

> Energy efficiency upgrades include adding insulation, sealing ducts, and installing more efficient windows, heaters, and cooling systems—and are lowering energy bills for lowincome families across the country, supporting economic growth and creating jobs. Weatherized homes saves the average household $400 in lower heating and cooling bills in the first year alone by reducing energy consumption by up to 35 percent.

Investments in Alternative Transportation Will Save Oil, Create Jobs

We must also invest in alternatives to oil. Plug-in hybrids and all electric vehicles consume little or no gasoline. During their first year, the combined sales of the plug-in hybrid Chevrolet Volt and the all-electric Nissan Leaf were twice as large as the now-familiar Toyota Prius and Honda Insight hybrids during their first year. It took fifteen years after its introduction for the Toyota Prius to become the third best-selling car in the world today. In March, Chevrolet sold more Volts than in any previous month. Sales in the emerging plug-in electric car market rose 323 percent while auto sales rose 13.4 percent in the quarter overall.

The Volt and other innovative American oil-savings technologies require enhanced infrastructure to speed their adoption. There is a long history of government support for the infrastructure that is essential to grow pioneering technologies, from FM radio to telephones. Electric vehicles would likewise benefit from such assistance with recharging infrastructure. The Electric Drive Vehicle Deployment Act of 2011, H.R. 1685, sponsored by Reps. Judy Biggert (R-IL) and Ed Markey (D-MA), would provide financial assistance to states for the deployment of electric vehicles.

In addition to making more sophisticated electric-fueled vehicles, the United States is investing in the advanced batteries necessary to power them. In 2009 the United States had only two factories manufacturing advanced vehicle batteries, producing less than 2 percent of the worldwide share. Due to investments made under the Recovery Act, battery and parts manufacturers are building 30 factories. As of January 2012 the battery program has created and saved more than 1,800 jobs not including construction jobs according to a ProPublica analysis.

Protect the Public from Pollution

Our use of coal and oil provide many essential economic and lifestyle benefits. These fuels have powered the United States to be the world's largest economy. At the same time, our reliance on coal and oil has a huge, hidden public health and economic price tag. The National Academy of Sciences concluded that combustion of these two fuels causes $120 billion annually in economic damage due to premature deaths, asthma attacks, hospitalizations, and lost productivity. Most vulnerable to acid rain, smog, toxics, and carbon pollution are children, seniors, and the infirm. Fortunately, it is possible to use these fuels while reducing the pollution that incurs these human and economic harms. The Clean Air Act of 1990 provides the administration with tools to protect the public from these deadly air pollutants.

The Environmental Protection Agency has recently finalized rules to reduce major pollutants from power plants. In 2011 it finished the "Cross-State Air Pollution Rule," designed to protect downwind states from acid rain or smog pollution from upwind states. It requires cuts in sulfur dioxide and nitrogen oxide pollution, the ingredients of acid rain and smog. This rule will prevent up to 34,000 premature deaths and avoid 858,000 other health problems annually, including 400,000 cases of aggravated asthma. These air quality improvements will result in $120 billion to $280 billion in annual benefits.

Another long overdue rule the EPA recently promulgated would require coal-fired power plants to dramatically reduce the emission of mercury, lead, arsenic, and other toxic pollutants. These contaminants can cause birth defects, brain damage, cancer, and other serious ailments. The EPA predicts that these reductions, which don't take effect until 2015 or 2016, will save 11,000 lives annually and prevent more than 100,000 asthma cases and heart

attacks too. These health improvements will provide economic benefits of up to $90 billion every year.

More Domestic Production Will Not Lower Gasoline Prices

High oil prices are responsible for high gasoline prices. The Energy Information Administration estimates that the cost of crude oil was 66 percent of the cost of a gallon of gas in May 2012. And oil prices are set on the global market, which is controlled by the Organization of Petroleum Exporting Countries, a cartel. The Federal Trade Commission found that "OPEC attempts to maintain the price of oil by limiting output and assigning quotas."

Other nations that produce most of their oil also experienced high gasoline prices this year. For instance, Canada had high gasoline prices too. The *Edmonton Journal* on March 30 reported that "Canadians are paying some of the highest prices they ever have for gasoline." No president has much control over oil prices, as noted by the Cato Institute and a survey of economists by the University of Chicago. The *Wall Street Journal* noted that:

> Producing a lot of oil doesn't lower the price of gasoline in your country. According to the U.S. Energy Information Administration, Germans over the past three years have paid an average of $2.64 a gallon (excluding taxes), while Americans paid $2.69, even though the U.S. produced 5.4 million barrels of oil per day while Germany produced just 28,000. Big Oil and their political allies claim that the expansion of oil drilling would lower gasoline prices. The Associated Press tested this hypothesis by analyzing three decades' worth of monthly oil production and gasoline price data. AP determined that there is "no statistical correlation between how much oil comes out of U.S. wells and the price at the pump."

House of Representatives Ignores "All of the Above" Strategy?

This hearing is designed to examine whether the Obama administration has pursued an "all of the above" energy strategy. The record clearly shows that it has.

Unfortunately, the House of Representatives does not appear to have joined the administration in pursuit of that strategy. The House-passed fiscal year 2013 budget resolution, H. Con. Res. 112, sponsored by House Budget Committee Chairman Paul Ryan (R-WI) favors fossil fuels at the expense of cleaner, new renewable energy technologies. In addition, the House has passed numerous bills that would put children, senior citizens, and the infirm at risk by blocking or delaying long-overdue safeguards to protect them from pollution. Let's quickly look at the House's record on "all of the above" energy:

- The FY 2013 budget proposal calls for a $3 billion cut in energy programs in FY 2013 alone. From

2013 through 2017 the Ryan budget would spend a paltry total of $150 million over these five years on these programs, barely 20 percent of what was invested in 2012 alone.

- The proposal includes scant specifics about cuts in energy programs. Yet it explicitly calls for ending investments in programs that promote emerging technologies, which would include renewable, efficiency, advanced vehicle, and other emerging technologies: This budget would . . . [pare] back duplicative spending and non-core functions, such as applied and commercial research or development projects best left to the private sector. And it would immediately terminate all programs that allow government to play venture capitalist with taxpayers' money.

- These cuts in energy programs could include:

    Investments in the development of advanced batteries, essential for electric vehicles that use little or no oil.

    Loans to auto companies to help them build super-fuel-efficient vehicles. For instance, a program signed into law by President George W. Bush provided a $5.9 billion loan to Ford to help it build 2 million fuel-efficient vehicles annually while creating 33,000 jobs.

    Tax incentives to encourage investment in wind and solar energy deployment, which will create electricity with little or no pollution.

- The Ryan budget would slash investments in clean energy technologies. According to the Office of Management and Budget: Clean energy programs would be cut by 19 percent over the next decade, derailing efforts to put a million electric vehicles on the road by 2015, retrofit residential homes to save energy and consumers money, and make the commercial building sector 20 percent more efficient by 2022.

- The House budget retains $40 billion in tax breaks for Big Oil companies over the coming decade.

- In the first session of the 112th Congress, the House of Representatives held 209 votes to weaken public health safeguards or environmental protections, according to an analysis by Reps. Herny Waxman (D-CA), Ed Markey (D-MA), and Howard Berman (D-CA). There were 77 votes to weaken the Clean Air Act, including efforts to "block EPA regulation of toxic mercury and other harmful emissions from power plants" and other major sources of dangerous air pollution.

- The House has not extended the production tax credit for wind and other renewable energy sources even though the credit expires at the end of 2012. Rep. Dave Reichert (RWA) introduced the American Renewable Energy Production Tax Credit Extension Act, H.R. 3307, last November. Although it has 100 co-sponsors from both parties, it has not moved through the Ways and Means Committee or to the House floor.

- The Electric Drive Vehicle Deployment Act, H.R. 1685, sponsored by Reps. Judy Biggert (R-IL) and Ed Markey (D-MA) was introduced in May 2011. It would create a "race to the top" for communities that wanted to invest in recharging infrastructure for electric vehicles. It has not been acted on, either.

Conclusion

As stated at the beginning, an "all of the above" strategy includes increasing oil and gas production, reducing use, investing in new clean energy technologies, and protection of public health. My testimony is just a brief summary of the available evidence that conclusively demonstrates based on the record and not rhetoric that President Obama has successfully pursued an "all of the above" energy strategy.

Just as clearly, the House of Representatives has ignored oil use reductions, slashed investments for new clean energy technologies, and would eliminate or eviscerate public health protection from hazardous pollutants. In particular, the House budget's disinvestment in clean energy threatens industries and jobs in a new worldwide economy that other nations are racing to claim. Such policies wave the white flag of surrender by proposing to kill the public-private investments essential to compete with China, Germany, and other nations.

The record demonstrates that President Obama has successfully pursued an "all of the above" energy strategy that creates jobs, builds new industries, reduces families' energy spending, and cuts pollution. Despite its rhetoric, it seems that the House of Representatives has pursued an "oil above all" strategy that would benefit big oil companies at the expense of everyone.

Hopefully, the House of Representatives will pass bipartisan legislation to invest in clean technologies, as well as join President Obama in supporting "an all of the above" energy strategy.

DANIEL J. WEISS is a Senior Fellow at the Center for American Progress Action Fund.

EXPLORING THE ISSUE

Should We Continue to Rely on Fossil Fuels?

Critical Thinking and Reflection

1. Why should we continue to rely on fossil fuels?
2. Why should we *not* continue to rely on fossil fuels?
3. In what ways does the use of fossil fuels affect human well-being?
4. Daniel J. Weiss discusses public health protection in his essay. Mark J. Perry does not. What part should public health protection play in a federal energy policy?

Is There Common Ground?

Both Mark J. Perry and Daniel J. Weiss agree that we will continue to use fossil fuels for a time. They differ in part on whether fossil fuels or alternative energy sources should be subsidized. They also differ on whether fossil fuels pose serious risks to human health and well-being.

1. List four ways the use of fossil fuels can damage human health and well-being.
2. Describe how each of these impacts on human health and well-being may be countered.
3. Is it necessary to stop using fossil fuels in order to protect human health and well-being?

Additional Resources

Joel K. Bourne, Jr., "The Gulf of Oil: The Deep Dilemma," *National Geographic* (October 2010)

Robert L. Hirsch, Roger H. Bezdek, and Robert M. Wendling, "Peaking Oil Production: Sooner Rather than Later?" *Issues in Science and Technology* (Spring 2005)

Mason Inman, "How to Measure the True Cost of Fossil Fuels," *Scientific American* (April 2013)

Create Central

www.mhhe.com/createcentral

Internet References . . .

350.org

www.350.org/

Behind the Numbers on Energy Return on Investment

www.scientificamerican.com/article.cfm?id
=eroi-behind-numbers-energy-return-investment

The National Renewable Energy Laboratory

www.nrel.gov/

Selected, Edited, and with Issue Framing Material by:
Thomas Easton, *Thomas College*

ISSUE

Is Shale Gas the Solution to Our Energy Woes?

YES: Diane Katz, from "Shale Gas: A Reliable and Affordable Alternative to Costly 'Green' Schemes," *Fraser Forum* (July/August 2010)

NO: Deborah Weisberg, from "Fracking Our Rivers," *Fly Fisherman* (April/May 2010)

Learning Outcomes

After reading this issue, you will be able to:

- Explain how shale gas threatens water supplies and even human health.
- Discuss the relative merits of putting energy-related decision making in the hands of private investors or government.
- Explain how an influx of cheap fossil fuel will affect the development of alternative (e.g., wind and solar) energy systems.
- Discuss whether increasing prices through regulation of an industry, in order to protect public health, is justifiable.

ISSUE SUMMARY

YES: Diane Katz argues that new technology has made it possible to release vast amounts of natural gas from shale far underground. As a result, we should stop spending massive sums of public money to develop renewable energy sources. The "knowledge and wisdom of private investors" are more likely to solve energy problems than government policymakers.

NO: Deborah Weisberg argues that the huge amounts of water and chemicals involved in "fracking"—hydraulic fracturing of shale beds to release natural gas—pose tremendous risks to both ground and surface water, and hence to public health. There is a need for stronger regulation of the industry.

Fossil fuels have undeniable advantages. They are compact and easy to transport. In the form of petroleum and natural gas and their derivatives, they are well suited to powering automobiles, trucks, and airplanes. They are also abundant and relatively inexpensive, although the end of the era of oil abundance is in sight, and prices are rising. However, fossil fuels also have disadvantages, for their use puts carbon dioxide in the air, which threatens us with global warming. Oil is associated with disastrous oil spills such as the one that resulted from the failure of the British Petroleum Deepwater Horizon drilling rig in the Gulf of Mexico in 2010. Coal mining leaves enormous scars on the landscape, and coal burning emits pollutants that must be controlled. Natural gas alone seems relatively benign, for though it emits carbon dioxide when burned, it emits less than oil or coal; on the other hand, the amount of methane released when drilling for natural gas may mean the overall impact of natural gas on global warming is greater than that of coal; see Robert W. Howarth, Rence Santoro, and Anthony Ingraffea, "Methane and the Greenhouse-Gas Footprint of Natural Gas from Shale Formations," *Climatic Change Letters* (June 2011). It produces fewer air pollutants, it cannot be spilled (if released, it can cause explosions and fires, but outdoors it mixes with air and blows away), and obtaining it has not meant huge damage to the environment. Indeed, some see it as a valuable partner for renewable energy; see Saya Kitasei, "Powering the Low-Carbon Economy: The Once and Future Roles of Renewable Energy and Natural Gas," *Worldwatch Report,* vol. 184 (Worldwatch Institute, 2010). Much of the United States' demand for natural gas is met by domestic production, but demand is rising and imports—now at about 15 percent of demand—will have to rise to keep up, just as they have with oil. Lacking new sources of natural gas or a shift to coal, nuclear power, or alternatives such as wind and solar power, the nation must inevitably become more dependent on foreign energy suppliers.

It has long been known that large amounts of "unconventional" natural gas reside in deep layers of sedimentary rock such as shale. However, this gas could not be extracted with existing technology, at least not at a price that would permit a profit once the gas was sold. In recent years, this has changed, for drilling technology now allows drillers to bend drill holes horizontally to follow rock layers. Injecting millions of gallons of water and chemicals at extraordinarily high pressure can fracture (or "frack") the rock surrounding a drill hole and permit trapped gas to escape. See Richard A. Kerr, "Natural Gas from Shale Bursts onto the Scene," *Science* (June 25, 2010). Mark Fischetti, "The Drillers Are Coming," *Scientific American* (July 2010), notes that the Marcellus shale formation, which stretches from upstate New York through Pennsylvania to Tennessee, may contain enough gas to meet U.S. needs for 40 years. There are other shale formations in the United States, Canada, Europe, and China. The total U.S. supply may be enough to meet needs for a century; see Steve Levine, "Kaboom!" *New Republic* (May 13, 2010). See also Peter Heywood, "Fracking: Safer and Greener?" *The Chemical Engineer* (April 2012), and Tom Wilber, *Under the Surface: Fracking, Fortunes and the Fate of the Marcellus Shale* (Cornell University Press, 2012). The same technology is being applied to extracting oil from shale formations; see Edwin Dobb, "America Strikes Oil: The Promise and Risk of Fracking," *National Geographic* (March 2013), and supply projections are rising; much more carbon will thus be added to the atmosphere when we need to be adding less. Reinforcing his point are new projections that say we have more reserves of oil and gas, largely due to fracking of oil shale, and can continue burning fossil fuels even longer before we run out; see "Supply Shock from North American Oil Rippling through Global Markets," International Energy Agency press release (May 14, 2013) (http://iea.org/newsroomandevents/pressreleases/2013/may/name,38080,en.html).

Not surprisingly, many people are concerned about the environmental impacts of "fracking" and disposing of used water, chemicals, and drilling wastes. Richard A. Kerr describes threats to groundwater in "Not Under My Backyard, Thank You," *Science* (June 25, 2010); see also Sharon Kelly, "The Trouble with Fracking," *National Wildlife* (World Edition) (October/November 2011). But the industry insists that it will deal responsibly with its wastes and hastens to reassure people living near drilling sites. Alex Halperin, "Drill, Maybe Drill?" *American Prospect* (May 2010), describes the debate over shale-gas drilling in upstate New York. The area has suffered large job losses, something the shale-gas industry may remedy. Many landowners—including farmers—see the potential for huge boosts to their income. But the industry has reportedly persuaded people to lease drilling rights on their property by making promises that cannot be kept. Environmental impacts are a huge concern. Indeed, one recent study found amounts of methane in drinking water supplies near natural gas wells so high as to pose "a potential explosion hazard"; see Stephen G. Osborn, Avner Vengosh, Nathaniel R. Warner,

and Robert B. Jackson, "Methane Contamination of Drinking Water Accompanying Gas-Well Drilling and Hydraulic Fracturing," *Proceedings of the National Academy of Science (PNAS)* (www.nicholas.duke.edu/hydrofracking/Osborn%20et%20al%20%20Hydrofracking%202011.pdf) (May 17, 2011). See also Joyce Nelson, "A 'Big Fracking Problem': Natural Gas Industry's 'Fracking' Risks Causing Earthquakes," *CCPA Monitor* (February 2011). Linda Marsa, "Fracking Nation," *Discover* (May 2011), is concerned that fracking may release radioactive and other toxic materials from shale in the process of gas extraction. Chris Mooney, "The Truth about Fracking," *Scientific American* (November 2011), notes that a great deal of research into the safety and side-effects of fracking has not yet been done, but the problems may lie less in the technology itself than in carelessness in well drilling and waste disposal.

According to Brydon Ross, "The 'Fracking' Rules," *Capitol Ideas* (May/June 2012), in New York, a requirement for further environmental review has imposed a *de facto* moratorium on fracking. Vermont has an actual moratorium, lasting until 2015. Several states require that fracking companies reveal the makeup of the chemicals they use. And regulations for fracking wastewater treatment and disposal are proliferating. The potential problems are discussed in Brian Colleran, "The Drill's About to Drop," *E—The Environmental Magazine* (March/April 2010). James C. Morriss, III, and Christopher D. Smith, "The Shales and Shale-Nots: Environmental Regulation of Natural Gas Development," *Energy Litigation Journal* (Summer 2010), contend that if companies in the industry act to prevent problems before regulators require such action, this both demands a better understanding of the technology and prevents future litigation.

In 2004 an Environmental Protection Agency (EPA) study said that fracking was little or no threat to drinking water and Congress exempted fracking from federal regulation. Now, however, the EPA is conducting a $1.9 million study to reevaluate fracking technology; see Tom Zeller, Jr., "E.P.A. Considers Risks of Gas Extraction," *New York Times* (July 23, 2010). Preliminary results of the study are expected in 2012. The EPA's web page on fracking and this study is at www.epa.gov/ogwdw000/uic/wells_hydrofrac.html. At a May 5, 2011, hearing of the House Science, Space, and Technology Committee on whether additional studies are needed to determine whether fracking is safe, chairman Ralph Hall (R-TX) called this EPA study "yet another example of this administration's desire to stop domestic energy development through regulation." Scientists are more concerned about potential side-effects of the technology; see Rachel Ehrenberg, "The Facts Behind the Frack," *Science News* (September 8, 2012). Among the side-effects of fracking—indeed, of just using methane as a fuel—is accelerated global warming; see Marianne Lavelle, "Good Gas, Bad Gas," *National Geographic* (December 2012).

So far, the only new fracking regulations from the EPA address air pollution; see e.g., Jeff Johnson, "EPA Issues Fracking Rules," *Chemical & Engineering News* (April 23, 2012). President Barack Obama has set up an interagency oversight committee to streamline regulation and ensure the industry's safe and responsible development: see Ben Wolfgang, "Obama Issues Order to Coordinate Fracking Oversight," *Washington Times* (April 13, 2012). Regulations are not nearly as restrictive as Hall feared. Nor does it seem quite fair when Kevin D. Williamson, "The Truth about Fracking," *National Review* (February 20, 2012), says that "the opposition to fracking isn't at its heart environmental or economic or scientific. It's ideological, and that ideology is *nihilism*." Environmentalists, he says, are opposed to modern technological civilization and would like nothing better than to phase out the human species.

In the YES selection, Diane Katz argues that the new fracking technology has made it possible to release vast amounts of natural gas from deep shale deposits. As a result, we should stop spending massive sums of public money to develop renewable energy sources. The "knowledge and wisdom of private investors" are more likely to solve energy problems than government policymakers. In the NO selection, Deborah Weisberg argues that the huge amounts of water and chemicals involved in fracking pose tremendous risks to both groundwater and surface water, and hence to public health. There is a need for stronger regulation of the industry.

YES ↵

Diane Katz

Shale Gas: A Reliable and Affordable Alternative to Costly "Green" Schemes

Governments at every level across North America are collectively showering billions of tax dollars on "green energy" schemes in an effort to avert global warming and end our "dependence on foreign oil." But in the political arena, there is precious little attention being paid to a far more affordable alternative energy source with great potential to reduce both fossil fuel emissions and imports of Middle Eastern oil.

In contrast to government tax breaks, preferential loans, grants, and other forms of subsidies to wind and solar projects, private investors are moving capital into the production of "shale gas." Trapped within dense sedimentary rock, this "unconventional" natural gas was for decades considered too costly to retrieve. But advances in drilling technologies, along with the rising cost of conventional natural gas, have transformed the economics of shale gas extraction. Consequently, the vast stores of shale gas buried a thousand meters or more below the surface of North America (and beyond) have the potential to dramatically alter both environmental politics and geopolitics.

The actual volume of recoverable shale gas remains imprecise as supplies are still being mapped and evaluated. The National Energy Board estimates Canada's volume to be 1,000 trillion cubic feet, with similar reserves in the United States. Europe also may be home to nearly 200 trillion cubic feet of shale gas.

In Canada, there are major shale gas "plays" in the Horn River Basin and the Montney Formation, both in British Columbia. Major exploration for shale gas is also occurring in the Colorado Group in Alberta and Saskatchewan, the Utica Shale in Quebec, and the Horton Bluff Shale in New Brunswick and Nova Scotia.

When burned, shale gas emits just half the carbon dioxide of coal. Unlike wind and solar power, which produce power intermittently, natural gas is continuously available to produce the steam that powers turbines in the production of electricity. In addition, distribution networks for natural gas already exist, meaning that there is less need to build costly infrastructure. These and other advantages of shale gas call into question the massive public outlays for more problematic "renewable" power sources.

According to energy analyst Amy Myers Jaffe, shale gas "is likely to upend the economics of renewable energy. It may be a lot harder to persuade people to adopt green power that needs heavy subsidies when there's a cheap, plentiful fuel out there that's a lot cleaner than coal, even if [natural] gas isn't as politically popular as wind or solar."

That very dynamic stymied energy mogul T. Boone Pickens in his plan to build the world's largest wind farm in the Texas Panhandle. The plan called for the construction of a wind farm with 687 turbines, driving the production of 1,000 megawatts of electricity—the equivalent of a nuclear power plant.

Shortly after the debut of the project in 2008, natural gas prices declined, making wind energy not competitive enough to attract the $2 billion needed in financing. As Pickens told the *Dallas Morning News,* "You had them standing in line to finance you when natural gas was $9 [per million Btu] . . . Natural gas at $4 [per million Btu] doesn't have many people trying to finance you." The lack of a transmission line to move the wind power to urban centers also contributed to his decision to kill the project, Pickens said.

But governments across Canada have virtually unlimited financing at their disposal in the form of tax revenues, and thus are forcing taxpayers to subsidize costly "renewable" energy projects and transmission build-outs, even though more efficient alternatives exist. The government of Ontario, for example, is forcing utilities (read consumers) to buy "green" power at more than double the market rate for conventional electricity.

In the past, the fine grain of shale rock made tapping the natural gas within particularly difficult. The National Energy Board describes shale as "denser than concrete" and thus virtually impermeable. But from the tenacity of a lone Texan, a productive method to set the gas flowing has emerged. As the *Sunday Times* reports:

> It all began in 1981 when Mitchell Energy & Development, a Texas gas producer, was, quite literally, running out of gas. [George] Mitchell, who founded the firm, ordered his engineers to look into tapping shale, which drillers usually passed through to get to the oil and gas fields below them. . . . For years, [the shale] had been ignored, but Mitchell had a hunch about their potential. "I thought there had to be a way to get at it," he said. "My engineers were always adamant. They would say, 'Mitchell, you're wasting your money.' And I said, 'Let me.'" It took 12 years, more than 30 experimental wells and millions of dollars before he came up with the technical solution.

From *Fraser Forum,* July/August 2010, pp. 18–20. Copyright © 2010 by Fraser Institute. Reprinted by permission. www.fraserinstitute.org

That technical solution is known as "hydraulic fracturing" (or "fracking"), which involves injecting at high pressure a mixture of water, sand, and chemicals into the shale to fracture the rock and allow the release of the natural gas therein. In conjunction with fracking, horizontal drilling is used to maximize the surface area of the borehole through which the gas is collected.

Some environmentalists complain that the chemical compounds used in fracking threaten to pollute soil and groundwater, and they decry the volumes of water used in the production process. In addition, some global warming alarmists oppose the development of new stores of fossil fuel. But in many instances, fracking is conducted thousands of feet below aquifers, and the strata are separated by millions of tons of impermeable rock. Moreover, ever larger quantities of the water used in fracking are recycled. The industry also maintains that stringent regulatory standards are in place to protect the environment. And, as detailed in another article in this edition of *Fraser Forum*, all sources of energy—"renewables" included—involve environmental trade-offs.

Initially, fracking and horizontal drilling were too costly for widespread adoption. But as oil prices rose, these techniques became more cost-effective. Since then, economies of scale and technological innovations have "halved the production costs of shale gas, making it cheaper even than some conventional sources."

Energy analysts expect further cost reductions in shale gas production as major oil and gas companies invest in new technologies. For example, production costs have fallen to $3 per million Btu at the Haynesville Formation, which encompasses much of the US Gulf Coast, down from $5 or more at the Barnett Shale in the 1990s.

The turnabout in shale gas fortunes is all the more remarkable given predictions in the past decade that Canada and the United States were running low on natural gas. US Federal Reserve Chairman Alan Greenspan, for example, declared in 2003 that the United States would have to import liquid natural gas to meet demand.

Doing so would have increased reliance on supplies from Russia and Iran, hardly an appealing prospect for anyone intent on "energy independence." Before the shale gas boom, both countries were thought to control more than half of the known conventional gas reserves in the world. Now, however, Canada and the United States have access to huge domestic stores.

This could cause dramatic shifts in global petro-politics. As energy analyst Amy Myers Jaffe notes, "Consuming nations throughout Europe and Asia will be able to turn to major US oil companies and their own shale rock for cheap natural gas, and tell the Chavezes and Putins of the world where to stick their supplies—back in the ground."

The new accessibility to shale gas will also moderate the influence of OPEC and any potential natural gas cartel by providing affordable and reliable alternative sources of energy. Indeed, US production of natural gas in March hit an historical monthly high of 2.31 trillion cubic feet, topping Russia to become the largest producer in the world. Consequently, natural gas exports once headed to North America are instead heading to Europe, thereby forcing Russia to lower prices for its once-captive customers.

Illustrating the new political tectonics is the recent agreement between Chevron and Poland for natural gas development and production. According to Dr. Daniel Fine of the Mining and Minerals Resources Institute at MIT, "When Chevron announces that they have gas [in Poland], then Russia is shut out" from having a monopoly in Eastern Europe.

Canada will also feel the effects of the energy market shifts. For example, the expansion of US supplies means that Canada will need to find new export opportunities for its natural gas. However, this should not cause problems, analysts say, because supplies of conventional natural gas are declining elsewhere while fuel demands for transportation and electricity are growing.

The private sector is adept at adjusting to shifting trends. For example, a shipping terminal for natural gas imports to be built by Kitimat LNG Inc. was redesigned for exports to the Pacific Rim due to "increases in supply throughout North America—including in the US, Canada's traditional export market."

Unfortunately, federal and provincial governments remain wedded to energy policies that lack the knowledge and wisdom of private investors and fail to account for the dynamic nature of the market. Vast infusions of subsidies obscure the true costs of various energy sources, while disparate regulations and mandates inhibit the unfettered competition that would otherwise determine the most efficient and beneficial fuels. Policy makers and politicians could dramatically improve energy policy by releasing their ham-fisted grip on the energy market.

DIANE KATZ is Director of Risk, Environment, and Energy Policy for the Fraser Institute, an independent non-partisan research and educational organization based in Canada.

Deborah Weisberg   ➡ **NO**

Fracking Our Rivers

On Christmas Day 2007, George Watson returned home from a family dinner to find one of his prized Black Angus cows dead alongside Hargus Creek, a stream that runs through his southwestern Pennsylvania farm.

Over the next three months, Watson lost 16 more cattle—all of which had been bred—making it, as he said, "a double loss." Up to three in one day were found lying near the water. A series of calves died soon after birth.

"I've been raising cattle for 22 years and never had anything like that," said Watson, a Vietnam veteran, who also was having problems with discolored, sludgy well water. A local vet tested the dead cows, but failed to find anything abnormal. Looking back now, Watson wishes he'd had someone test the water in the creek.

Although natural gas wells were being developed all around him, rumors of illegal wastewater dumping in local streams, and a 43-mile fish kill on Dunkard Creek in the same Monongahela River watershed two years later, fueled his darkest fears.

"After my cows died, I suspected it was from brine and waste being dumped, although I can't prove it now," said Watson, who later leased the mineral rights on his farm to Range Resources for $3,000 an acre plus 15 percent production royalties. Drilling hadn't begun as of late last year.

Range is one of 40 companies driving the boom in hydraulic fracturing for natural gas in Pennsylvania, where 53,000 wells are turning pastures and woods into industrial sites. Although hundreds of thousands more have changed the landscape in at least 31 states, Pennsylvania and New York have an abundance of Marcellus Shale wells and, unlike out West, they are close to end users. While vertical drilling and "hydrofracking" for gas has existed for decades, new technologies enable extractors to go more than a mile deep and a mile horizontally to fracture the Marcellus—and release embedded gas—using millions of gallons of sandy, chemical-laden water.

Dunkard Fish Kill

CONSOL Energy's Morris Run borehole and other sources in the Dunkard watershed are under investigation by several federal and state agencies, including the Pennsylvania Attorney General's Office, over possible illegal discharges of hydraulic fracturing fluid, since the level of total dissolved solids, including chlorides, in Dunkard Creek was higher than anything previously associated with coal bed methane wastewater, the only discharge permitted at Morris Run.

"There's pretty strong evidence there was more than coal bed methane water going down that borehole," said Charlie Brethauer of Pennsylvania DEP's water management section. "As far as allegations of illegal activity, I think there's something to it, although to what extent, we don't have any idea yet. We haven't ruled out 'fracking' fluid."

Ed Pressley and his wife Verna live along Dunkard Creek in Brave, Pennsylvania, and watched in horror as fish began going belly up in September 2009 in what would become a massive loss of wildlife that continued for a month. The shells of rare mussels popped open, said Verna, and muskellunge and smallmouth bass bled to death from their gills.

"Kids were putting fish into buckets trying to save them—the tears were running down their cheeks—but there was nowhere to take the fish to," said Verna, a retired science teacher. "We counted 600 dead fish—the stench was overwhelming—just below our dam. It was one of the most devastating emotional experiences of my life."

What made it especially heartbreaking for the Pressleys is that their dream was to turn their property into a living classroom, where children could study kingfishers, blue herons, mudpuppies, turtles, and other forms of wildlife sustained by the water. They were negotiating a conservation easement agreement with the US Department of Agriculture that would protect their land against development for generations to come, and with American Rivers to have a relic industrial dam removed from their section of the stream.

"The folks at Agriculture and American Rivers say they're going to stick with it," Verna said as she stood along Dunkard and peered into the eerily empty water last fall. "But it's going to be years before you'll see fish in here again. I know it's not going to happen in my lifetime."

An EPA interim report about Dunkard's demise cites the presence of golden algae, a toxic organism indigenous to southern U.S. coastal waters, but never before documented in Pennsylvania. Whether it got to Dunkard on migratory birds' feet, drilling equipment that originated in Texas, or by some other means may never be known, but the EPA confirmed that excessive levels of total dissolved solids turned Dunkard so salty the algae were able to thrive.

Golden algae was later found on Whitely Creek, a stocked trout fishery in the same watershed, said Brethauer, who indicated it is likely to spread to other streams.

While the gas drilling industry touts hydraulic fracturing as America's path to energy independence—the Natural Gas Supply Association claims there are enough reserves to meet the nation's needs for a century—some watchdogs say weak regulations and poor enforcement are fueling an environmental nightmare.

The 2005 Energy Policy Act exempts injection of hydraulic fracturing fluids from a key provision in the Safe Drinking Water Act, and federal regulations governing wastewater disposal are limited, according to Deborah Goldberg of Earthjustice, a nonprofit environmental law firm. "Gas wastewater treatment is mostly left to states to regulate and monitor, and most states are way behind the curve."

Ron Bishop, a biochemistry lecturer at SUNY College at Oneonta and a nationally certified chemical hazards management expert, put it this way: "You have to go through more permitting hoops to put a new garage on your property than to drill for gas."

Pennsylvania is in the process of tightening limits on total dissolved solids that can be discharged in rivers and streams, and New York is considering new permitting requirements—generating a de facto moratorium on drilling—although many environmental stakeholders, including New York City, say they aren't strong enough to protect watersheds such as the Delaware River, which provides drinking water to 17 million people.

"Government has to ramp up its regulations tremendously," said Jeff Zimmerman, an attorney for Damascus Citizens for Sustainability and Friends of the Upper Delaware River, groups which formally have protested the New York proposal. "Until an environmentally infallible extraction system can be assured without qualification, the gas drilling industry should not be allowed to operate. It must be failsafe. A single mistake or uncontrolled accident can wipe out, for years and years, important resources, such as those of Dunkard Creek."

Federal lawmakers are also considering legislation—the Fracturing Responsibility and Awareness of Chemicals (FRAC) Act—that would reverse the Clean Water Act exemption and force industry to disclose the names of all of the hundreds of chemicals used in the hydrofracking process, including those traditionally guarded as proprietary information. Pennsylvania makes the names of chemicals available, but not the proportions.

"Some of them are really nasty, like toluene and benzene, which are known to cause cancer," said Bishop. "Others are harmful to wildlife. DB-NPA is a biocide commonly added to fracking water to kill bacteria and algae. Even in amounts too tiny to show up on chemical tests, it's lethal to bay oysters, water fleas, and brown trout."

The staggering volume of fracking fluid used in each horizontal well—up to 6 million gallons of water and 50,000 pounds of chemicals—means environmental impacts can occur on a massive scale, Bishop said. Spills at drill sites and well casing failures—the two most common problems associated with hydrofracking—can cause escaping fluids to contaminate ground and surface water, and gas to migrate underground.

Violations

PADEP cited drillers for more than 450 violations last year. Cabot Oil Co. was charged with a series of spills that polluted a wetlands and killed fish in Stevens Creek, a Susquehanna River tributary in northeast Pennsylvania.

In a separate matter, Cabot is being sued by 15 Dimock residents who claim drilling operations contaminated their drinking water and caused them to suffer neurological and gastrointestinal ills. They are seeking a halt to drilling plus establishment of a trust fund to cover their medical care.

[On Jan. 9, 2010, PADEP also announced it had fined **Atlas Resources $85,000 for violations at 13 different well sites in Greene, Fayette, and Washington counties. The violations included failure to restore well sites after drilling, failure to prevent discharges of silt-laden runoff, and for discharging industrial waste including production fluids onto the ground at 7 of the 13 sites. THE EDITOR.**]

Among the many environmental threats or impacts associated with hydrofracking—including huge withdrawals of water from lakes and streams and erosion and sedimentation from truck traffic on rural roads—one of the more concerning is disposal of wastewater, since about half the liquid used in fracking flows back with additional toxins, including brine six times saltier than ocean water, Bishop said. "This hazardous, industrial waste must be disposed of, but there's no good answer as to how or where. Texas and Oklahoma allow deep well injection, but it doesn't work in Pennsylvania and New York because our rock 10,000 feet down isn't porous enough to absorb the waste."

Before water can be discharged into streams it must be strained, desalinated and restored to an acceptable pH level, but few sewage treatment plants are equipped to deal with the volume and chemical composition of fracking water, and many streams have reached their capacity for assimilating more total dissolved solids, Bishop said. "It's a gigantic problem."

Chris Tucker of Energy In-Depth, a coalition of trade groups managed by the Independent Petroleum Association of America, agrees wastewater disposal is one of the industry's biggest bugaboos.

"Everyone knows we have to get on top of it," he said. "Producers are taking a lot of the wastewater from Pennsylvania into Ohio for deep-well injection. The industry is also looking at mobile recycling facilities, but they're getting quoted one cent a gallon. Consider what that would cost when you're dealing with 3 or 4 million gallons of water."

Although Range Resources' CEO John Pinkerton insists that his company's wastewater poses no threat to freshwater streams, Range has turned to recycling in Pennsylvania, where one-acre impoundments and miles of above-ground pipes circulate frack wastewater among several wells. Range also is exploring additional technologies, including crystallization and evaporation—essentially boiling wastewater and skimming off the salt which could be sold for road de-icing.

"We don't know how much Marcellus play there will be but wastewater disposal will keep pace. If it doesn't, the drilling will cease," Pinkerton said. "We have millions of dollars invested in each well. We have to know where every gallon coming out of the ground will go. It's in our best interest to do it right. To do otherwise would be business suicide."

As a fly fisher, Pinkerton considers himself an environmentalist, and he said natural gas extraction is the only practical alternative to foreign oil and coal. "The idea that we can go to 100 percent renewables before you and I pass away is ludicrous. We need a portfolio of energy solutions—a balanced energy policy—so if oil goes to $300 a barrel, we're not stuck. If we don't figure this out, we're dead meat."

He said every industry has risks and impacts— "you've got to cut down trees to print your magazine," he said—"but temporary inconveniences are necessary for tremendous, long-term gain, unless we all want to walk or ride horses to work."

Both Pinkerton and Tucker decry direct EPA permitting, which the FRAC Act would require. "It wouldn't just slow us down, it would bring us to a stop for four or five years," said Tucker, who points to the job growth he claims his industry has spawned. "We put 48,000 people to work in Pennsylvania and zero in New York because of the de facto moratorium. Where I come from, Wilkes-Barre/Scranton, gas is a godsend for folks who are economically depressed."

Fly fishing guide Glenn McConnell said he felt better about leasing the mineral rights to his land in the Pennsylvania Wilds after Range agreed to address Trout Unlimited.

"The drillers are just as concerned about the environment as you and me," McConnell said. "They don't want to make a bad name for themselves. If something isn't right, they'll correct it immediately."

PA Council Trout Unlimited environmental chair Greg Grabowicz is more focused on problem prevention. "We want assurances that operations will be fail-safe. Our immediate concern is whether DEP can enforce even existing regulations, with such a small staff and so many wells," said Grabowicz, a professional forester. "There's no doubt Pennsylvania's watersheds will change dramatically over the next 30 years from new roads and pipelines, but only time will tell if drillers run into problems that cause catastrophes."

Others, though, already have seen impacts to their favorite coldwater fisheries, including Sam Harper, the DEP water management program chief monitoring Dunkard, who has a camp in the Allegheny National Forest. "There's been a dramatic change in the South Branch of Tionesta Creek, where I fish," he said. "We're seeing a lot fewer brook trout and a lot more roads leading to wells."

And there are likely to be more impacts to woodland streams as ozone from diesel-powered trucks and drilling equipment cause leaf burn and deforestation, according to Al Appleton, a former New York City Department of Environmental Protection commissioner, who serves as technical advisor to Damascus Citizens.

Appleton said too little is also made of the millions of gallons of water sucked from lakes and streams for each hydrofracking operation.

"They may not impact flow during certain times of the year, but drilling isn't a seasonal business," he said. "These companies are withdrawing significant amounts of water constantly."

While PADEP raised drilling permit fees last year to help pay for more site inspections, it also streamlined the permit approval process to 28 days with completion of a basic application—even though the agency admits the need to put more teeth into existing regulations. "Environmentalists focus on wastewater, but the biggest issue for us is what happens at the site," said PADEP spokesman Tom Rathbun. "Is the well 'cased' properly? Are the water pipes built properly?

What about how trucks are crossing streams? That's where our focus needs to be."

In the meantime, lawmakers expect to hold hearings on hydrofracking and to request an EPA study on its effects on the environment, according to Kristopher Eisenla, an aide to FRAC Act co-sponsor Congresswoman Diana DeGette (D-Colorado). "The industry has had a free ride for so long, if greater oversight costs it a few more bucks, in the interest of public health, it's worth it."

[In the Sept. 2009 issue, John Randolph in his page 2 article "The Threats Posed by Marcellus Drilling" identified "the single largest threat to Pennsylvania (also new York, West Virginia, and Ohio) wild-trout streams since the coal/steel era of the Industrial Revolution." After that issue went to subscribers, a "total" fish kill on 43 miles of Dunkard Creek in Pennsylvania raised the question again: "Is the Commonwealth of Pennsylvania protecting its waterways?"   THE EDITOR.]

DEBORAH WEISBERG is an award-winning journalist whose work appears in the *Pittsburgh Post-Gazette, New York Times,* and other publications.

EXPLORING THE ISSUE

Is Shale Gas the Solution to Our Energy Woes?

Critical Thinking and Reflection

1. Do we need energy so badly that we should ignore risks to water supply and human health?
2. In what sense is the knowledge and wisdom of private investors preferable to that of government policymakers when it comes to deciding what to do about energy?
3. How will ample supplies of cheap natural gas affect development of renewable energy supplies such as wind and solar power?
4. Is it true that if greater oversight of an industry (such as the shale gas industry) costs a few more bucks, in the interest of public health, it's worth it?

Is There Common Ground?

Even if the proponents of unrestrained exploitation of shale gas by fracking are right when they say it solves our energy problems, the supply of shale gas will not last forever. The public—and its health—will remain a concern, as will concern over carbon emissions and the need for ample amounts of energy to run our civilization.

1. Should we, as suggested by Diane Katz, stop investing public money in developing alternative energy sources? If not, why?
2. Is government regulation essential to protect public health? Visit the Public Health Service at www.usphs.gov/aboutus/mission.aspx to explore one agency's approach.
3. Another kind of fossil fuel we have not yet tapped in any major way is shale oil. (See the Bureau of Land Management's oil shale information at http://ostseis.anl.gov/guide/oilshale/.) Discuss the potential benefits (and environmental costs) of exploiting this resource.

Create Central

www.mhhe.com/createcentral

Additional Resources

Mark Fischetti, "The Drillers Are Coming," *Scientific American* (July 2010)

Sharon Kelly, "The Trouble with Fracking," *National Wildlife* (World Edition) (October/November 2011)

Marianne Lavelle, "Good Gas, Bad Gas," *National Geographic* (December 2012)

Chris Mooney, "The Truth about Fracking," *Scientific American* (November 2011)

Internet References . . .

Energy from Shale

www.energyfromshale.org/hydraulic-fracturing/shale-gas

Fracking at ProPublica

www.propublica.org/series/fracking

National Oil Shale Association

oilshaleassoc.org/

Selected, Edited, and with Issue Framing Material by:
Thomas Easton, *Thomas College*

ISSUE

Is Renewable Energy Green?

YES: Andrea Larson, from "Growing U.S. Trade in Green Technology," testimony before the U.S. House Committee on Energy and Commerce, Subcommittee on Commerce, Trade and Consumer Protection (October 7, 2009)

NO: Lamar Alexander, from "The Perils of 'Energy Sprawl,'" remarks given to Resources for the Future Policy Leadership Forum (October 5, 2009)

Learning Outcomes
After reading this issue, you will be able to:
• Explain what "green" means.
• Describe the environmental impacts of renewable energy technologies.
• Explain how renewable energy technologies benefit the economy.
• Explain why nuclear power may be preferable to renewable energy technologies.

ISSUE SUMMARY

YES: Professor Andrea Larson argues that "green" technologies include, among other things, renewable energy technologies and these technologies are essential to future U.S. domestic economic growth and to international competitiveness.

NO: Senator Lamar Alexander (R-TN) argues that the land use requirements of solar and wind power threaten the environment. We must therefore be very careful in how we implement these "green" energy technologies. He also believes the best way to address climate change (by cutting carbon emissions) is with nuclear power.

"**G**reen" has long been understood to mean environmentally friendly. A "green" energy technology is sustainable. It is not based on fossil fuels—which is not quite the same as saying it is based on renewable energy sources—so it does not add carbon dioxide or other greenhouse gases to the atmosphere. It does not pollute air or water. It does not diminish biodiversity or harm ecosystems.

Commonly cited examples of "green" energy technologies include solar power, wind power, biomass power, hydropower, tidal power, geothermal power, and wave power. The first two—solar and wind—are the ones most frequently discussed in magazines and newspapers, perhaps because they can be installed in units small enough to fit in a family's backyard. John Gulland and Wendy Milne describe in "Choosing Renewable Energy," *Mother Earth News* (April/May 2008), the process of converting their home to run mostly on solar and wind power, with a wood stove (biomass power) for heat. "[U]sing renewables," they write, "has increased our independence and

sense of security, and lessened our carbon emissions . . . another benefit is in knowing that we are contributing, even in a small way, to the health and sustainability of the local community." Gulland and Milne also note that going green wasn't always easy. There are trade-offs, both at their very local scale and at much larger scales. Yekang Ko, Derek K. Schubert, and Randolph T. Hester, "A Conflict of Greens: Green Development versus Habitat Preservation—The Case of Incheon, South Korea," *Environment* (May/June 2011), describe the conflict between tidal power and wetland preservation.

"Green" energy technologies can also be scaled up to provide much larger amounts of electricity. An Australian study says they could provide 60 percent of Australia's electricity by 2040 and reduce associated carbon dioxide emissions by 78 percent; at present, Australia has more than 50 wind farms with about a thousand wind turbines. In Iceland, "green" energy technologies account for three quarters of the installed electrical generation capacity. In California the corresponding figure is one quarter, in Sweden, one third, in Norway, one half. Wind alone meets a fifth of Denmark's needs. See Rachel Sullivan and Mary-Lou Considine, "Hastening Slowly in the Global Renewables Race," *Ecos* (April–May 2010). See also the special section on "Scaling Up Alternative Energy," *Science* (August 13, 2010), which notes among other things that "there is potentially more than enough energy available from the Sun, wind, and other renewable sources to replace fossil fuels."

In the United States, "green" energy technologies play much less of a role. Hydropower, at 7 percent (77,000 MW) of national electrical generating capacity, is an important contributor; unfortunately, potential growth is very limited. Wind power is growing rapidly (see *Vital Signs 2010*, Worldwatch, 2010) and the potential is huge but at the end of 2009, it provided only 34,863 MW of electrical generating capacity, about 3 percent of the total (and only about 0.2 percent of actual electricity produced). Solar power plays an even smaller role, with just over 2,000 MW installed; it too is growing rapidly. Wind and solar thus promise continued growth in manufacturing of equipment, jobs for installation, and displacement of fossil fuels.

Unfortunately, wind and solar power take up a lot of land. Wind farms cover miles of high or windy terrain with towers and spinning blades. One major trend is to move them offshore, where they are out of sight to most people (and the wind is steadier). Solar electric power requires large expanses of solar panels or concentrators, preferably in locations where the sky is rarely cloudy, meaning arid lands or deserts. There is in fact a plan afoot to develop solar power in the Sahara desert to supply· Europe with up to 15 percent of its electricity as early as 2015. See Ashley Seager, "Solar Power from Sahara a Step Closer," *The Guardian* (November 1, 2009; www.guardian.co.uk/business/2009/nov/01/solar-power-sahara-europe-desertec), and Daniel Clery, "Sending African Sunlight to Europe, Special Delivery," *Science* (August 13, 2010). It has been noted that if just 1 percent of the Sahara were covered with solar concentrators, it could meet the electrical needs of the entire world.

Power lines can be run across or under the Mediterranean Sea to bring electricity from the Sahara to Europe. Getting it to the rest of the world is a more difficult problem. It thus seems sensible to install "green" energy technologies closer to where the energy will be used. In Ontario, Canada, the provincial government has embarked on a major effort to increase the "green" component of its energy supply. As one component of the effort, the Hay Solar company is offering to provide farmers with free barns—whose roofs are covered with solar cells. Hay calculates that the barns will provide a large area for solar energy collection and—despite many more cloudy days in Ontario than in the Sahara—pay for themselves over 20 years, after which they belong to the farmers. See Chris Sorensen, "Absolute Power?" *Maclean's* (June 14, 2010). Rooftop solar power has been discussed in the United States as well; it has a number of advantages—

it does not interfere with other uses of land, has minimal red tape, and is close to end users. Southern California Edison has proposed generating 500 MW of electricity by scattering solar power units on rooftops in the region (see David Anthony, "Where Will Solar Power Plants Be Built—Deserts or Rooftops?" *Greentechmedia* (February 22, 2010; www.greentechmedia.com/articles/read/where-will-solar-power-plants-be-built-deserts-or-rooftops/). Sara C. Bronin, "Curbing Energy Sprawl with Microgrids," *Connecticut Law Review* (December 2010), notes that solar power is not the only thing that can be localized—small wind power systems, geothermal wells, and fuel cells can also contribute to a decentralized, efficient energy system.

It thus seems clear that "green" energy technologies can be deployed without necessarily having huge environmental impacts. The real question is whether we can deploy *enough* "green" energy technology to replace fossil fuels, and one major obstacle is the sheer scale of the need for energy; see Richard A. Kerr, "Do We Have the Energy for the Next Transition?" *Science* (August 13, 2010). Many people contend that nuclear power is much better suited to meeting future needs; see Charles Forsberg, "The Real Path to Green Energy: Hybrid Nuclear-Renewable Power," *Bulletin of the Atomic Scientists* (November/December 2009), and J. B. Bradley, "The Nuclear Revivalist," *Popular Science* (July 2010). Allan Sloan and Marilyn Adamo, "If You Believe in Magic, Green Energy Will Be Our Salvation," *Fortune* (July 26, 2010), add that part of the problem in shifting to "green" energy is that we must start from such a small base. Another part is that we use so much energy for transportation. Jesse Ausubel, "Renewable and Nuclear Heresies," *International Journal of Nuclear Governance, Economy and Ecology* (vol. 1, no. 3, 2007), argues that renewable energy technologies are not really green, largely because when developed to a scale at which they might contribute meaningfully to society's energy requirements, they will cause serious environmental harm. He considers nuclear power a much "greener" way to meet society's energy needs. Stewart Brand, *Whole Earth Discipline: An Ecopragmatist Manifesto* (Viking, 2009), finds renewable energy environmentally unacceptable for similar reasons. Robert I. McDonald, et al., "Energy Sprawl or Energy Efficiency: Climate Policy Impacts on Natural Habitat for the United States of America," *PLoS One* (vol. 4, no. 8, 2009) (www.plosone.org/article/info:doi/10.1371/journal.pone.0006802), conclude, "The land-use intensity of different energy production techniques varies over three orders of magnitude, from 1.9–2.8 km^2/TW [terawatt] hr/yr for nuclear power to 788–1000 km^2/TW hr/yr for biodiesel from soy. In all scenarios, temperate deciduous forests and temperate grasslands will be most impacted by future energy development, although the magnitude of impact by wind, biomass, and coal to different habitat types is policy-specific. . . . The possibility of widespread energy sprawl increases the need for energy conservation, appropriate siting, sustainable production practices, and compensatory mitigation offsets."

Amory B. Lovins, "Renewable Energy's 'Footprint' Myth," *Electricity Journal* (June 2011), energetically criticizes claims that renewable energy has a large footprint and nuclear power is to be preferred, as well as the idea that power plants (such as wind and solar) must be able to generate power a large percentage of the time. Because the power grid interconnects thousands of power plants, says Lovins, deficits in single power plants are amply compensated, and this will remain true as solar and wind (etc.) play larger roles.

Testifying at the October 7, 2009, hearing on "Growing U.S. Trade in Green Technology" before the U.S. House Committee on Energy and Commerce, Subcommittee on Commerce, Trade and Consumer Protection, Mary Saunders, Acting Assistant Secretary for Manufacturing and Services, International Trade Administration, U.S. Department of Commerce, says she sees increasing demand for green energy technologies providing great export opportunities and adds, "Policies that support the early development and commercialization of green technologies are critical to the competitiveness of U.S. firms and improve their competitive edge in the global marketplace." If we do not support such technologies, other nations will claim the lead and the economic benefits.

The current rapid growth of wind and solar power supports, in the YES selection, Professor Andrea Larson's argument that "green" technologies—including, among other things, renewable energy technologies such as wind and solar power—are essential to future U.S. domestic economic growth and to international competitiveness. In the NO selection, Senator Lamar Alexander (R-TN) argues that the land use requirements of solar and wind power threaten the environment. We must therefore be very careful in how we implement these "green" energy technologies. He also believes the best way to address climate change (by cutting carbon emissions) is with nuclear power.

YES ⬅

Andrea Larson

Growing U.S. Trade in Green Technology

Green technology and clean commerce are the future. Green technology has become, and will increasingly be, a major economic growth area for the U.S. and world trade. There is no reason the U.S. cannot be a world leader through export of clean technology and clean commerce innovation, and U.S. leadership should be a strategic goal. Why? Because:

1. Investing in clean energy and clean materials is essential for intelligent economic development, human health protection, and ecosystem preservation.
2. U.S. leadership in clean energy and materials (green technology) creates jobs, stimulates innovation, drives exports, and differentiates U.S. technology, education, and skills in global markets.
3. The U.S. could have an advantage in world trade, but on the current path the U.S. will continue to fall behind.

Green Tech and Clean Commerce Is the Future

Population and economic development pressures are colliding with the ability of nature to deliver clean air, water, and soil. Yet the design of the industrial system that brought us to this point in history was based on assumptions of limitless resources and limitless capacity for natural system regeneration, even in the face of our waste streams. Responding to climate change and green tech opportunities are just the beginning of a major shift in this century for business. New design for business is imperative because the forces of change are accelerating.

It is not just the current economic downturn that confounds us. We face unacceptable income and opportunity disparities at home and poverty worldwide as global population grows from 6.5 to 9 billion in the next few decades. Worldwide over 2 billion people are moving rapidly into the middle class, and they will want all the opportunities and material wealth that the richest populations in western societies now view as normal. Today we concurrently face an economic downturn, a climate crisis, an energy security crisis, energy price volatility, new environmental health challenges, and ecological systems in dramatic decline.

If that were not enough, the U.S. also faces a competitiveness crisis as it loses ground to other countries that are already strategically committed to mobilizing state resources behind domestic businesses that will produce solutions to these problems. Other countries have mounted national efforts to reach clean commerce goals (e.g. renewable energy, domestic "green" companies, dramatic efficiencies, accelerating advances in PV solar design innovation, advancing clean public transportation, protecting consumers from toxic materials, and providing subsidies and incentives to advance their industries in global markets).

The larger picture shows capitalism as currently designed is at a crossroads. It must deliver on its promise of broad prosperity, yet its very design appears to undermine the ecological systems and healthy communities on which it depends. It needs an overhaul: clean energy and materials provide an answer. The U.S. should be leading this change, not following. . . .

Defining Green Technology

Green technology is one term of several used today to encompass a range of activity and innovation to simultaneously address economic development needs, health protection, and preservation of ecosystem services (e.g. the natural systems that provide us with clean air, water, soil, and food). Other terms include sustainability, clean commerce, cleantech, sustainable business, and sustainability innovation. The activities these terms reference challenge existing ways of designing and delivering not just energy, but the entire set of interdependent systems and supply chains that provide food, shelter, consumer products, and transportation modes.

We will use the abbreviation GT/CC throughout this testimony to refer to green tech and clean commerce, two terms that represent the ideas under discussion.

GT/CC refers to technology innovation, but also non-technical innovation, the latter represented by innovative supply chain management or innovative financing mechanisms to install urban PV solar installations that pay residents to sell excess electricity back to the grid. The non-technical innovative frontier must also be a focus for green tech and clean commerce innovation and U.S. competitiveness.

Furthermore, GT/CC is not just about energy. The fundamental basis of commerce and trade is energy AND materials. Both must be managed and designed to meet human needs and optimize ecological system functions.

U.S. House of Representatives, October 7, 2009.

Thus green chemistry and green engineering practices are equally as important to green tech and clean commerce (GT/CC) as renewable energy technologies. PV solar systems that expose their production workers to toxins, are thrown away in landfills after use, then pollute water supplies, are not the solutions we need. "Fresh" vegetables and fruit grown with agricultural chemicals, processed, and transported thousands of miles and lacking fundamental nutrients that urban garden-grown food provides are not the solutions we need. More efficient lighting replacements that create mercury waste may save energy but are still poor designs. In other words, poorly thought out, so-called green technology improvements focused on today's hot topics (climate and energy are the focus today) are common. But a deeper design perspective is needed. First, a systems view is required. One that understands every "green" energy solution, in fact every energy AND product selection by a company or a consumer, reflects materials choices and embedded energy decisions that must be made visible, examined and evaluated for their life cycle implications. Fortunately this is now happening, led by innovative entrepreneurs. But it must be expanded and accelerated.

Nor is green technology just about efficiency. It is about that, but more importantly it is about innovation. Efficiency just allows us to do the same old things at lower cost and using less energy and fewer materials. A laudable improvement, but not the solution. Innovation creates fundamentally new solutions, preferably systems-oriented solutions that prevent and eliminate the problems we face now with climate alteration and unsafe products.

The concept that ties together innovation and both clean energy and materials is the notion of cradle to cradle design. Our current commercial practices extract raw materials, make products, generate waste streams that impact air and water, expose production workers, sell to consumers who use the products and throw them away, and leave the materials to decompose and contaminate our air and water from the landfill, incinerator or Third World country dumping destination. Think about how the costs and benefits are allocated in this linear system. This is called a cradle to grave product life cycle. The alternative is cradle to cradle design derived from systems thinking, that reduces or eliminates energy and material inputs, including toxicity *by design from the outset* to avoid employee, user/consumer, and ecosystem contamination. Under a cradle to cradle design, selected materials can be safely returned to the earth or maintained within closed recycling systems that use waste from one production and use process, as the feedstock for another.

The "greentech" issues or what I am calling the green technology and clean commerce issues (GT/CC) constitute a central challenge for governments. Providing ever growing volumes of products and services (under current design parameters) to support economic development also gives us pollution and costs that are externalized (and inequitably so) onto the population in one form or another (higher taxes for regulation, disease, and more expensive health insurance for chronic illnesses). Examples are air pollution (excessive concentrations of toxins in the air contributing to the asthma epidemic, among other respiratory problems), unsafe foods (linked to diabetes, obesity, and food contamination), excessive carbon dioxide concentration in the atmosphere (climate change and volatility), and water supply threats and shortages due to industrial contamination.

As world population rises to 9 billion in the next few decades and capitalism as currently designed stumbles in its promise of greater prosperity and results instead in wealth creation accompanied by income disparities, climate change, and waste streams increasingly tied to chronic human health challenges, a clean commerce solution is emerging. This is an alternative approach to business that we call green technology and clean commerce. This movement is obvious in the current emphasis on clean energy alternatives in response to climate change. . . .

U.S. Competitiveness

Transformation in the next decade to an alternative mindset about energy and materials is key to U.S. competitiveness and mandatory if global society is to handle the challenges of population growth, energy demands, and material throughput volumes required to provide prosperity for billions more people. We can choose to let others lead or we can mobilize and combine all the elements we have in this country to lead.

This discussion acknowledges that the U.S. has declared 25% renewable energy goals by 2025 with the February 2009 ARRA legislation. The clean technology stimulus accounts for about $66 billion, just ahead of China's stimulus investment. The important fact, nonetheless, is that we come to the table late. By way of example, according to the U.S. International Trade Commission, "Denmark, Germany, India, Japan, and Spain accounted for a combined 91 percent of global exports of wind-powered generating sets in 2008."

Globally, investments in GT/CC have been growing rapidly. For instance, new investments in sustainable energy increased between 25% and 73% annually from 2002 to 2007, until growth fell to only 5% in 2008 following the 2007–08 recession. Nonetheless, even in 2008, total investments in sustainable energy projects and companies reached $155 billion, with wind power representing the largest share at $51.8 billion. Meanwhile, the world's 12 major economic stimulus packages proposed to invest another $180 billion collectively in coming years. Also in 2008, sustainability-focused companies as identified by the Dow Jones Sustainability Index or Goldman Sachs SUSTAIN list outperformed their industries by 15% over a six-month period. Longer horizon analyses indicate companies screened for sustainability factors match or exceed the performance of conventional firms. These are companies that focus not only on renewable energy sources but also energy conservation, environmentally safer products, and improved corporate governance.

Despite being a leader in some areas, however, the U.S. was not an overall leader in GT/CC. From 2000 to 2008, venture capital investments in U.S.-based renewable energy companies increased from 0.6% of all VC investments to 11.84%, and in 2008, venture capital and private equity made new investments in energy efficiency and renewable energy worth $7.72 billion in North America and $3.05 billion in Europe. Moreover, the U.S. had the most GT/CC business incubators in 2008, with 56. The UK was next in incubators with 21, and 16 were in Germany. Yet Europe as whole was home to 46% of the global total of incubators, versus 40% for the U.S. Furthermore, North American investments in sustainable energy shrank 8% in 2008 to $30.1 billion, while in Europe they increased 2% to $49.7 billion. Many other major emerging economies also saw investments in their renewable energy sectors increase: Brazil's increased 76% to $10.8 billion (mainly due to ethanol), China's increased 18% to $15.6 billion, and India's increased 12% to $3.7 billion. Even in Spain investments reached $17.4 billion in 2008, or $430 per capita compared to North America's $57 per capita. For investments specifically in publically traded renewable energy and efficiency companies, Chinese companies led in 2008 with $2.8 billion, followed by Portugal ($2.6 billion), the U.S. ($2.1 billion), and Germany ($1.5 billion). In fact, in 2008, China became the world's largest manufacturer of photovoltaic panels, with 95% of them destined for export. This output means China may soon surpass both German and American manufacturers.

Indeed, China has recently made massive moves toward a CT/CC economy. For instance, China now has 60% of the total global capacity for solar thermal water heaters. Even such a relatively minor innovation saved 3 million tons of oil equivalent in 2006 according to the International Energy Agency. China is also nurturing and protecting its domestic wind power producers, reserving contracts for them and restricting foreign firms. The size of China's market for GT/CC creates significant opportunities for development of domestic innovators and mass producers. Nonetheless, China has a way to go: other countries have put themselves into leadership positions over the past two decades through a series of policies. Those world leaders have been Japan, Denmark, Spain, and Germany.

In 1996, Japan set a target by 2010 of using 3% (roughly 19 gigaliters oil equivalent) of primary energy supply from renewable sources excluding hydropower and geothermal energy. In 2008, the target was amended to represent an upper bound while 15.1 Gl was established as a lower bound. That goal plus grants for residential solar PV installations allowed Japan to lead the world in installed solar capacity from 1999 to 2005, which also allowed Japanese companies such as Sharp to gain an early manufacturing lead. Sharp and other Japanese companies remain competitive in the U.S. market to this day, even though

Germany overtook Japan in installed capacity in 2006. In 2007, Japan established Renewable Portfolio Standards that required utilities to use renewable sources of electricity generation, to reach 16 TWh by 2014. The RPS also set prices for solar PV rates, and in December 2008, Japan allocated another $9 billion for solar subsidies, which is less than California's current solar subsidy program but reaches more eligible people. Japan continues to invest in solar research, including space-based solar energy.

Denmark began to shape its lead in GT/CC in 1976, when its Energy Research Program granted generous subsidies to renewable energies. Danish renewable energy companies turned heavily toward wind power, selling that technology domestically and abroad, especially in California. In 1989, new laws required utilities to buy electricity from renewable sources and co-generation plants, and a series of subsidies and other government support boosted GT/CC through the 1990s. By 2003, Denmark dominated the global market for wind-power generator sets, selling $966 million or 79.5% of the market. Denmark still gets a larger share of its energy from wind than any other country and sold $1.2 billion worth of generator sets in 2008, or 23.4% of the global market. Meanwhile, Danish Vestas controls 17.8% of the wind turbine market, putting Danish companies behind Germany and ahead of the U.S., Spain, and China in that field. In 2008, the Danish government's Agreement on Energy Policy set goals of 20% of gross energy consumption from renewable sources by 2011, with incentives for de-centralized production, research, and other activity.

On the other side of Europe, Spain had a mere 979 GWh of renewable energy generation, almost all of it hydro-electric, in 1990. Yet in 2007, that same generation had risen 33-fold to 32,714 GWh, with wind accounting for about two-thirds of the total. A series of steps similar to those in Japan and Denmark led to this rapid rise, which has ultimately left Spain a major force in the world's solar and wind energy markets. Spain's 1980 Law for the Conservation of Energy first established subsidies for renewable energy sources feeding into grid. In 1997, the Law of the Electricity Sector guaranteed grid access for renewable sources and later laws set prices as well as targets, such as 12% of energy from renewable sources by 2010. With this support, Spain ranked third globally in 2008 in installed wind capacity with 16.8 GW and controlled 8.8% of the market for wind generator sets and 14.9% for turbines. It has also been a leader in solar thermal plants, building Europe's first in 2007 and continuing to develop others.

Germany, finally, has achieved some of the broadest, most profound changes en route to a GT/CC economy. It reached its Kyoto Protocol emissions target of a 20% reduction of GHG emissions from 1990 levels in 2007, a year early. A series of policies has enabled this progress, such as the 1991 Feed-in Tariff Act that required utilities to purchase electricity from any supplier on the grid. Later laws, such as the 2000 Renewable Energies Act and its subsequent updates, have guaranteed prices for renewable energies and set broad environmental targets. Germany in 2009 set even more ambitious plans for reducing overall emissions and dependence on fossil fuels. . . .

In 2008 in Germany, revenue from construction of renewable energy facilities was 13.1 billion Euros (approximately $19.7 billion) and from operation was 15.7 billion Euros ($23.6 billion), representing approximately 278,000 jobs in all. The total revenue from these two activities increased 188% relative to 2003. Meanwhile, the German government's Market Incentive Program, through grants and other incentives, encourages renewable energies by direct funding, which attracts additional investment. From 2000 to 2008, 1.2 billion Euros of direct funding attracted an additional 8.6 billion Euros of outside investment, with government funding for renewable energy R&D directed mainly to solar and wind. The results have been a near quintupling of electricity generated from renewable sources since 1990. In contrast, U.S. government subsidies totaled $29 billion from 2002–2008 for renewable energies, more than half for corn ethanol, which paled in comparison to $72 billion in subsidies for fossil fuels.

What you see when reviewing different countries' strategies is policy variation customized to local conditions but built upon a consistent pattern of core features that includes protections to control consumer costs and mitigation for windfall profits to any players. Simplicity is important to keep public administration costs low and company and individual transaction costs minimal. Consistent policies, gradual amendments to update, and stable supports (whether direct investments or tax incentives) are essential to encourage equipment manufacturers to innovate and to mass produce. Clear and consistent signals also reassure investors that markets will be relatively predictable within adequate time frames for generating returns. In summary, successful government policies appear to include key stakeholders and set ambitious targets, and then address concerns about price-gouging and the factors that typically drive innovators and companies away: instability, uncertainty, and inconsistency.

The U.S. can catch up, but when other countries are working from 20 year-plus guaranteed grid access for renewable energy producers in Spain and Germany (starting in 1991 in Germany) and well-established Spanish Feed-In Tariffs (TIFs) that built on German and Danish examples established well over a decade ago, it suggests the magnitude of the catch up challenge. These countries jumped in early, learned and adapted, and can now act faster and more effectively to build their CT/CC going forward. For the huge and rapidly growing markets for GT/CC in India and China, the U.S. faces governments quickly moving to protect and support fledgling industries that will produce clean cars and public transportation technologies to address pollution impacts, clean energy production (to offset reliance on dirty coal), and the state of the art green components and systems to address the many development and pollution/health problems they know they must solve.

Final Thoughts

The economic growth paradigm and accompanying common knowledge that told us growth had to come first, followed only much later by investment in environmental and health protection (the path of western industrialized societies) will not be sufficient for India and China. I tell my MBA students that given the pace of innovation in those countries around clean commerce goals, the U.S. will be buying most of its clean technology solutions from Indian and Chinese companies in 10 years.

I would also suggest that the U.S.'s geopolitical decline, should it come to pass, will be reflected in our unwillingness to step up to the GT/CC challenge that current population, resource, pollution, and technology development conditions impose.

I am not an advocate of government regulation unless the private sector lacks the ability to provide for the public good. Unfortunately, companies trying to move toward GT/CC, while admirable, are in a race against the cumulative decisions of firms and individuals that continue to erode the commons that is our ultimate source of all wealth, social and financial.

We tend to think of the commons as natural systems (air, water, or land); we might want to consider adding our children's bodies to that collective commons. The Centers for Disease Control [and Prevention] extensive research on contaminants in human blood, immune, and reproductive systems suggest that this century long industrial experiment that clearly has had decisive negative influences on our ecological systems and atmosphere, is also at work on the human body and children's health. Are we surprised?

The last thing I want to see is unnecessary regulation. I work with private sector innovators and emphasize the amazing capacity of markets and entrepreneurial forces in society to create the changes we need to see. But this activity must be framed with enabling and supporting policy that sets the rules and provides consistent and intelligent guidance so that markets and human ingenuity can do the rest.

In addition, let us keep in mind, in the polarized and ideologically laced discussions that pass for policy debate, that there are no purists. State subsidies and consistent long-term government support for fossil fuels played a large part in giving us the energy and materials system we live with today. Subsidies, just in recent years alone, explain why GT/CC activities remain vulnerable and investment capital moves slowly.

Can the U.S. build a GT/CC strategy? Through insufficient investment and lack of policy leadership the U.S. continues to lose ground in its learning pace and its domestic experience to countries willing to back their companies with capital and create mutually reinforcing incentives to mobilize citizen behavior, corporate investment, education, and state decision making. While the hesitancy of the U.S. to create industrial policy to lead in GT/CC is historically understandable, other countries without our political and ideological history (and gridlock) have put policies in place. First we must get our own house in order. It is only then that we will have built the necessary platform for leadership in world trade.

The challenge is straightforward, if ambitious. Future prosperity depends on economic development solutions that address poverty and extreme disparities in income distribution while simultaneously delivering on job creation, skill development, and education for the future. Industrial and commercial activity that fails to actively support provision of clean, healthy products, and clean air, water, shelter, transport, and food, by definition undermines that prosperity. Fortunately the know-how and tools are now available in the form of GT/CC practices and innovation. . . .

ANDREA LARSON is a Professor in the Darden School of Business, University of Virginia. Her research has focused on innovation and entrepreneurship, strategy, and sustainability.

Lamar Alexander

➡ **NO**

The Perils of "Energy Sprawl"

. . . **I** believe that a new Nature Conservancy scientific paper titled "Energy Sprawl or Energy Efficiency: Climate Policy Impacts on Natural Habitat for the United States of America" will one day occupy a place among the pioneering actions that we honor in the conservation movement. The paper warns that during the next 20 years new energy production, especially biofuels and wind power, will consume a land mass larger than the state of Nebraska. This "energy sprawl," as the authors termed it, will be the result of government cap and trade and renewable mandate proposals designed to deal with climate change. The paper should serve as a Paul Revere ride for the coming renewable energy sprawl. There are negative consequences, as well as positive effects, from producing energy from the sun, the wind, and the earth. And, unless we are as wise in our response as the authors have been in their analysis, our nation runs the risk of damaging the environment in the name of saving the environment.

The first insight in the *Nature Conservancy* paper is in describing the sheer size of the sprawl. The second insight is in carefully estimating the widely varying amounts of land consumed by different kinds of energy production. Finally, the paper suggests four ways to reduce carbon emissions while minimizing the side effects of energy sprawl on the landscape and wildlife habitat. The first recommendation is energy conservation. Second is generating electricity on already developed sites, as when solar panels are put on rooftops or when a chemical company uses byproducts from its production processes to make heat and power. The third recommendation is to make carbon regulation flexible enough to allow for coal plants that recapture carbon or nuclear power plants that produce no carbon or for international offsets. Fourth, the paper suggests careful site selection.

This makes me think of my own experience as Governor 25 years ago when Tennessee banned new billboards and junkyards on a highway over which 2 million visitors travel each year to the Great Smoky Mountains National Park. Then, that decision attracted little attention. Today, it helps to preserve one of the most attractive gateways to any national park. But, as all of us know, if the billboards had gone up then it would be almost impossible today to take them down today. The same will be true with wind turbines, solar thermal plants, and other new forms of energy production.

My purpose here today is to challenge you and the organizations who have traditionally protected our landscape, air and water, and wildlife habitat to do the same with the threat of energy sprawl. To ask you, first, to suggest to governments and policy makers and landowners before it is too late the best choices and the most appropriate sites for low-carbon or carbon-free energy production. And, second, I want to ask you to do something that gives many conservationists a stomachache whenever it is mentioned—and that is to rethink nuclear power, because as the *Nature Conservancy*'s paper details, nuclear power in several ways produces the largest amounts of carbon-free electricity with the least impact.

I learned a long time ago that it helps an audience to know where a speaker is coming from. Well, I grew up hiking and camping in the Great Smoky Mountains where I still live two miles from the park. As a Senator I have fought for strict emission standards for sulfur, nitrogen, and mercury because many of us still breathe air that is too polluted. I have introduced legislation to cap carbon from coal plants because I believe that human production of carbon contributes to global warming. I have helped to create 10,000 acres of conservation easements adjacent to the Smokies because it preserves views of the mountains and wildlife needs the space. I drive one of the first plug-in electric hybrid cars because I believe electrifying half our cars and trucks is the quickest way to clean the air, keep fuel prices down, reduce foreign oil use, and help deal with climate change. And I object to 50-story wind turbines along the Appalachian Trail, for the same reason I am co-sponsor of legislation to end the coal mining practice called mountaintop removal—not because I am opposed to coal plants or wind power in appropriate places, but because I want to save our mountaintops.

* * *

Let me offer a few examples to paint a clearer picture of what this energy sprawl might look like in 20 years.

As the *Nature Conservancy* paper notes, most new renewable electricity production will come from wind power which today provides about 1.5 percent of our country's electricity. Hydroelectric dams produce about 7 percent of our electricity and some of them are being dismantled. Solar and all other forms of renewable electricity produce less than 1 percent today. President Bush first suggested that wind power could grow from 1.5 percent to 20 percent by 2030 and President Obama has set out enthusiastically to get this done. In fact, the combination of presidential rhetoric, taxpayer subsidies and mandates have very nearly turned our national electricity policy into a national windmill policy.

To produce 20 percent of America's electricity from wind turbines would require erecting 186,000 1.5-megawatt wind turbines covering an area the size of West Virginia. According to the American Wind Energy Association, one megawatt of wind requires about 60 acres of land, or in other words, that's one 1.5 megawatt wind turbine every 90 acres. These are not your grandmother's windmills. They are 50 stories high. Or, if you are a sports fan, they are three times as tall as the skyboxes at the University of Tennessee football stadium. The turbines themselves are the length of a football field, they are noisy and their flashing lights can be seen for up to twenty miles. In the eastern U.S., where the wind blows less, turbines would work best along scenic ridge tops and coastlines. National Academy of Sciences says up to 19,000 miles of new high voltage transmission lines would be needed to carry electricity from 186,000 wind turbines in remote areas to and through population centers.

So many wind turbines can create real threats to wildlife. The Governor of Wyoming has expressed concern about protecting the Sage Grouse's diminishing population in his state as a result of possible habitat destruction from wind farms. The American Bird Conservancy estimates that each wind turbine in this country kills as many as seven or eight birds each year. Multiply that by 186,000 wind turbines and you could predict the annual death of close to 1.4 million birds per year. Then there are the solar thermal plants, which use big mirrors to heat a fluid and which can spread over many square miles. Secretary of Interior Ken Salazar recently announced plans to cover 1,000 square miles of federally owned land in Nevada, Arizona, California, Colorado and New Mexico and Utah with such solar collectors to generate electricity. Senator Dianne Feinstein of California, who has spent most of her career trying to make the Mojave desert a national monument, strongly objected to a solar thermal plant in the desert on federal land just outside the Mojave National Preserve that would have covered an area 3 miles by 3 miles. Plans for the plant were recently canceled.

The only wind farm in the southeastern United States is on the 3,300 foot tall Buffalo Mountain in Tennessee. The wind there blows less than 20 percent of the time making the project a commercial failure. Because of the unavailability of wind power, renewable energy advocates suggest that we southeasterners use biomass, a sort of controlled bonfire that burns wood products to make electricity. Biomass has promise, to a point. Paper mills can burn wood byproducts to make energy. And clearing forests of dead wood and then burning it not only produces energy but can help to avoid forest fires. According

to the Conservancy's paper, biofuels and biomass burning of energy crops for electricity take the most space per unit of energy produced. For example, the Southern Company is building a new 100 megawatt biomass plant in Georgia. Southern estimates it will keep 180 trucks a day busy hauling a million tons of wood a year to the plant. One hundred megawatts is less than one-tenth the production of a nuclear plant which will fit on one square mile. To produce the same amount of energy as one nuclear plant would require continuously foresting an area one-third larger than the 550,000 acre Great Smoky Mountain National Park. You can make your own estimate of the number of trucks it would take to haul that much wood.

That is the second important insight of the *Nature Conservancy* report: a careful estimate of the widely different amounts of land each energy-producing technique requires. The gold standard for land usage is nuclear power. You can get a million megawatt hours of electricity a year—that's the standard unit the authors chose—per square mile, using nuclear power. The second most compact form of renewable energy is geothermal energy. To generate the same amount of power, coal requires four square miles, taking into account all the land required for mining and extraction. Solar thermal takes six square miles. Natural gas takes seven square miles and petroleum seventeen. Photovoltaic cells that turn sunlight directly into electricity require 14 square miles and wind is even more dilute, taking 28 square miles to produce the same unit of electricity.

These differences in land use are pronounced even though the paper's analysis is conservative. The authors include upstream inputs and waste disposal as part of their estimate of an energy producer's footprint. They add uranium mining and Yucca Mountain's 220 square miles to the area our 104 nuclear reactors actually occupy. If one were to consider only each energy plant's footprint, to produce 20 percent of U.S. electricity would take 100 nuclear reactors on 100 square miles or 186,000 wind turbines on 25,000 square miles. Visualize the difference this way. Thru hikers regularly travel the 2,178 miles of the Appalachian Trail from Springer Mountain in Georgia to Mount Katahdin in Maine. A row of fifty story wind turbines along that entire 2,178-mile trail would produce the same amount of electricity produced by 4 nuclear reactors on four square miles.

So, because of these wide differences, policy makers have the opportunity to choose carefully among the various forms of producing carbon-free electricity as well as to think about where such energy production should or should not go.

These are the four ways that the *Nature Conservancy* suggests we approach those decisions:

First, focus on energy conservation. This is the paper's preferred alternative to energy sprawl—and it is hard to see how anyone could disagree. To cite just one example, my home state of Tennessee leads the nation in residential per capita electricity use. If Tennesseans simply used electricity at the national average, the amount of electricity we would save each year would equal that amount produced by two nuclear power plants. Oak Ridge National Laboratory scientists have said that fuel efficiency standards are the single most important step our country can take to reduce carbon emissions.

The second recommendation for dealing with energy sprawl is end-use generation of electricity which usually occurs on already developed sites. One example of this is the co-generation that occurs at a paper factory that uses waste product to produce electricity and heat to run its facility. The most promising example is likely to be solar power on rooftops. In other words, since rooftops already exist, covering them with hundreds of square miles of solar panels would create no additional sprawl. There still are obstacles to the widespread use of solar power. In the southeast, solar still costs 4–5 times what TVA pays on the average for other electricity. There is the obstacle of aesthetics. But companies are now producing solar film embedded within attractive roofing materials—although this costs more. And there still is the problem that solar power is only available when the sun shines and, like wind, it can't be stored in large quantities. But unlike wind, which often blows at night when there is plenty of unused electricity, the sun shines when most people are at their peak power use. As former Energy Secretary James Schlesinger wrote in the *Washington Post*, because of their intermittency, wind and solar systems have to be backed up by other forms of electricity generation—which adds to cost and land usage.

The third recommendation is to make carbon regulation flexible, allowing for carbon recapture at coal plants, for nuclear power and for international offsets. So far the sponsors of climate and energy bills in the Congress haven't heeded this advice. In fact, both the Waxman Markey bill in the House and the Bingaman energy bill in the Senate contain very narrowly defined "renewable energy" mandates. Instead of allowing states to choose their methods of producing the required amount of carbon-free electricity, the legislation heavily tilts toward requiring wind power. For example, the legislation allows existing and new wind turbines, but only new hydroelectric. It does not count nuclear power, municipal solid waste, or landfill gas as "renewable." In the same way, 75 percent of the so-called "renewable electricity" subsidies enacted since 1978 have gone to wind developers. A study by the Energy Information Administration shows that wind gets a subsidy 31 times that of all other renewables combined. These policies have created a heavy bias toward the form of renewable electricity—wind power—that would consume our treasured mountaintops and can be very destructive to wildlife. And a national policy that also encourages wind power in the southeast where the wind barely blows makes as much sense as mandating new hydroelectric dams in the western desert where there is no water. It is my opinion that if we are truly seeking to reduce our carbon output, the policy that would create

the least energy sprawl would be a "carbon-free electricity standard," allowing for the maximum flexibility for those renewable electricity techniques that consume less land and require fewer new transmission lines.

Finally, the *Nature Conservancy* suggests paying attention to site selection for new energy projects. This is where those of you who represent organizations who have spent a century protecting wildlife and treasured landscapes could be of the greatest help in asking the right questions and providing wise answers. For example, should energy projects be placed in National Parks? In National Forests? If so, which forests and which energy projects? Should there be generous taxpayer subsidies for renewable energy projects within 20 miles of the Grand Tetons or along the Appalachian Trail? What about the large amounts of water needed for solar thermal plants or nuclear plants? Should turbines be concentrated in shallow waters 20 miles or more offshore where they can't be seen from the coast and transmission lines run underwater? Couldn't turbines be located in the center of Lake Michigan instead of along its shoreline? Should there be renewable energy zones, such as the solar zones Secretary Salazar is planning, where most new projects are placed—and where are the most appropriate locations for those zones and their transmission lines? In a recent op-ed in the *New York Times*, the Massachusetts Secretary of Energy and Environmental Affairs asked, wouldn't it make more sense to place wind turbines in the Atlantic and run transmission lines underwater than to build new transmission lines to carry wind power from the Great Plains to Boston? Should the subsidies for cellulosic ethanol be larger than those for corn ethanol, or should there be no subsidies at all? Should there be a special effort to encourage conservation easements on private lands that protect treasured viewscapes and habitats? According to the Wall Street Journal, on August 13 Exxon Mobil pleaded guilty in federal court to killing 85 birds that had come into contact with crude oil or other pollutants in uncovered tanks of waste-water facilities on its properties. The birds were protected by the Migratory Bird Treaty Act, which dates back to 1918. The company paid $600,000 in fines and fees.

Should the Migratory Bird law be enforced against developers of other energy projects—for example, renewable electricity and transmission lines? One wind farm near Oakland California estimates that its turbines kill 80 golden eagles a year. The American Bird Conservancy estimates that the 25,000 wind turbines in the United States kill between 75,000 and 275,000 birds per year. "Somebody is getting a get out of jail card free," Michael Fry of the Bird Conservancy told the Journal. And what would be the fine for the almost 1.4 million birds that 186,000 turbines might kill?

This raises the question of whether there should be some parity among all energy companies in the application of laws and policies. For example, oil and gas companies receive taxpayer subsidies but they bid to lease and drill on federal land or waters and then pay a royalty for the privilege. Should taxpayer subsidized renewable energy companies also be required to pay a royalty for the privilege of producing electricity on federal lands or waters? And, if so, could this be a source of permanent funding for the Land and Water Conservation fund or other conservation projects on the theory that if the law allows an environmental burden it ought to require an environmental benefit? Based on estimates from the Joint Committee on Taxation and the Congressional Budget Office, taxpayers will pay wind developers a total of $29 billion in federal subsidies over the next 10 years to increase windpower production from 1.5 to 4 percent of our total electricity. . . .

This brings me to my last point, which is to ask you to rethink nuclear power.

In our country, fears about safety, proliferation, and waste disposal have stymied the "atoms for peace" dream of large amounts of low-cost clean reliable energy from nuclear power. Twelve states even have moratoria against building new nuclear plants. Still, the 104 U.S. reactors built between 1970 and 1990 produce 19 percent of America's electricity and, as I have said, 70 percent of our carbon-free electricity. I believe that what Americans should most fear about nuclear power is this: the rest of the world will use it to create low-cost, carbon-free electricity while we—who invented the technology—will not. That would send our jobs overseas looking for their cheap energy. And it would deprive us of the technology most likely to produce large amounts of carbon free electricity to help deal with climate change—and to do it in the way least likely to harm the landscape and wildlife habitats.

Look at what the rest of the world is doing. Of the top five emitters, who together produce 55 percent of the carbon in the world, only the U.S. has no new nuclear plants under construction. China, the world's largest carbon emitter, recently upped its goal for nuclear reactors to 132. Russia, the number three emitter, plans two new reactors every year until 2030. Of the next two emitters, India has six reactors under construction and ten more planned. Japan already has 55 reactors, gets 35 percent of its electricity from nuclear, has two under construction and plans for ten more by 2018.

According to the International Atomic Energy Agency (IAEA), worldwide there are 53 reactors under construction in 11 countries, mostly in Asia. South Korea gets nearly 40 percent of its electricity from nuclear and plans another eight reactors by 2015. Taiwan gets 18 percent of its power from nuclear and is building two new reactors.

In the West, France gets 80 percent of its electricity from nuclear and has among the lowest electricity rates and carbon emissions in Western Europe (behind Sweden and Switzerland which are both half nuclear.) Great Britain has hired the French electric company EDF to help build reactors. Italy has announced it will go back to nuclear.

So where does this leave the United States? Well, we still know how to run reactors better than anyone else; we just don't build them anymore. Our fleet of 104 plants is up and running 90 percent of the time. We have 17 appli-

cations for new reactors pending before the Nuclear Regulatory Commission but haven't started construction on any new ones—and the 104 we currently have in operation will begin to grow too old to operate in twenty years.

That is why I believe the U.S. should build 100 new nuclear plants in 20 years. This would bring our nuclear-produced electricity to more than 40 percent of our total generation. Add 10 percent for hydroelectric dams, 7–8 percent for wind and solar (now 1.5 percent), 25 percent for natural gas (which is low-carbon) and you begin to get a real clean—and low-cost—electricity policy.

According to the National Academy of Sciences, construction costs for 100 nuclear plants are about the same as for 186,000 wind turbines. New reactors could be located mostly on sites with existing reactors. There would be little need for new transmission lines. Taxpayer subsidies for nuclear would be one-tenth what taxpayers would pay wind developers over 10 years. As for so called "green jobs," building 100 nuclear plants would provide four times as many construction jobs as building 186,000 turbines. And, of course, nuclear is a base load source of power operating 90 percent of the time, the kind of reliable power that a country that uses 25 percent of the energy in the world must have. Wind and solar are useful supplements but they are only available, on average, about one-third of the time and can't be stored in large amounts.

And what about the lingering fears of nuclear? Obama Administration Energy Secretary Dr. Steven Chu, the Nobel Prize winning physicist, says nuclear plants are safe and he wouldn't mind living near one. That view is echoed by the thousands of U.S. Navy personnel who have lived literally on top of nuclear reactors in submarines and Navy ships for 50 years without incident. The Nuclear Regulatory Commission agrees and its painstaking supervision and application process is intended to do everything humanly possible to keep our commercial fleet of reactors safe.

On the issue of waste, Dr. Chu says there is a two step solution. Step one is store the waste on site for 40 to 60 years. The Nuclear Regulatory Commission agrees this can be done safely, perhaps even for 100 years. Step two is research and development to find the best way to recycle fuel so that its mass is reduced by 97 percent, pure plutonium is never created, and the waste is only radioactive for 300 years instead of 1 million years.

That kind of recycling would take care of both the waste and the third fear of nuclear power, the threat that other countries might somehow use plutonium to build a bomb. One could argue that because the U.S. failed to lead in developing the safe use of nuclear technology for the last 30 years, we may have made it easier for North Korea and Pakistan to steal or buy nuclear secrets from rogue countries.

Now, let me conclude with this prediction: taking into account these energy sprawl concerns, I believe the best way to reach the necessary carbon goals for climate change with the least damage to our environment and to our economy will prove to be (1) building 100 new nuclear plants in 20 years, (2) electrifying half the cars and trucks in 20 years; we probably have enough unused electricity to plug these vehicles in at night without building one new power plant; (3) putting solar panels on our rooftops. To make this happen, the government should launch mini-Manhattan projects like the one we had in World War II: for recycling used nuclear fuel, for better batteries for electric vehicles, to make solar panels cost competitive, and in addition, to recapture carbon from coal plants. This plan should produce the largest amount of electricity with the smallest amount of carbon at the lowest possible cost thereby avoiding the pain and suffering that comes when high-cost energy pushes jobs overseas and makes it hard for low-income Americans to afford their heating and cooling bills.

My fellow Tennessean Al Gore won a Nobel Prize for arguing that global warming is the inconvenient problem. If you believe he is right, and if you are also concerned about energy sprawl, then I would suggest that nuclear power is the inconvenient solution.

Lamar Alexander, Republican Senator from Tennessee, has been a Tennessee Governor (1979–1987) and a United States Secretary of Education (1991–1993).

EXPLORING THE ISSUE

Is Renewable Energy Green?

Critical Thinking and Reflection

1. What does it mean to be "green"?
2. Which is the better reason for promoting the development of an energy technology—impact on the environment or impact on domestic economic growth?
3. How could society diminish its energy needs to make the problem of supplying enough energy more manageable?
4. If we choose to build huge arrays of solar power collectors, where should we put them, in deserts, on rooftops, or at sides of skyscrapers?

Is There Common Ground?

In the debate over whether renewable energy technologies are truly "green," both sides agree that we need to replace fossil fuels. Unfortunately, there is no easy way to do that, for every energy technology has environmental impacts. But solar and wind power are only two of several renewable energy technologies. Investigate geothermal power (in this book), wave power (www.energysavers.gov/renewable_energy/ocean/index.cfm/mytopic=50009), and Ocean Thermal Energy Conversion (OTEC; see www.nrel.gov/otec/what.html) and answer the following questions:

1. Which of these three energy technologies has the smallest "footprint" on the landscape?
2. In what ways are these three energy technologies likely to interfere with human activities?

3. How large a contribution to society's energy needs can these three energy technologies make?

Create Central

www.mhhe.com/createcentral

Additional Resources

Jesse Ausubel, "Renewable and Nuclear Heresies," *International Journal of Nuclear Governance, Economy and Ecology* (vol. 1, no. 3, 2007)

Stewart Brand, *Whole Earth Discipline: An Ecopragmatist Manifesto* (Viking, 2009)

Amory B. Lovins, "Renewable Energy's 'Footprint' Myth," *Electricity Journal* (June 2011)

Internet References . . .

The National Renewable Energy Laboratory

www.nrel.gov/

Renewable Energy World

www.renewableenergyworld.com/rea/home

U.S. Department of Energy: Energy Efficiency and Renewable Energy

www.eere.energy.gov/

Selected, Edited, and with Issue Framing Material by:
Thomas Easton, *Thomas College*

ISSUE

Are Biofuels a Reasonable Substitute for Fossil Fuels?

YES: Keith Kline et al., from "In Defense of Biofuels, Done Right," *Issues in Science and Technology* (Spring 2009)

NO: David Pimentel et al., from "Food Versus Biofuels: Environmental and Economic Costs," *Human Ecology* (February 2009)

Learning Outcomes
After reading this issue, you will be able to:
• Explain why there is so much interest in using biofuels to replace fossil fuels for powering vehicles. • Describe how biofuels production affects food supply and prices. • Describe how biofuels production affects the environment. • Explain why some kinds of biofuels are more desirable than others.

ISSUE SUMMARY

YES: Keith Kline, Virginia H. Dale, Russell Lee, and Paul Leiby argue that the impact of biofuels production on food prices is much less than alarmists claim. If biofuels development focused on converting biowastes and fast-growing trees and grasses into fuels, the overall impact would be even better, with a host of benefits in reduced fossil fuel use and greenhouse gas emissions, increased employment, enhanced wildlife habitat, improved soil and water quality, and more stable land use.

NO: David Pimentel, Alison Marklein, Megan A. Tuth, Marissa N. Karpoff, Gillian S. Paul, Robert McCormack, Joanna Kyriazis, and Tim Krueger argue that it is not possible to replace more than a small fraction of fossil fuels with biofuels. Furthermore, producing biofuels consumes more energy (as fossil fuels) than it makes available, and because biofuels compete with food production for land, water, fertilizer, and other resources, they necessarily drive up the price of food, which disproportionately harms the world's poor. It must also damage the environment in numerous ways.

The threat of global warming has spurred a great deal of interest in finding new sources of energy that do not add to the amount of carbon dioxide in the air. Among other things, this has meant a search for alternatives to fossil fuels, which modern civilization uses to generate electricity, heat homes, and power transportation. Finding alternatives for electricity generation (which relies much more on coal than on oil or natural gas) or home heating (which relies more on oil and natural gas) is easier than finding alternatives for transportation (which relies on oil, refined into gasoline and diesel oil). In addition, the transportation infrastructure, consisting of refineries, pipelines, tank trucks, gas stations, and an immense number of cars and trucks that will be on the road for many years, is well designed for handling liquid fuels. It is not surprising that industry and government would like to find non-fossil liquid fuels for cars and trucks (as well as ships and airplanes).

There are many suitably flammable liquids. Among them are the so-called biofuels or renewable fuels, plant oils and alcohols that can be distilled from plant sugars. According to Daniel M. Kammen, "The Rise of Renewable Energy," *Scientific American* (September 2006), the chief biofuel in the United States so far is ethanol, distilled from corn and blended with gasoline. Production is subsidized with $2 billion of federal funds, and "when all the inputs and outputs were correctly factored in, we found that ethanol" contains about 25 percent more energy (to be used when it is burned as fuel) than was used to produce it. At least one study says the "net energy" is actually less than the energy used to produce ethanol from corn; see Dan Charles, "Corn-Based Ethanol Flunks Key Test," *Science* (May 1, 2009). If other sources, such as cellulose-rich switchgrass or cornstalks, can be used, the "net energy" is supposedly much better. However, generating ethanol requires first converting cellulose to fermentable sugars, which is so far an expensive process (although many people are working on making the process cheaper; see e.g., Jennifer Chu, "Reinventing Cellulosic Ethanol Production," *Technology Review* online, June 10, 2009, www.technologyreview.com/energy/22774/?nlid=2091, George W. Huber and Bruce E. Dale, "Grassoline at the Pump," *Scientific American* (July 2009), and Rachel Ehrenberg, "The Biofuel Future," *Science News* (August 1, 2009). A significant additional concern is the amount of land needed for growing crops to be turned into biofuels; in a world where hunger is widespread, this means land is taken out of food production. If additional land is cleared to grow biofuel crops, this must mean loss of forests and wildlife habitat, increased erosion, and other environmental problems. Lars Hein and Rik Leemans, "The Impact of First-Generation Biofuels on the Depletion of the Global Phosphorus Reserve," *AMBIO* (June 2012), note that the need of biofuels for fertilizer will deplete supplies of the essential plant nutrient, phosphorus, to the point that future food production will be impaired. Sustainability is thus a major concern; see Barry D. Solomon, "Biofuels and Sustainability," *Annals of the New York Academy of Sciences* (January 2010). One study says that greenhouse-gas emissions may be significantly less; see Kevin Bullis, "Do Biofuels Reduce Greenhouse Gases?" *Technology Review* online (May 20 2011) (www.technologyreview.com/energy/37609/?mod=chfeatured), but Jussi Lankoski and Markku Ollikainen ask "Biofuel Policies and the Environment: Do Climate Benefits Warrant Increased Production from Biofuel Foodstocks?" *Ecological Economics* (February 2011), and conclude that the case for biofuels "is not as evident as has been generally thought."

Under the Energy Policy Act of 2005, the U.S. Environmental Protection Agency (EPA) requires that gasoline sold in the United States contain a minimum volume of renewable fuel. The Energy Independence and Security Act (EISA) of 2007 expanded the Renewable Fuel Standard (RFS) program to include diesel, in addition to gasoline, and to increase the volume of renewable fuel required to be blended into transportation fuel to 36 billion gallons by 2022. The aim of the program is to reduce greenhouse gas emissions, reduce imported petroleum, and encourage the development and expansion of our nation's renewable fuels sector. However, some think it is premature and even dangerous to put so much emphasis on biofuels. Robbin S. Johnson and C. Ford Runge, "Ethanol: Train Wreck Ahead," *Issues in Science and Technology* (Fall 2007), argue that the U.S. government's bias in favor of corn-based ethanol rigs the market against more efficient alternatives. It also leads to rising food prices, which particularly affects the world's poor. Pat Thomas, introducing *The Ecologist*'s special report on biofuels in the March 2007 issue, wrote that "the science is far from complete, the energy savings far from convincing and, although many see biofuels as a way to avoid the kind of resource wars currently raging in the Middle East and elsewhere, going down that road may in the end provoke a wider series of resource wars—

this time over food, water and habitable land." William Tucker, "Food Riots Made in the USA," *The Weekly Standard* (April 28, 2008), blames the food riots seen in many countries in 2008 on price increases due in large part to the shift in agricultural production from food grains to biofuels crops. Donald Mitchell argues in "A Note on Rising Food Prices," The World Bank Development Prospects Group (July 2008), that although many factors contributed to the increase in internationally traded food prices from January 2002 to June 2008, the most important single factor—accounting for as much as 70 percent of the rise in food prices—was the large increase in biofuels production from grains and oilseeds in the United States and EU. Without these increases, global wheat and maize stocks would not have declined appreciably and price increases due to other factors would have been moderate. See also Robert Bryce, "A Promise Not Kept: The Case against Ethanol," *The Conference Board Review* (May/June 2008). Jikun Huang, et al., "Biofuels and the Poor: Global Impact Pathways of Biofuels on Agricultural Markets," *Food Policy* (August 2012), find that by increasing farm income biofuels may help the world's poor, despite higher food prices.

There *are* problems with biofuels, and those problems are getting a great deal of attention. Robin Maynard, "Against the Grain," *The Ecologist* (March 2007), stresses that when food and fuel compete for farmland, food prices will rise, perhaps drastically. The poor will suffer, as will rainforests. Renton Righelato, "Forests or Fuel," *The Ecologist* (March 2007), reminds us that when forests are cleared, they no longer serve as "carbon sinks"; deforestation thus adds to the global warming problem, and it may take a century for the benefit of biofuels to show itself. David Pimentel, "Biofuel Food Disasters and Cellulosic Ethanol Problems," *Bulletin of Science, Technology & Society* (June 2009), says that because using 20 percent of the U.S. corn crop displaces a mere 1 percent of oil consumption, corn ethanol is a disaster, while using crop wastes and other biological materials poses its own problems. Robert F. Service, "Another Biofuels Drawback: The Demand for Irrigation," *Science* (October 23, 2009), notes that biofuel crops can compete for irrigation water, leading to both water supply and water quality problems. Gernot Stoeglehner and Michael Narodoslawsky, "How Sustainable Are Biofuels? Answers and Further Questions Arising from an Ecological Footprint Perspective," *Bioresource Technology* (August 2009), note that the sustainability of biofuels production depends very much on regional context. Alena Buyx and Joyce Tait, "Ethical Framework for Biofuels," *Science* (April 29, 2011), lay out five ethical principles for biofuels development, including sustainability, human rights, greenhouse-gas reductions, and equitable benefits and note a lack of incentives to support such development. See also Joyce Tait, "The Ethics of Biofuels," *GCB Bioenergy* (June 2011).

Laura Venderkam, "Biofuels or Bio-Fools?" *American: A Magazine of Ideas* (May/June 2007), describes the huge amounts of money being invested in companies planning to bring biofuels to market. A great deal of research is also going on, including efforts to use genetic engineering to produce enzymes that can cheaply and efficiently break cellulose into its component sugars (see Matthew L. Wald, "Is Ethanol for the Long Haul?" *Scientific American*, January 2007, and Michael E. Himmel, et al., "Biomass Recalcitrance: Engineering Plants and Enzymes for Biofuels Production," *Science*, February 9, 2007), make bacteria or yeast that can turn a greater proportion of sugar into alcohol (see Francois Torney, et al., "Genetic Engineering Approaches to Improve Bioethanol Production from Maize," *Current Opinion in Biotechnology*, June 2007), and even make bacteria that can convert sugar or cellulose into hydrocarbons that can easily be turned into gasoline or diesel fuel (see Neil Savage, "Building Better Biofuels," *Technology Review*, July/August 2007). If these efforts succeed, the price of biofuels may drop drastically, leading investors to abandon the field. Such a price drop would, of course, benefit the consumer and lead to wider use of biofuels. It would also, say C. Ford Runge and Benjamin Senauer, "How Biofuels Could Starve the Poor," *Foreign Affairs* (May/June 2007), ease the impact on food supply. However, David Biello, "The False Promise of Biofuels," *Scientific American* (August 2011), notes that progress in developing alternatives to corn-based ethanol (which competes directly with food production) is very slow and biofuels are not yet commercially competitive; some producers are already shifting from making relatively low-profit biofuels to making high-profit specialty chemicals.

D. A. Walker, "Biofuels—for Better or Worse?" *Annals of Applied Biology* (May 2010), views much biofuels advocacy as based on misinformation. Peter Rosset, "Agrofuels, Food Sovereignty, and the Contemporary Food Crisis," *Bulletin of Science, Technology & Society* (June 2009), says that biofuels are not a prime cause of the 2008 food crisis but they are "clearly contraindicated." On the other hand, Jose C. Escobar, et al., "Biofuels: Environment, Technology and Food Security," *Renewable & Sustainable Energy Reviews* (August 2009), consider that increased reliance on biofuels is inevitable due to "the imminent decline of the world's oil production, its high market prices and environmental impacts." According to Amela Ajanovic, "Biofuels versus Food Production: Does Biofuels Production Increase Food Prices," *Energy* (April 2011), biofuels and food production can coexist, especially for second-generation biofuels, but

it will never be possible to replace with biofuels all the fossil fuel used for transportation. On the other hand, Kevin Bullis, "Record Food Prices Linked to Biofuels," *Technology Review* online (www.technologyreview.com/energy/37848/) (June 17, 2011), reports that both the World Trade Organization and the United States Department of Agriculture have linked subsidies for biofuels production to high and volatile food prices; the WTO has called for governments to cut support for biofuels. See also "Price Volatility in Food and Agricultural Markets: Policy Responses," Policy Report including contributions by FAO, IFAD, IMF, OECD, UNCTAD, WFP, the World Bank, the WTO, IFPRI, and the UN HLTF (June 2, 2011).

Suzanne Hunt, "Biofuels, Neither Saviour nor Scam: The Case for a Selective Strategy," *World Policy Journal* (Spring 2008), argues that all the concerns are real, but the larger problem is that "our current agricultural, energy, and transport systems are failing." John Ohlrogge, et al., "Driving on Biomass," *Science* (May 22, 2009), point to a potential fix for the failure of the transportation system when they show that converting biomass to biofuels is much less efficient than burning it to produce electricity and using the electricity to power electric cars, which are in fact projected to "gain substantial market share in the coming years."

In the YES selection, Keith Kline, Virginia H. Dale, Russell Lee, and Paul Leiby argue that the impact of biofuels production on food prices is much less than alarmists claim. If biofuels development focused on converting biowastes and fast-growing trees and grasses into fuels, the overall impact would be even better, with a host of benefits in reduced fossil fuel use and greenhouse gas emissions, increased employment, enhanced wildlife habitat, improved soil and water quality, and more stable land use. In the NO selection, David Pimentel, Alison Marklein, Megan A. Tuth, Marissa N. Karpoff, Gillian S. Paul, Robert McCormack, Joanna Kyriazis, and Tim Krueger argue that it is not possible to replace more than a small fraction of fossil fuels with biofuels. Furthermore, producing biofuels consumes more energy (as fossil fuels) than it makes available, and because biofuels compete with food production for land, water, fertilizer, and other resources, they necessarily drive up the price of food, which disproportionately harms the world's poor. It must also damage the environment in numerous ways.

YES ↩

Keith Kline et al.

In Defense of Biofuels, Done Right

Biofuels have been getting bad press, not always for good reasons. Certainly important concerns have been raised, but preliminary studies have been misinterpreted as a definitive condemnation of biofuels. One recent magazine article, for example, illustrated what it called "Ethanol USA" with a photo of a car wreck in a corn field. In particular, many criticisms converge around grain-based biofuel, traditional farming practices, and claims of a causal link between U.S. land use and land-use changes elsewhere, including tropical deforestation.

Focusing only on such issues, however, distracts attention from a promising opportunity to invest in domestic energy production using biowastes, fast-growing trees, and grasses. When biofuel crops are grown in appropriate places and under sustainable conditions, they offer a host of benefits: reduced fossil fuel use; diversified fuel supplies; increased employment; decreased greenhouse gas emissions; enhanced habitat for wildlife; improved soil and water quality; and more stable global land use, thereby reducing pressure to clear new land.

Not only have many criticisms of biofuels been alarmist, many have been simply inaccurate. In 2007 and early 2008, for example, a bumper crop of media articles blamed sharply higher food prices worldwide on the production of biofuels, particularly ethanol from corn, in the United States. Subsequent studies, however, have shown that the increases in food prices were primarily due to many other interacting factors: increased demand in emerging economies, soaring energy prices, drought in food-exporting countries, cut-offs in grain exports by major suppliers, market-distorting subsidies, a tumbling U.S. dollar, and speculation in commodities markets.

Although ethanol production indeed contributes to higher corn prices, it is not a major factor in world food costs. The U.S. Department of Agriculture (USDA) calculated that biofuel production contributed only 5% of the 45% increase in global food costs that occurred between April 2007 and April 2008. A Texas A&M University study concluded that energy prices were the primary cause of food price increases, noting that between January 2006 and January 2008, the prices of fuel and fertilizer, both major inputs to agricultural production, increased by 37% and 45%, respectively. And the International Monetary Fund has documented that since their peak in July 2008, oil prices declined by 69% as of December 2008, and global food prices declined by 33% during the same

period, while U.S. corn production has remained at 12 billion bushels a month, one-third of which is still used for ethanol production.

In another line of critique, some argue that the potential benefits of biofuel might be offset by indirect effects. But large uncertainties and postulations underlie the debate about the indirect land-use effects of biofuels on tropical deforestation, the critical implication being that use of U.S. farmland for energy crops necessarily causes new land-clearing elsewhere. Concerns are particularly strong about the loss of tropical forests and natural grasslands. The basic argument is that biofuel production in the United States sets in motion a necessary scenario of deforestation.

According to this argument, if U.S. farm production is used for fuel instead of food, food prices rise and farmers in developing countries respond by growing more food. This response requires clearing new land and burning native vegetation and, hence, releasing carbon. This "induced deforestation" hypothesis is based on questionable data and modeling assumptions about available land and yields, rather than on empirical evidence. The argument assumes that the supply of previously cleared land is inelastic (that is, agricultural land for expansion is unavailable without new deforestation). It also assumes that agricultural commodity prices are a major driving force behind deforestation and that yields decline with expansion. The calculations for carbon emissions assume that land in a stable, natural state is suddenly converted to agriculture as a result of biofuels. Finally, the assertions assume that it is possible to measure with some precision the areas that will be cleared in response to these price signals.

A review of the issues reveals, however, that these assumptions about the availability of land, the role of biofuels in causing deforestation, and the ability to relate crop prices to areas of land clearance are unsound. Among our findings:

First, sufficient suitably productive land is available for multiple uses, including the production of biofuels. Assertions that U.S. biofuel production will cause large indirect land-use changes rely on limited data sets and unverified assumptions about global land cover and land use. Calculations of land-use change begin by assuming that global land falls into discrete classes suitable for agriculture—cropland, pastures and grasslands, and forests—and results depend on estimates of the extent, use, and productivity of these lands, as well as presumed future interactions among land-use classes. But several major organizations, including the Food and Agriculture Organization (FAO), a primary data clearinghouse, have documented significant inconsistencies surrounding global land-cover estimates. For example, the three most recent FAO Forest Resource Assessments, for periods ending in 1990, 2000, and 2005, provide estimates of the world's total forest cover in 1990 that vary by as much as 470 million acres, or 21% of the original estimate.

Cropland data face similar discrepancies, and even more challenging issues arise when pasture areas are considered. Estimates for land used for crop production range from 3.8 billion acres (calculated by the FAO) to 9 billion acres (calculated by the Millennium Ecosystem Assessment, an international effort spearheaded by the United Nations). In a recent study attempting to reconcile cropland use circa 2000, scientists at the University of Wisconsin-Madison and McGill University estimated that there were 3.7 billion acres of cropland, of which 3.2 billion were actively cropped or harvested. Land-use studies consistently acknowledge serious data limitations and uncertainties, noting that a majority of global crop lands are constantly shifting the location of cultivation, leaving at any time large areas fallow or idle that may not be captured in statistics. Estimates of idle croplands, prone to confusion with pasture and grassland, range from 520 million acres to 4.9 billion acres globally. The differences illustrate one of many uncertainties that hamper global land-use change calculations. To put these numbers in perspective, USDA has estimated that in 2007, about 21 million acres were used worldwide to produce biofuel feedstocks, an area that would occupy somewhere between 0.4% and 4% of the world's estimated idle cropland.

Diverse studies of global land cover and potential productivity suggest that anywhere from 600 million to more than 7 billion additional acres of underutilized rural lands are available for expanding rain-fed crop production around the world, after excluding the 4 billion acres of cropland currently in use, as well as the world's supply of closed forests, nature reserves, and urban lands. Hence, on a global scale, land per se is not an immediate limitation for agriculture and biofuels.

In the United States, the federal government, through the multiagency Biomass Research and Development Initiative (BRDI), has examined the land and market implications of reaching the nation's biofuel target, which calls for producing 36 billion gallons by 2022. BRDI estimated that a slight net reduction in total U.S. active cropland area would result by 2022 in most scenarios, when compared with a scenario developed from USDA's so-called "baseline" projections. BRDI also found that growing biofuel crops efficiently in the United States would require shifts in the intensity of use of about 5% of pasture lands to more intensive hay, forage, and bioenergy crops (25 million out of 456 million acres) in order to accommodate dedicated energy crops, along with using a combination of wastes, forest residues, and crop residues. BRDI's estimate assumes that the total area allocated to USDA's Conservation Reserve Program (CRP) remains constant at about 33 million acres but allows about 3 million acres of the CRP land on high-quality soils in the Midwest to be offset by new CRP additions in other regions. In practice, additional areas of former cropland that are now in the CRP could be managed for biofuel feedstock production in a way that maintains positive impacts on wildlife, water, and land conservation goals, but this option was not included among the scenarios considered.

Yields are important. They vary widely from place to place within the United States and around the world. USDA projects that corn yields will rise by 20 bushels per acre by 2017; this represents an increase in corn output equivalent to adding 12.5 million acres as compared with 2006, and over triple that area as compared with average yields in many less-developed nations. And there is the possibility that yields will increase more quickly than projected in the USDA baseline, as seed companies aim to exceed 200 bushels per acre by 2020. The potential to increase yields in developing countries offers tremendous opportunities to improve welfare and expand production while reducing or maintaining the area harvested. These improvements are consistent with U.S. trends during the past half century showing agricultural output growth averaging 2% per year while cropland use fell by an average of 0.7% per year. Even without large yield increases, cropland requirements to meet biofuel production targets may not be nearly as great as assumed.

Concerns over induced deforestation are based on a theory of land displacement that is not supported by data. U.S. ethanol production shot up by more than 3 billion gallons (150%) between 2001 and 2006, and corn production increased 11%, while total U.S. harvested cropland fell by about 2% in the same period. Indeed, the harvested area for "coarse grains" fell by 4% as corn, with an average yield of 150 bushels per acre, replaced other feed grains such as sorghum (averaging 60 bushels per acre). Such statistics defy modeling projections by demonstrating an ability to supply feedstock to a burgeoning ethanol industry while simultaneously maintaining exports and using substantially less land. So although models may assume that increased use of U.S. land for biofuels will lead to more land being cleared for agriculture in other parts of the world, evidence is lacking to support those claims.

Second, there is little evidence that biofuels cause deforestation, and much evidence for alternative causes. Recent scientific papers that blame biofuels for deforestation are based on models that presume that new land conversion can be simulated as a predominantly market-driven choice. The models assume that land is a privately owned asset managed in response to global price signals within a stable rule-based economy—perhaps a reasonable assumption for developed nations.

However, this scenario is far from the reality in the smoke-filled frontier zones of deforestation in less-developed countries, where the models assume biofuel-induced land conversion takes place. The regions of the world that are experiencing first-time land conversion are characterized by market isolation, lawlessness, insecurity, instability, and lack of land tenure. And nearly all of the forests are publicly owned. Indeed, land-clearing is a key step in a long process of trying to stake a claim for eventual tenure. A cycle involving incremental degradation, repeated and extensive fires, and shifting small plots for subsistence tends to occur long before any consideration of crop choices influenced by global market prices.

The causes of deforestation have been extensively studied, and it is clear from the empirical evidence that forces other than biofuel use are responsible for the trends of increasing forest loss in the tropics. Numerous case studies document that the factors driving deforestation are a complex expression of cultural, technological, biophysical, political, economic, and demographic interactions. Solutions and measures to slow deforestation have also been analyzed and tested, and the results show that it is critical to improve governance, land tenure, incomes, and security to slow the pace of new land conversion in these frontier regions.

Selected studies based on interpretations of satellite imagery have been used to support the claims that U.S. biofuels induce deforestation in the Amazon, but satellite images cannot be used to determine causes of land-use change. In practice, deforestation is a site-specific process. How it is perceived will vary greatly by site and also by the temporal and spatial lens through which it is observed. Cause-and-effect relationships are complex, and the many small changes that enable larger future conversion cannot be captured by satellite imagery. Although it is possible to classify an image to show that forest in one period changed to cropland in another, cataloguing changes in discrete classes over time does not explain why these changes occur. Most studies asserting that the production and use of biofuels cause tropical deforestation point to land cover at some point after large-scale forest degradation and clearing have taken place. But the key events leading to the primary conversion of forests often proceed for decades before they can be detected by satellite imagery. The imagery does not show how the forest was used to sustain livelihoods before conversion, nor the degrees of continual degradation that occurred over time before the classification changed. When remote sensing is supported by a ground-truth process, it typically attempts to narrow the uncertainties of land-cover classifications rather than research the history of occupation, prior and current use, and the forces behind the land-use decisions that led to the current land cover.

First-time conversion is enabled by political, as well as physical, access. Southeast Asia provides one example where forest conversion has been facilitated by political access, which can include such diverse things as government-sponsored development and colonization programs in previously undisturbed areas and the distribution of large timber and mineral concessions and land allotments to friends, families, and sponsors of people in power. Critics have raised valid concerns about high rates of deforestation in the region, and they often point an accusing finger at palm oil and biofuels.

Palm oil has been produced in the region since 1911, and plantation expansion boomed in the 1970s with growth rates of more than 20% per year. Biodiesel represents a tiny fraction of palm oil consumption. In 2008, less than 2% of crude palm oil output was processed for biofuel in Indonesia and Malaysia, the world's largest producers and exporters. Based on land-cover statistics alone, it is impossible to determine the degree of attribution that oil palm may share with other causes of forest conversion in Southeast Asia. What is clear is that oil palm is not the only factor and that palm plantations are established after a process of degradation and deforestation has transpired. Deforestation data may offer a tool for estimating the ceiling for attribution, however. In Indonesia, for example, 28.1 million hectares were deforested between 1990 and 2005, and oil palm expansion in those areas was estimated to be between 1.7 million and 3 million hectares, or between 6% and 10% of the forest loss, during the same period.

Initial clearing in the tropics is often driven more by waves of illegitimate land speculation than agricultural production. In many Latin American frontier zones, if there is native forest on the land, it is up for grabs, as there is no legal tenure of the land. The majority of land-clearing in the Amazon has been blamed on livestock because, in part, there is no alternative for classifying the recent clearings and, in part, because land holders must keep it "in production" to maintain claims and avoid invasions. The result has been the frequent burning and the creation of extensive cattle ranches. For centuries, disenfranchised groups have been pushed into the forests and marginal lands where they do what they can to survive. This settlement process often includes serving as low-cost labor to clear land for the next wave of better-connected colonists. Unless significant structural changes occur to remove or modify enabling factors, the forest-clearing that was occurring before this decade is expected to continue along predictable paths.

Testing the hypothesis that U.S. biofuel policy causes deforestation elsewhere depends on models that can incorporate the processes underlying initial land-use change. Current models attempt to predict future land-use change based on changes in commodity prices. As conceived thus far, the computational general equilibrium models designed for economic trade do not adequately incorporate the processes of land-use change. Although crop prices may influence short-term land-use decisions, they are not a dominant factor in global patterns of first-time conversion, the land-clearing of chief concern in relating biofuels to deforestation. The highest deforestation rates observed and estimated globally occurred in the 1990s. During that period, there was a surplus of commodities on world markets and consistently depressed prices.

Third, many studies omit the larger problem of widespread global mismanagement of land. The recent arguments focusing on the possible deforestation attributable to biofuels use idealized representations of crop and land markets, omitting what may be larger issues of concern. Clearly, the causes of global deforestation are complex and are not driven merely by a single crop market. Additionally, land mismanagement, involving both initial clearing and maintaining previously cleared land, is widespread and leads to a process of soil degradation and environmental damage that is especially prevalent in the frontier zones. Reports by the FAO and the Millennium Ecosystem Assessment describe the environmental consequences of repeated fires in these areas. Estimates of global burning vary annually, ranging from 490 million to 980 million acres per year between 2000 and 2004. The vast majority of fires in the tropics occur in Africa and the Amazon in what were previously cleared, nonforest lands. In a detailed study, the Amazon Institute of Environmental Research and Woods Hole Research Center found that 73% of burned area in the Amazon was on previously cleared land, and that was during the 1990s, when overall deforestation rates were high.

Fire is the cheapest and easiest tool supporting shifting subsistence cultivation. Repeated and extensive burning is a manifestation of the lack of tenure, lack of access to markets, and severe poverty in these areas. When people or communities have few or no assets to protect from fire and no incentive to invest in more sustainable production, they also have no reason to limit the extent of burning. The repeated fires modify ecosystem structure, penetrate ever deeper into forest margins, affect large areas of understory vegetation (which is not detected by remote sensing), and take an ever greater cumulative toil on soil quality and its ability to sequester carbon. Profitable biofuel markets, by contributing to improved incentives to grow cash crops, could reduce the use of fire and the pressures on the agricultural frontier. Biofuels done right, with attention to best practices for sustained production, can make significant contributions to social and economic development as well as environmental protection.

Furthermore, current literature calculates the impacts from an assumed agricultural expansion by attributing the carbon emissions from clearing intact ecosystems to biofuels. If emission analyses consider empirical data reflecting the progressive degradation that occurs (often over decades) before and independently of agriculture market signals for land use, as well as changes in the frequency and extent of fire in areas that biofuels help bring into more stable market economies, then the resulting carbon emission estimates would be worlds apart.

Brazil provides a good case in point, because it holds the globe's largest remaining area of tropical forests, is the world's second-largest producer of biofuel (after the United States), and is the world's leading supplier of biofuel for global trade. Brazil also has relatively low production costs and a growing focus on environmental stewardship. As a matter of policy, the Brazilian government has supported the development of biofuels since launching a National Ethanol Program called Proálcool in 1975. Brazil's ethanol industry began its current phase of growth after Proálcool was phased out in 1999 and the government's role shifted from subsidies and regulations toward increased collaboration with the private sector in R&D. The government helps stabilize markets by supporting variable rates of blending ethanol with gasoline and planning for industry expansion, pipelines, ports, and logistics. The government also facilitates access to global markets; develops improved varieties of sugarcane, harvest equipment, and conversion; and supports improvements in environmental performance.

New sugarcane fields in Brazil nearly always replace pasture land or less valuable crops and are concentrated around production facilities in the developed southeastern region, far from the Amazon. Nearly all production is rain-fed and relies on low input rates of fertilizers and agrochemicals, as compared with other major crops. New projects are reviewed under the Brazilian legal framework of Environmental Impact Assessment and Environmental Licensing. Together, these policies have contributed to the restoration or protection of reserves and riparian areas and increased forest cover, in tandem with an expansion of sugarcane production in the most important producing state, Sao Paulo.

Yet natural forest in Brazil is being lost, with nearly 37 million acres lost between May 2000 and August 2006, and a total of 150 million acres lost since 1970. Some observers have suggested that the increase in U.S. corn production for biofuel led to reduced soybean output and higher soybean prices, and that these changes led, in turn, to new deforestation in Brazil. However, total deforestation rates in Brazil appear to fall in tandem with rising soybean prices. This co-occurrence illustrates a lack of connection between commodity prices and initial land clearing. This phenomenon has been observed around the globe and suggests an alternate hypothesis: Higher global commodity prices focus production and investment where it can be used most efficiently, in the plentiful previously cleared and underutilized lands around the world. In times of falling prices and incomes, people return to forest frontiers, with all of their characteristic tribulations, for lack of better options.

Biofuels Done Right

With the right policy framework, cellulosic biofuel crops could offer an alternative that diversifies and boosts rural incomes based on perennials. Such a scenario would create incentives to reduce intentional burning that currently affects millions of acres worldwide each year. Perennial biofuel crops can help stabilize land cover, enhance soil carbon sequestration, provide habitat to support biodiversity, and improve soil and water quality. Furthermore, they can reduce pressure to clear new land via improved incomes and yields. Developing countries have huge opportunities to increase crop yield and thereby grow more food on less land, given that cereal yields in less developed nations are 30% of those in North America. Hence, policies supporting biofuel production may actually help stop the extensive slash-and-burn agricultural cycle that contributes to greenhouse gas emissions, deforestation, land degradation, and a lifestyle that fails to support farmers and their families.

Biofuels alone are not the solution, however. Governments in the United States and elsewhere will have to develop and support a number of programs designed to support sustainable development. The operation and rules of such programs must be transparent, so that everyone can understand them and see that fair play is ensured. Among other attributes, the programs must offer economic incentives for sustainable production, and they must provide for secure land tenure and participatory land-use planning. In this regard, pilot biofuel projects in Africa and Brazil are showing promise in addressing the vexing and difficult challenges of sustainable land use and development. Biofuels also are uniting diverse stakeholders in a global movement to develop sustainability metrics and certification methods applicable to the broader agricultural sector.

Given a priority to protect biodiversity and ecosystem services, it is important to further explore the drivers for the conversion of land at the frontier and to consider the effects, positive and negative, that U.S. biofuel policies could have in these areas. This means it is critical to distinguish between valid concerns calling for caution and alarmist criticisms that attribute complex problems solely to biofuels.

Still, based on the analyses that we and others have done, we believe that biofuels, developed in an economically and environmentally sensible way, can contribute significantly to the nation's—indeed, the world's—energy security while providing a host of benefits for many people in many regions.

KEITH KLINE, VIRGINIA H. DALE, RUSSELL LEE, AND PAUL LEIBY are Scientists in the Center for BioEnergy Sustainability at the Oak Ridge National Laboratory, Oak Ridge, Tennessee.

David Pimentel et al.   → **NO**

Food Versus Biofuels: Environmental and Economic Costs

Introduction

With global shortages of fossil energy, especially oil and natural gas, and heavy biomass energy consumption occurring, a major focus has developed worldwide on biofuel production. Emphasis on biofuels as renewable energy sources has developed globally, including those made from crops such as corn, sugarcane, and soybean. Wood and crop residues also are being used as fuel. Though it may seem beneficial to use renewable plant materials for biofuel, the use of crop residues and other biomass for biofuels raises many environmental and ethical concerns.

Diverse conflicts exist in the use of land, water, energy and other environmental resources for food and biofuel production. Food and biofuels are dependent on the same resources for production: land, water, and energy. In the USA, about 19% of all fossil energy is utilized in the food system: about 7% for agricultural production, 7% for processing and packaging foods, and about 5% for distribution and preparation of food. In developing countries, about 50% of wood energy is used primarily for cooking in the food system.

The objective of this article is to analyze: (1) the uses and interdependencies among land, water, and fossil energy resources in food versus biofuel production and (2) the characteristics of the environmental impacts caused by food and biofuel production.

Food and Malnourishment

The Food and Agricultural Organization (FAO) of the United Nations confirms that worldwide food available per capita has been declining *continuously* based on availability of cereal grains during the past 23 years. Cereal grains make up an alarming 80% of the world's food supply. Although grain yields per hectare in both developed and developing countries are still gradually increasing, the rate of increase is slowing, while the world population and its food needs are rising. For example, from 1950 to 1980, US grain yields increased at about 3% per year. Since 1980, the annual rate of increase for corn and other grains is only approximately 1%. Worldwide the rate of increase in grain production is not keeping up with the rapid rate of world population growth of 1.1%.

The resulting decrease in food supply results in widespread malnutrition. There are more deaths from malnutrition than any other cause of death in the world today. The World Health Organization reports more than 3.7 billion people (56% of the global population) are currently malnourished, and that number is steadily increasing. Although much of the land worldwide is occupied by grains and other crops, malnutrition is still globally prevalent.

World Cropland and Water Resources

More than 99.7% of human food comes from the terrestrial environment, while less than 0.3% comes from the oceans and other aquatic ecosystems. Worldwide, of the total 13 billion hectares of land area, the percentages in use are: cropland, 11%; pasture land, 27%; forest land, 32%; urban, 9%; and other 21%. Most of the remaining land area (21%) is unsuitable for crops, pasture, and/or forests because the soil is too infertile or shallow to support plant growth or the climate and region are too harsh, too cold, dry, steep, stony, or wet. Most of the suitable cropland is already in use.

As the human population continues to increase rapidly, there has been an expansion of diverse human activities that have dramatically reduced cropland and pasture land. Much vital cropland and pastureland has been covered by transportation systems and urbanization. In the USA, about 0.4 ha (one acre) of land per person is covered with urbanization and highways. In 1960, when the world population numbered only three billion, approximately 0.5 ha was available per person for the production of a diverse, nutritious diet of plant and animal products. It is widely agreed that 0.5 ha is essential for a healthy diet. China's recent explosion in development provides an example of rapid declines in the availability of per capita cropland. The current available cropland in China is only 0.08 ha per capita. This relatively small amount of cropland provides the people in China with a predominantly vegetarian diet, which requires less energy, land, and biomass than the typical American diet.

In addition to land, water is a vital controlling factor in crop production. The production of 9 t/ha of corn requires about seven million liters of water (about 700,000 gallons of water per acre). Other crops also require large amounts of water. Irrigation provides much of the water for world food production. For example, 17% of the crops that are irrigated

From *Human Ecology,* February 2009, pp. 1–7, 9. Copyright © 2009 by Springer Science and Business Media. Reprinted by permission via Rightslink.

worldwide provide 40% of the world food supply. A major concern is that world-wide availability of irrigation water is projected to decline further because of global warming.

Energy Resources and Use

Since the industrial revolution of the 1850s, the rate of energy use from all sources has been growing even faster than the world population. For example, from 1970 to 1995, energy use increased at a rate of 2.5% per year (doubling every 30 years) compared with the worldwide population growth of 1.7% per year (doubling every 40 to 60 years). Developed countries annually consume about 70% of the fossil energy worldwide, while the developing nations, which have about 75% of the world population, use only 30% of world fossil energy.

Although about 50% of all the solar energy captured worldwide by photosynthesis is used by humans for food, forest products, and other systems, it is still inadequate to meet all human food production needs. To make up for this shortfall, about 473 quads (one quad = 1×10^{15} BTU) of fossil energy—mainly oil, gas, coal, and a small amount of nuclear—are utilized worldwide each year. Of these 473 quads, about 100 quads (or about 22%) of the world's total energy are utilized just in the United States, which has only 4.5% of the world's population.

Each year, the USA population uses three times more fossil energy than the total solar energy captured by all harvested US crops, forests, and grasses. Industry, transportation, home heating and cooling, and food production account for most of the fossil energy consumed in the United States. Per capita use of fossil energy in the United States per year amounts to about 9,500 l of oil equivalents—more than seven times per capita use in China. In China, most fossil energy is used by industry, although approximately 25% is now used for agriculture and in the food production system.

Worldwide, the earth's natural gas supply is considered adequate for about 40 years and that of coal for about 100 years. In the USA, natural gas is already in short supply: it is projected that the USA will deplete its natural gas resources in about 20 years. Many agree that the world reached peak oil and natural gas in 2007; from this point, these energy resources are declining slowly and continuously, until they run out altogether.

Youngquist reports that earlier estimates of the amount of oil and gas new exploration drilling would provide were very optimistic as to the amount of these resources to be found in the United States. Both the US oil production rate and existing reserves have continued to decline. Domestic oil and natural gas production has been decreasing for more than 30 years and are projected to continue to decline. Approximately 90% of US oil resources have already been exploited. At present, the United States is importing more than 63% of its oil, which puts its economy at risk due to fluctuating oil prices and difficult international political situations, as was seen

previously during the 1973 oil crisis, the 1991 Gulf War, and the current Iraq War.

Biomass Resources

The total sustainable world biomass energy potential has been estimated to be about 92 quads (10^{15} BTU) per year, which represents 19% of total global energy use. The total forest biomass produced worldwide is 38 quads per year, which represents 8% of total energy use. In the USA, only 1% to 2% home heating is achieved with wood.

Global forest area removed each year totals 15 million ha. Global forest biomass harvested is just over 1,431 billion kg per year, of which 60% is industrial roundwood and 40% is fuelwood. About 90% of the fuelwood is utilized in developing countries. A significant portion (26%) of all forest wood is converted into charcoal. Production of charcoal causes between 30% and 50% of the wood energy to be lost and produces large quantities of smoke. On the other hand, charcoal is cleaner burning and thus produces less smoke than burning wood fuel directly; it is dirty to handle but lightweight.

Worldwide, most biomass is burned for cooking and heating. In developing countries, about 2 kcal of wood are utilized in cooking 1 kcal of food. Thus, more biomass and more land and water are needed to produce the biofuel for cooking than are needed to produce the food. However, biomass can also be converted into electricity. Assuming that optimal yield globally of three dry metric tons (t/ha) per year of woody biomass can be harvested sustainably, this would provide a gross energy yield of 13.5 million kcal/ha. Harvesting this wood biomass requires an energy expenditure of approximately 30 l of diesel fuel per hectare, plus the embodied energy for cutting and collecting wood for transport to an electric power plant. Thus, the energy input per output ratio for such a system is calculated to be 1:25.

Per capita consumption of woody biomass for heat in the USA amounts to 625 kg per year. The diverse biomass resources (wood, crop residues, and dung) used in developing nations averages about 630 kg per capita per year.

Woody biomass has the capacity to supply the USA with about five quads (1.5×10^{12} kWh thermal) of its total gross energy supply by the year 2050, provided that the amount of forest-land stays constant. A city of 100,000 people using the biomass from a sustainable forest (3 t/ha per year) for electricity requires approximately 200,000 ha of forest area, based on an average electrical demand of slightly more than one billion kilowatt-hours (electrical energy [e]) (860 kcal = 1 kWh).

Air quality impacts from burning biomass are less harmful than those associated with coal, but more harmful than those associated with natural gas. Biomass combustion releases more than 200 different chemical pollutants, including 14 carcinogens and four cocarcinogens, into the atmosphere. As a result, approximately four billion people globally suffer from continuous exposure to smoke. In the USA, wood smoke kills 30,000 people each year, although many of the pollutants from electric plants that use wood and other biomass can be mitigated. These controls include the same scrubbers that are frequently installed on coal-fired plants.

An estimated 2.0 billion tons of biomass is produced per year on US land area. This translates into about 32 quads of energy, which means that the solar energy captured by all the plants in the USA per year equates to only 32% of the energy currently consumed as fossil energy. There is insufficient US biomass for ethanol and biodiesel production to make the USA oil independent.

Of the total world land area in cropland, pasture, and forest, about 38% is cropland and pasture and about 30% is forests. Devoting a portion of this cropland and forest land to biofuels will stress both managed ecosystems and will not be sufficient to solve the world fuel problem.

Corn Ethanol

In the United States, ethanol constitutes 99% of all biofuels. For capital expenditures, new plant construction costs from $1.05 to $3.00 per gallon of ethanol. Fermenting and distilling corn ethanol requires large amounts of water. The corn is finely ground and approximately 15 l of water are added per 2.69 kg of ground corn. After fermentation, to obtain a liter of 95% pure ethanol from the 10% ethanol and 90% water mixture, 1 l of ethanol must be extracted from the approximately 10 l of the ethanol/water mixture. To be mixed with gasoline, the 95% ethanol must be further processed and more water must be removed, requiring additional fossil energy inputs to achieve 99.5% pure ethanol. Thus, a total of about 12 l of wastewater must be removed per liter of ethanol produced, and this relatively large amount of sewage effluent has to be disposed of at energy, economic, and environmental costs.

Manufacture of a liter of 99.5% ethanol uses 46% more fossil energy than it produces and costs $1.05 per liter ($3.97 per gallon). The corn feedstock alone requires more than 33% of the total energy input. The largest energy inputs in corn-ethanol production are for producing the corn feedstock plus the steam energy and electricity used in the fermentation/distillation process. The total energy input to produce a liter of ethanol is 7,474 kcal. However, a liter of ethanol has an energy value of only 5,130 kcal. Based on a net energy loss of 2,344 kcal of ethanol produced, 46% more fossil energy is expended than is produced as ethanol. The total cost, including the energy inputs for the fermentation/distillation process and the apportioned energy costs of the stainless steel tanks and other industrial materials, is $1,045 per 1,000 l of ethanol produced.

Subsidies for corn ethanol total more than $6 billion per year. This means that the subsidies per liter of ethanol are 60 times greater than the subsidies per liter of gasoline. In 2006, nearly 19 billion liters of ethanol were produced on 20% of US corn acreage. This 19 billion liters represents only 1% of total US petroleum use.

However, even if we completely ignore corn ethanol's negative energy balance and high economic cost, we still find that it is absolutely not feasible to use ethanol as a replacement for US oil. If all 341 billion kilograms of corn produced annually in the USA were converted into ethanol at the current rate of 2.69 kg per liter of ethanol, then 129 billion liters of ethanol could be produced. This would provide only 5% of total oil consumption in the USA. And of course, in this situation there would be no corn available for livestock and other needs.

In addition, the environmental impacts of corn ethanol are enormous:

1. Corn production causes more soil erosion than any other crop grown.
2. Corn production uses more nitrogen fertilizer than any other crop grown and is the prime cause of the dead zone in the Gulf of Mexico. In 2006, approximately 4.7 million tons of nitrogen was used in US corn production. Natural gas is required to produce nitrogen fertilizer. The USA now imports more than half of its nitrogen fertilizer. In addition, about 1.7 million tons of phosphorus was used in the USA.
3. Corn production uses more insecticides than any other crop grown.
4. Corn production uses more herbicides than any other crop grown.
5. More than 1,700 gallons of water are required to produce one gallon of ethanol.
6. Enormous quantities of carbon dioxide are produced. This is due to the large quantity of fossil energy used in production, and the immense amounts of carbon dioxide released during fermentation and soil tillage. All this speeds global warming.
7. Air pollution is a significant problem. Burning ethanol emits pollutants into air such as peroxyacetyl nitrate (PAN), acetaldhyde, alkylates, and nitrous oxide. These can have significant detrimental human health effects as well as impact other organisms and ecosystems.

In addition to corn ethanol's intensive environmental degradation and inefficient use of food-related resources, the production of corn ethanol also has a great effect on world food prices. For instance, the use of corn for ethanol production has increased the prices of US beef, chicken, pork, eggs, breads, cereals, and milk by 10% to 20%. Corn prices have more than doubled during the past year.

Grass and Cellulosic Ethanol

Tilman et al. suggest that all 235 million hectares of grassland available in the USA plus crop residues can be converted into cellulosic ethanol. This suggestion causes concerns among scientists. Tilman et al. recommend that crop residues, like corn stover, can be harvested and utilized as a fuel source. This would be a disaster for the agricultural ecosystem because crop residues are vital for protecting topsoil. Leaving the soil unprotected would intensify soil erosion by tenfold or more and may increase soil loss as much as 100-fold. Furthermore, even a partial removal of the stover can result in increased CO_2 emissions and intensify acidification and eutrophication due to increased runoff. Already, the US crop system is losing soil ten times faster than the sustainable rate. Soil formation rates at less than 1 t ha^{-1} year^{-1}, are extremely slow. Increased soil erosion caused by the removal of crop residues for use as biofuels facilitates soil–carbon oxidation and contributes to the greenhouse emissions problem.

Tilman et al. assume about 1,032 l of ethanol can be produced through the conversion of the 4 t ha^{-1} year^{-1} of grasses harvested. However, Pimentel and Patzek report a negative 68% return in ethanol produced compared with the fossil energy inputs in switchgrass conversion. Furthermore, converting all 235 million hectares of US grassland into ethanol at the optimistic rate suggested by Tilman et al. would still provide only 12% of annual US consumption of oil. Verified data, however, confirm that the output in ethanol would require 1.5 l of oil equivalents to produce 1 l of ethanol.

To achieve the production of this much ethanol, US farmers would have to displace the 100 million cattle, seven million sheep, and four million horses that are now grazing on 324 million hectares of US grassland and rangeland. Already, overgrazing is a serious problem on US grassland and a similar problem exists worldwide. Thus, the assessment of the quantity of ethanol that can be produced on US and world grasslands by Tilman et al. appears to be unduly optimistic.

Converting switchgrass into ethanol results in a negative energy return. The negative energy return is 68% or a slightly more negative energy return than corn ethanol production. The cost of producing a liter of ethanol using switchgrass was 93¢.

Several problems exist [in] the conversion of cellulosic biomass into ethanol. First, it takes from two to five times more cellulosic biomass to achieve the same quantity of starches and sugars as are found in the same quantity of corn grain. Thus, two to five times more cellulosic material must be produced and handled compared with corn grain. In addition, the starches and sugars are tightly held in lignin in the cellulosic biomass. They can be released using a strong acid to dissolve the lignin. Once the lignin is dissolved, the acid action is stopped with an alkali. Now the solution of lignin, starches, and sugars can be fermented.

Some claim that the lignin can be used as a fuel. Clearly, it cannot when dissolved in water. Usually less than 25% of the lignin can be extracted from the water mixture using various energy intensive technologies.

Soybean Biodiesel

Processed vegetable oils from soybean, sunflower, rapeseed, oil palm, and other oil plants can be used as fuel in diesel engines. Unfortunately, producing vegetable oils for use in diesel engines is costly in terms of economics and energy. A slight net return on energy from soybean oil is possible only if the soybeans are grown without commercial nitrogen fertilizer. The soybean, since it is a legume, will under favorable conditions produce its own nitrogen. Still soy has a 63% net fossil energy loss.

The USA provides $500 million in subsidies for the production of 850 million liters of biodiesel, which is 74 times greater than the subsidies per liter of diesel fuel. The environmental impacts of producing soybean biodiesel are second only to that of corn ethanol:

1. Soybean production causes significant soil erosion, second only to corn production.
2. Soybean production uses large quantities of herbicides, second only to corn production. These herbicides cause major pollution problems with natural biota in the soybean production areas.
3. The USDA reports a soybean yield worldwide to be 2.2 tons per hectare.

With an average oil extraction efficiency of 18%, the average oil yield per year would be approximately 0.4 tons per hectare. This converts into 454 l of oil per hectare. Based on current US diesel consumption of 227 billion liters/year, this would require more than 500 million hectares of land in soybeans or more than half the total area of the USA planted just for soybeans! All 71 billion tons of soybeans produced in the USA could only supply 2.6% of total US oil consumption. . . .

Algae for Oil Production

Some cultures of algae consist of 30% to 50% oil. Thus, there is growing interest using algae to increase US oil supply based on the theoretical claims that 47,000 to 308,000 l ha^{-1} year^{-1} (5,000 to 33,000 gallons/acre) of oil could be produced using algae. The calculated cost per barrel would be $15. Currently, oil in the US market is selling for over $50 per barrel. If the above estimated production and price of oil produced from algae were exact, US annual oil needs could theoretically be met if 100% of all US land were in algal culture.

Despite all the algae-related research and claims dating back to 1970s, none of the projected algae and oil yields have been achieved. To the contrary, one calculated estimate based on all the included costs using algae would be $800 per barrel, not $15 per barrel, as quoted above. Algae, like all plants, require large quantities of nitrogen and water in addition to significant fossil energy inputs for the production system.

Conclusion

A rapidly growing world population and rising consumption of fossil fuels is increasing demand for both food and biofuels. That will exaggerate both food and fuel shortages. Producing biofuels requires huge amounts of both fossil energy and food resources, which will intensify conflicts among these resources.

Using food crops to produce ethanol raises major nutritional and ethical concerns. Nearly 60% of humans in the world are currently malnourished, so the need for grains and other basic foods is critical. Growing crops for fuel squanders land, water, and energy resources vital for the production of food for people. Using food and feed crops for ethanol production has brought increases in the prices of US beef, chicken, pork, eggs, breads, cereals, and milk of 10% to 20%. In addition, Jacques Diouf, Director General of the UN Food and Agriculture Organization reports that using food grains to produce biofuels already is causing food shortages for the poor of the world. Growing crops for biofuel not only ignores the need to reduce natural resource consumption, but exacerbates the problem of malnourishment worldwide by turning food grain into biofuel.

Recent policy decisions have mandated increased production of biofuels in the United States and worldwide. For instance, in the Energy Independence and Security Act of 2007, President Bush set "a mandatory renewable fuel standard (RFS) requiring fuel producers to use at least 36 billion gallons of biofuel in 2022." This would require 1.6 billion tons of biomass harvested per year and would require harvesting 80% of all biomass in the USA, including all agricultural crops, grasses, and forests. With nearly total biomass harvested, biodiversity and food supplies in the USA would be decimated.

Increased biofuel production also has the capability to impact the quality of food plants in crop systems. The release of large quantities of carbon dioxide associated with the planting and processing of plant materials for biofuels is reported to reduce the nutritional quality of major world foods, including wheat, rice, barley, potatoes, and soybeans. When crops are grown under high levels of carbon dioxide, protein levels may be reduced as much as 15%.

Many problems associated with biofuels have been ignored by some scientists and policy-makers. The production of biofuels that are being created in order to diminish dependence on fossil fuels actually depends on fossil fuels. In most cases, more fossil energy is required to produce a unit of biofuel compared with the energy that it produces. Furthermore, the USA is importing oil and natural gas to produce biofuels, which is making the USA [less] oil independent. Publications promoting biofuels have used incomplete or insufficient data to support their claims.

For instance, claims that cellulosic ethanol provides net energy have not been experimentally verified because most of their calculations are *theoretical*. Finally, environmental problems, including water pollution from fertilizers and pesticides, global warming, soil erosion, and air pollution are intensifying with biofuel production. There is simply not enough land, water, and energy to produce biofuels.

Most conversions of biomass into ethanol and biodiesel result in a negative energy return based on careful up-to-date analysis of all the fossil energy inputs. Four of the negative energy returns are: corn ethanol at minus 46%; switchgrass at minus 68%; soybean biodiesel at minus 63%; and rapeseed at minus 58%. Even palm oil production in Thailand results in a minus 8% net energy return, when the methanol requirement for transesterification is considered in the equation.

Increased use of biofuels further damages the global environment and especially the world food system.

DAVID PIMENTEL is Professor Emeritus in the Departments of Entomology and Ecology and Evolutionary Biology at Cornell University in Ithaca, New York.

ALISON MARKLEIN is a Graduate Student in Earth Systems Ecology and Biogeochemistry at University of California, Davis.

MEGAN A. TUTH is a Graduate Student in Environmental Studies at the University of Oregon in Eugene, Oregon.

MARISSA N. KARPOFF is currently pursuing an MD at Tulane Medical School in New Orleans.

GILLIAN S. PAUL has a Masters of Forest Science from the Yale School of Forestry and Environmental Studies and now works for the Environmental Leadership and Training Initiative (ELTI), a global training and capacity building organization that is a joint initiative of Yale University and the Smithsonian Tropical Research Institute (STRI).

ROBERT MCCORMACK is currently completing a Masters in Environmental Law and Policy at Vermont Law School.

JOANNA KYRIAZIS is currently enrolled in the Law School at the University of Toronto.

TIM KRUEGER is working for Policy Matters, Ohio, an economic research institute.

EXPLORING THE ISSUE

Are Biofuels a Reasonable Substitute for Fossil Fuels?

Critical Thinking and Reflection

1. Why is there so much interest in using ethanol as a portion of automobile fuel?
2. Why does the production of biofuels affect the food supply and food prices?
3. Is using biofuels to replace fossil fuels compatible with feeding the world's population?
4. What alternatives exist to using liquid fuels to power automobiles?
5. Which biofuels are likely to have the least undesirable impact on the environment and food supply?

Is There Common Ground?

There is agreement that using prime agricultural land and expensive fertilizers and processing to grow biofuel crops necessarily reduces resources needed to grow food and unsurprisingly has effects on food supply and prices. However, there are biofuel crops that are less dependent on agricultural resources. Investigate what is happening with "cellulosic" biofuels, algae-based biofuels, and bacteria-based biofuels and answer the following questions:

1. Can these alternatives be implemented in the near future?
2. Will their prices be competitive?
3. Can they meet demand?
4. Will their production compete with food production?

Create Central

www.mhhe.com/createcentral

Additional Resources

David Biello, "The False Promise of Biofuels," *Scientific American* (August 2011)

Robert Bryce, "A Promise Not Kept: The Case against Ethanol," *The Conference Board Review* (May/June 2008)

Suzanne Hunt, "Biofuels, Neither Saviour nor Scam: The Case for a Selective Strategy," *World Policy Journal* (Spring 2008)

Daniel M. Kammen, "The Rise of Renewable Energy," *Scientific American* (September 2006)

Internet References . . .

International Energy Agency

www.iea.org/topics/biofuels/

National Biofuels Energy Laboratory

www.eng.wayne.edu/page.php?id=4765

The National Renewable Energy Laboratory

www.nrel.gov/

Selected, Edited, and with Issue Framing Material by:
Thomas Easton, *Thomas College*

ISSUE

Is Hydropower a Sound Choice for Renewable Energy?

YES: Steve Blankinship, from "Hydroelectricity: The Versatile Renewable," *Power Engineering* (June 2009)

NO: Mike Ives, from "Dam Bad," *Earth Island Journal* (Autumn 2011)

Learning Outcomes

After reading this issue, you will be able to:

- Describe three advantages offered by hydroelectric power.
- Describe three environmental drawbacks of hydroelectric power.
- Explain how energy can be obtained from water without building dams.
- Describe how water can be used to store energy (as "pumped storage").

ISSUE SUMMARY

YES: Steve Blankinship argues that hydroelectric power is efficient, cheap, reliable, and flexible. It can serve as baseload electricity, backup for wind farms, and even as energy storage, and there is significant room for expansion, including using new technology that does not require dams. It is therefore drawing increasing interest as a way of dealing with rising demand and ever more expensive fossil fuels.

NO: Mike Ives argues that hydroelectric dams such as one proposed for the Mekong River in Laos pose flooding risks, threaten the livelihoods of farmers and fishers, and may be vulnerable to earthquakes. Decisions to build them (or not) are guided by politics, not environmental and social impacts.

Dams have long been an icon of civilization. Building dams of all sizes, from those that hold back village mill ponds to the giant Hoover Dam, was a crucial step in the settling and development of America. They supplied mills with the mechanical power generated when flowing water spun waterwheels. Combined with locks and canals, dams improved the navigability of waterways in the days before railroads. They provided water for irrigation, reduced flooding, and generated electricity. In other parts of the world, building dams for these benefits has been an important step in moving from "undeveloped" to "developing" to "developed" status. See e.g., Ibrahim Yuksel, "Hydropower in Turkey for a Clean and Sustainable Energy Future," *Renewable & Sustainable Energy Reviews* (August 2008). However, hydropower does have its critics. American Rivers, a nonprofit organization dedicated to the protection and restoration of North America's rivers, argues in "Hydropower: Not the Answer to Preventing Climate Change" (2007) that suggesting that hydropower is the answer to global warming hurts opportunities for alternative renewable energy technologies such as solar and wind and distracts from the most promising solution, energy

efficiency. Aviva Imhof and Guy R. Lanza, "Greenwashing Hydropower," *World Watch* (January/February 2010), stress hydropower's drawbacks and argue that "the industry's attempt to repackage hydropower as a green, renewable technology is both misleading and unsupported by the facts, and alternatives are often preferable. In general, the cheapest, cleanest, and fastest solution is to invest in energy efficiency." In the United States, the Army Corps of Engineers (ACE) built almost all hydroelectric power systems until the late 1970s. Today, according to the ACE pamphlet, "Hydropower: Value to the Nation" (www.vtn.iwr .usace.army.mil/pdfs/Hydropower.pdf), the ACE is the single largest owner and operator of hydroelectric power plants in the country and one of the largest in the world. Its 75 power plants produce nearly 100 billion kilowatt-hours of electricity per year. Hydroelectric power is renewable, efficient, and clean. It does not generate air or water pollution, and it emits no greenhouse gases to contribute to global warming. According to Matt Lucky ("Global Hydropower Installed Capacity and Use Increase," in Vital Signs 2012 [Island Press, 2012]), global use of hydropower has been steadily increasing for years. In 2010, it "accounted for 16.1 percent of electricity use and 3.4 percent of energy use worldwide." North America accounts for about a fifth of global hydropower production, the Asia-Pacific region almost a third, Eurasia a quarter, and South and Central America a fifth. R. Bakis, "The Current Status and Future Opportunities of Hydroelectricity," *Energy Sources: Part B: Economics, Planning, and Policy* (vol. 2, no. 3, 2007), finds that hydropower has unique benefits (some of which are associated with reservoir development), is clean and affordable, and has an important role to play in the future.

In the United States, most good sites for large hydropower plants have already been developed. In the rest of the world, the best sites, such as China's Three Gorges project, are rapidly being developed. Unfortunately, many people see problems in these projects. Mara Hvistendahl's "China's Three Gorges Dam: An Environmental Catastrophe?" *Scientific American* online (March 25, 2008) (www.sciam.com/article.cfm?id=chinas-three-gorges-dam-disaster) is indicative. In Chile, the need for electricity versus the risks to the environment is still being debated; see Gaia Vice, "Dams for Patagonia," *Science* (July 23, 2010); Patrick Symmes, "The Beautiful & the Dammed," *Outside* (June 2010), finds the Patagonian landscape far too beautiful to spoil with dams and power lines (as well as too vulnerable to volcanic eruptions and earthquakes), but the CEO of HidroAysen—the company that proposes to build five large dams on Patagonian rivers—says other renewables (solar and wind) just can't meet the energy demand; see Aaron Nelsen, "CEO of Chilean Energy Company Defends Project to Dam Patagonia," *Global Spin* (a *Time* blog) (May 20, 2011) (http://globalspin.blogs.time.com/2011/05/20/ceo-of-chilean-energy-company-defends-project-to-dam-patagonia/). In Southeast Asia, at issue is the Mekong River and its tributaries, which offer huge potential for hydropower. Unfortunately, as David Fullbrook, "Dams It Is!" *The World Today* (June 2008), notes, millions of farmers and fishers depend on the Mekong for food and livelihoods, and they are threatened by proposals to meet energy demand—both locally and in neighboring countries such as China—with hydropower. See also Richard Stone, "Mayhem on the Mekong," *Science* (August 12, 2011). In 2012, Laos approved construction of the Xayaburi dam on the lower Mekong ("Megadam Gets Green Light," *Science*, November 9, 2012).

Because dams flood the land behind them, they destroy forests, farmland, and villages and displace thousands—even millions—of people. Species decline and vanish. Sediment trapped behind the dam no longer reaches the sea to nourish fisheries. River flow is altered. Diseases change their patterns. When a dam breaks, the resulting sudden, immense flood can do colossal damage downstream. The destruction of forests also means that even though hydropower itself does not release greenhouse gases, displaced forests are no longer removing carbon dioxide from the air. According to the International Hydropower Association (http://hydropower.org), the lakes behind dams also emit greenhouse gases, and lake sediments can store large amounts of organic matter. The real issue is the *net effect*, which is ignored by those who wish to discredit hydroelectric power, who also focus on the worst cases, which tend to be in tropical environments. The net emissions of greenhouse gases from hydropower impoundments is an area of ongoing research. See Alain Tremblay, Louis Varfalvy, Charlotte Roehm, and Michelle Garneau, *Greenhouse Gas Emissions—Fluxes and Processes: Hydroelectric Reservoirs and Natural Environments* (Springer, 2005). The tension between the real benefits and the real drawbacks of hydropower is discussed by R. Sternberg in "Hydropower: Dimensions of Social and Environmental Coexistence," *Renewable & Sustainable Energy Reviews* (August 2008).

Fortunately, it is possible to get energy from water without building traditional dam-and-reservoir hydroelectric systems. Tidal power requires dams across estuaries; at high tide, water flows upstream through turbines; at low tide, the turbines are reversed to extract energy as the water flows back to the sea. Turbines can also be placed on the sea floor to take advantage of tidal (and other) currents and in rivers to use "run of the river" flows. See Jonathon Porritt, "Catch the Tide," *Green Futures* (January 2008), and David Kerr, "Marine Energy," *Philosophical Transactions: Mathematical, Physical & Engineering Sciences* (April 2007). In North America, a number of tidal power projects are being started in the Bay of Fundy; see Colin Woodard, "On US Border, a Surge in Tidal-Power Projects," *Christian Science Monitor* (August 15, 2007). Zafer Defne et al., "National Geodatabase of Tidal Stream Power Resource in USA," *Renewable & Sustainable Energy Reviews* (June 2012), find the greatest numbers of tidal power "hotspots" in Alaska and Maine. See also the U.S. Department of Energy's Marine and Hydrokinetic Technology Database at www1.eere.energy.gov/water/hydrokinetic/default.aspx.

There is also wave power, which uses the motion of waves to work special arrangements of pistons, levers, and air chambers to extract energy. See Ewen Callaway, "To Catch a Wave," *Nature* (November 8, 2007), David C. Holzman, "Blue Power: Turning Tides into Electricity," *Environmental Health Perspectives* (December 2007), and Elisabeth Jeffries, "Ocean Motion Power," *World Watch* (July/August 2008). Yet even this can draw attention from environmentalists, who fear that the equipment may interfere with marine animals and that associated electromagnetic fields may harm sensitive species such as sharks and salmon; see Stiv J. Wilson, "Wave Power," *E Magazine* (May/June 2008). However, Urban Henfridsson et al., "Wave Energy Potential in the Baltic Sea and the Danish Part of the North Sea, with Reflections on the Skagerrak," *Renewable Energy: An International Journal* (October 2007), note that the potential contribution of wave energy to civilization's needs is large. To take advantage of that potential requires "Sound engineering, in combination with producer, consumer and broad societal perspective." See also Kester Gunn and Clym Stock-Williams, "Quantifying the Global Wave Power Resource," *Renewable Energy: An International Journal* (August 2012).

In the following selections, Steve Blankinship argues that hydroelectric power is efficient, cheap, reliable, and flexible. It can serve as baseload electricity, backup for wind farms, and even as energy storage, and there is significant room for expansion, including using new technology that does not require dams. It is therefore drawing increasing interest as a way of dealing with rising demand and ever more expensive fossil fuels. Mike Ives argues that hydroelectric dams such as one proposed for the Mekong River in Laos pose flooding risks, threaten the livelihoods of farmers and fishers, and may be vulnerable to earthquakes. Decisions to build them (or not) are guided by politics, not environmental and social impacts.

YES

Steve Blankinship

Hydroelectricity: The Versatile Renewable

As one of the earliest and most elementary forms of power generation, hydropower remains by far the largest source of renewable energy in the world, including in North America.

In the early 1900s, hydropower was the dominant source of U.S. electric generation and as recently as the 1940s accounted for more than 40 percent of electricity production. By the 1950s, developers had tapped the hydro potential of the most mountainous regions in the U.S.—many in the Northwest—where steep inclines supply the strongest river flows and permit the most cost-efficient projects.

Hydropower supplies almost two-thirds of Canada's power and makes it the world's largest hydropower producer, representing 13 percent of global output. It's also the world's second largest exporter of hydro (after France). Altogether, its roughly 450 hydro plants, half of which produce less than 10 MW, account for 72,660 MW with another 1,800 MW currently under construction and an additional 12,000 MW under consideration, according to the Canadian Hydropower Association.

Today, hydro represents about 8 percent of all power in the U.S. and more than 90 of all the renewable power generated in the nation. Hydro provides more than 16 times as much energy as wind and solar power combined.

And hydro's use is increasing, both through updates to older hydro generation technology and through new technologies. Utilities are proposing more than 70 projects that would boost U.S. hydroelectric capacity by at least 11,000 MW over the next decade.

Driving a new wave of hydropower development is unprecedented demand for renewable energy and rising fossil fuel costs. The American Recovery and Reinvestment Act and other programs provide tax provisions to attract investment in incremental hydropower, hydro at non-powered dams, ocean, tidal and in-stream hydrokinetic technologies.

Upgrades and New Builds

AMP-Ohio, which owns and operates power production facilities for 126 member entities in Ohio, Pennsylvania, Michigan, Virginia, West Virginia and Kentucky, is developing multiple hydroelectric projects, representing one of the largest deployments of hydroelectric generation in the U.S. The projects are run-of-the-river hydroelectric facilities to be installed on existing dams on the Ohio River and the New River in West Virginia. Combined, these six projects would add more than 380 MW of new generation at an estimated construction cost of just over $1.5 billion.

As part of the project, AMP-Ohio signed a contract worth more than $300 million with Voith Siemens to manufacture turbines and generators for the first three of these projects at the Smithland, Cannelton and Willow Island locks and dams. A fourth Ohio River project will be at the Meldahl locks and dam. In addition, AMP-Ohio is pursuing a project at the R.C. Byrd locks and dam on the Ohio River and performing a feasibility study for a project at the Bluestone dam on the New River.

PPL Corp. recently filed a request with the Federal Energy Regulatory Commission for a 125 MW expansion at its Holtwood hydroelectric plant on the Susquehanna River in Pennsylvania. Holtwood currently is rated at 108 MW and has generated power since 1910. The estimated $440 million project is subject to availability of stimulus funding. Construction could start early next year.

"The hydro business is so robust right now that the contractors only go after big projects, leaving lots of room for smaller players to stay busy with medium and smaller projects," said Norm Bishop, vice president for the hydro dams market sector of MWH, a Colorado-based firm that helps develop new hydro projects and update existing ones.

In addition to the demand for renewables and the rising costs of fossil fuels, Bishop cites hydro's flexibility that allows it to meet today's power market demands, including ancillary grid support, which is especially critical in places with increasingly high penetrations of wind farms.

And the potential to make cheap power from water has barely been tapped. Of the existing dams in the U.S., 3 percent (or around 2,400) are equipped to produce power and annually generate 270,000 GWh, according to the U.S. Department of Energy. DOE estimates another 30,000 MW of capacity could be developed, including 17,000 MW at existing dams.

Pump It Up

After decades of little or no development, U.S. hydro pumped storage (by which water is pumped uphill during off peak demand periods and released to flow down hill to spin generating turbines on peak) is seeing renewed attention. In the past two years, FERC has approved 21 preliminary permits for pumped-storage projects totaling more than 12,000 MW.

Earlier this year, Energy Secretary Steven Chu said hydro pumped storage must be a part of a national plan to expand clean-energy resources and integrate variable renewable energy resources into the transmission grid. Chu said the U.S. has limited existing resources for storing energy and most of what it does have comes from the 20,355 MW of pumped-storage capacity now in service.

National Hydropower Association (NHA) Executive Director Linda Church-Ciocci said that expanding hydro pumped storage capacity will be a high priority for her association's new pumped storage council.

"Right now, the federal government has no program to spur expansion of U.S. pumped storage," she said. "We advocate investment tax credits or other similar measures that can incentivize pumped storage development immediately."

A benefit could be changes to the licensing process, an initiative NHA has worked on for a number of years. The new process focuses on collaboration among agencies, which should reduce the amount of time required for a new or renewed license from 15 years to as little as three or four years.

Relicensing is hot right now as owners hope to eke even 2 or 3 percent improvements for a price tag that can be as low as $200/kW in some cases.

"There's a tremendous opportunity to repower and upgrade the mechanical aspects of existing facilities to increase output," said Don Erpenbeck, another MWH vice president. He's particularly upbeat with some new technologies that include ultra low head hydro and emerging technologies such as hydrokinetic.

"If a project is 20 years old there's a good chance today's technology can eke out more power at a very small cost per kW."

Water to Wire = Ultimate Efficiency

Hydro power has always had high availability and quick ramp rates. It also enjoys an overall efficiency unmatched by any other power source. No fuel is needed, just the volume and motion of the water. Mechanical efficiency is high and the only true efficiency losses are limited to line loss.

"Availability is pushing 90 percent with hydro and on the mechanical side we hit 95 percent efficiency," said Erpenbeck. But some plants have lost as much as 10 percent of their efficiency due to the age of their turbine/generators. New technology can not only reclaim that efficiency, but increase output above previous levels.

"You could be looking at up to 20 percent efficiency increases if the existing machines are in bad shape," he said.

Hydro's ability to ramp quickly enhances its attractiveness as a power portfolio asset. New technology can expand that flexibility. "We can make the efficiency curve flatter," said Erpenbeck, "so hydro is more efficient running off peak. We can now run with even greater flexibility and

respond to market conditions across a wider range of megawatts in terms of cycling, load following and turn down."

Erpenbeck said hydropower can routinely operate at 55 to 100 percent of rated load and back off to 20 to 40 percent as needed.

Increases to operating range provide prime quality spinning reserve for grid support, which is more important today than ever before. The increases are achieved through the ability to run in condensing mode where the generator is synchronized and motoring while the turbine spins air, or synchronized at low power (20 to 40 percent of rated load) going to full power in seconds. "With all the wind power coming onto the grid this is being used much more," said Bishop. "Modern hydro's ramp rate is fast."

Hydro also provides high-quality ancillary grid support. For example, Grand Coulee Dam on the Columbia River in Washington State can go from low load to full load (about 800 MW) in matter of seconds.

Technology Improvements

Improvements to conventional hydro technology provide a variety of upgrades that help hydropower remain low cost while offering environmental benefits.

Grant County Public Utility District, part of the Mid-Columbia System, is installing 10, $15 million fish-friendly turbines at Wanapum Dam and plans to replace another 10 turbines at Priest Rapids. The old turbines are being replaced with models that use six smaller blades instead of five. When completed in 2012, the project is expected to improve each turbines' efficiency by 3 percent and the facility's overall capacity by 15 percent.

Recent upgrades to the Sacramento Municipal Utility District's Jaybird and Loon Lake hydroelectric powerhouses have led to still more efficiency gains. Installing new computerized controllers to better regulate water flow to the turbines has increased output by 15 MW for the same amount of water when running at low power levels. The new governor control system automatically regulates the amount of water that shoots out of six high-pressure nozzles and onto a wheel that spins the generator.

With the old equipment, the controller opened all six needles at once, boosting water flow to the turbine as electricity demand rose. When the unit was at low load, it required less water. But this fanned out of the needles similar to a garden hose set to a wide spray pattern and caused most of the water to miss the turbine wheel.

The new equipment opens two needles initially and adds others as demand for power rises. By moving the same volume of water through two needles instead of six, the water stream is more tightly focused and hits the turbine wheel more directly. This results in significant water savings for the same amount of power generation. Based on current short-term power price forecasts, the utility estimates the equipment will save it $130,000 a year.

Canada's two largest hydro utilities—Ontario Hydro and Hydro Quebec—continue to expand capacity. Hydro

Quebec will announce this summer a revised hydro expansion schedule. But current expansion plans include completing the last generating units at the Péribonka development and the first units at Chute-Allard and Rapides-des-Coeurs.

Work also proceeded at the Eastmain-1-A/Sarcelle/Rupert jobsite. The project will divert a portion of the flow from the Rupert River watershed into the Eastmain River watershed. That will involve four dams, a spillway on the Rupert River, 74 dikes, two diversion bays, construction of a 1.8-mile-long tunnel and a network of canals and hydraulic structures on the Rupert River to maintain post-diversion water levels along half of the river's length.

Ontario Hydro's Niagara tunnel project will increase the amount of water flowing to turbines at the Sir Adam Beck generating stations at Niagara Falls, allowing them to better use available water. Upon completion of the 6.5-mile-long tunnel, average annual generation from the Beck stations is expected to increase by about 1.6 TWh. Ontario Power also recently completed a 12.5 MW hydroelectric generating station on the English River. The new Lac Seul generating station uses most of the spill currently passing the existing Ear Falls generating station, thus increasing overall efficiency, capacity and energy generated from the plant. The project is expected to be inservice later this year.

Ontario Power is proceeding with the definition phase for a 450 MW development on the Lower Mattagami River, including replacing the Smoky Falls station and expanding the Little Long, Harmon and Kipling stations. The company also approved redeveloping four existing stations, which otherwise would have been removed from service.

New Wave for Hydropower

The tremendous force of moving water is obvious to anyone who has stood in breaking ocean waves or swum against a river's current. Hydrokinetic—or kinetic hydro—technologies generate electricity from waves or directly from the flow of water in ocean currents, tides or inland waterways and is gaining increased attention.

Hydrokinetic technology uses stream flow to make power and requires a steady three to five knots of flow to operate. Hydrokinetic water turbines (HUTs) can be placed where there is no dam; for example, they may be attached to bridges or to frames on the river bottom. Hydrokinetic boosts potential capacity far beyond conventional hydro power. As one example, thousands of miles of canals in California are designed primarily for irrigation, but could also host kinetic turbines.

HUTs are smaller than wind turbines because water is about 800 times denser than air. Ocean tidal current can deliver predictable 20 hours per day energy and a HUT can produce up to four times more energy than a wind turbine on a good day. Venturi and centrifugal designs can accelerate water speed through the turbine and double the

energy produced. Current project proposals suggest that kinetic hydro produced by U.S. waves, tides and rivers, could produce 13,000 MW by 2025.

FERC has approved the nation's first commercially-operational hydrokinetic power station in Hastings, Minn. The 4.4 MW run-of-river hydropower plant will use two hydrokinetic units, each with a nameplate capacity of 100 kW.

Near and off-shore ocean waves might have the greatest hydrokinetic potential. Extracting just 15 percent of the energy in U.S. coastal waves would generate as much electricity as is currently produced at conventional hydro dams. Much of this wave potential is along the Pacific Coast and close to population centers.

Beyond the sheer size of the resource, hydrokinetics are attractive for their predictability. Wave patterns can be predicted days in advance. Since the kinetic energy held in a stream is related to its speed cubed, extracting the most electricity from each hydrokinetic project will depend heavily on site selection. Energy output increases eight times with only twice as much water current speed.

State and federal policymakers across the U.S. have taken notice of the potential of hydrokinetic energy and have begun to support its development through legislative and monetary means. Ocean energy is eligible for credit under renewable electricity standards in 16 states and for federal renewable production tax credits, as expanded in the Energy Policy Act of 2005. Furthermore, hydrokinetic energy development was marked for increased research funding appropriations in the 2007 Energy Independence and Security Act.

Overcoming Environmental Opposition

Hydro has all but disappeared from the renewable energy options usually cited by renewable energy advocates. Many environmentalists have long opposed hydroelectric power and do not consider it "green" or renewable. Much of the opposition is based on the water diversions required by traditional hydroelectric projects and the effects on land and wildlife. Fish killed as a result of passing hydro turbines has also led to a substantial amount of environmental concern.

Because of this environmental opposition, some states restrict the extent to which hydroelectric projects may qualify under renewable portfolio standards.

"Policy makers at the federal and state level have a difficult task of designing regulations and incentives that recognize the fact that an existing renewable source like hydropower can be further developed with the right incentives," said Michael Cutter, vice president of engineering and development for Brookfield Renewable Power (BRP). The company has developed, owned and operated hydro power facilities for more than 100 years and has 100 hydropower facilities totaling nearly 2,000 MW in nine U.S. states.

Cutter said opportunities exist throughout the U.S. for continued development of hydro electric generation. "Recent studies show the amount of hydropower could double from the current amount of installed hydro generating capacity by 2030 if the country could upgrade existing hydropower, add hydropower at non power dams and develop some of the new technologies," he said. "To reach hydropower's potential it is important to continue to strengthen federal and state energy policies and educate the public on hydropower's role as an indigenous, renewable energy source."

Steve Blankinship is an Associate Editor of *Power Engineering* magazine.

Mike Ives

 NO

Dam Bad

Laos' Plans to Dam the Mekong Could Open the Floodgates to Further Dams on the River

Sathian Megboon is a DJ for 94.5 FM, a radio station in northeast Thailand. It's a fun gig, he says, because the station broadcasts across the Mekong River to Vientiane, the tiny capital of the Lao People's Democratic Republic, which gives him the chance to take requests from listeners in two countries. Last year, listeners started calling more than usual, but not to ask for the folk songs Sathian likes to play. They wanted to know what he thought of reports that a Thai company was planning to build a $3.5 billion dam a few hundred kilometers upstream on the Lao–Thai border in a remote and impoverished mountain area.

Sathian's Vientiane listeners may have read about the proposed Xayaburi dam in *The Vientiane Times*, a mouthpiece for the secretive Lao Communist Party. But they wanted the real story: Namely, how would a hydropower dam in Xayaburi Province affect them?

The DJ said he wasn't sure, because he isn't a politician or hydropower expert. But building dams on the Mekong—the world's tenth-longest river—seemed to him a terrible idea. China, which borders northern Laos, built four hydroelectric dams on the upper Mekong between 1986 and 2009. Sathian and his neighbors say those dams have changed the Mekong's "flood pulse," the seasonal ebb and flow that regulates agriculture and fishing and feeds the Lower Mekong Basin's 60 million residents. According to elder farmers who grew up beside the Mekong, erratic flow patterns that appeared in the 1990s have made it harder to grow staple crops such as chili peppers, eggplants, and corn. They say they cannot imagine how additional dams would improve the situation.

Sathian and his neighbors claim that, in 2007, Chinese dams triggered violent flooding. "It happened very fast, and we didn't have any warning," Sathian, who is 63 and generously tattooed, told me on a lethargic mid-April afternoon. "It wasn't raining, but the river flooded for seven days! If they build a new dam in Xayaburi, I'm afraid the next flood could be like a tsunami."

We were standing on the banks of the Mekong under a white banner that said NO MEKONG DAMS in English. Looking across the river, I saw monks in orange robes bathing against the backdrop of Vientiane's unassuming skyline. Fishing skiffs were gliding downstream, and the air felt sticky and stagnant. It was hard to imagine that a planned hydropower dam hundreds of miles upstream, in the mountains of landlocked Laos, had caused such fear. But a few weeks earlier, 263 nongovernmental organizations from 51 countries had written to the Lao prime minister and urged him to cancel the project. Apparently Xayaburi was kind of a big deal.

Hydropower dams are common in Southeast Asia and have already been constructed on some of Mekong's tributaries. But the 1,280-megawatt Xayaburi dam, which Laos proposed last September, would be the first of 11 dams planned for the river's mainstream. Nine would be sited in Laos, the other two in Cambodia. Along with a proposed "river diversion" project in Laos, the dams could supply about 65 terawatt hours of electricity per year—up to 8 percent of the Mekong region's projected 2025 electricity demand and slightly less energy than Americans use each year to power their televisions. About 90 percent of the power would go to Vietnam and Thailand.

Environmental activists and civil society groups across Southeast Asia say that if the Xayaburi dam is built, it will lay the political groundwork for the other dams, which they fear would have devastating cumulative impacts on ecosystems and livelihoods. They also worry that hydropower developers are ignoring climate change, which, according to scientists, will affect Mekong hydrology this century. The activists have rallied behind a 2010 study by the International Centre for Environmental Management (ICEM) that recommended a ten-year moratorium on new Mekong dams because the dams, if built, would provoke "permanent and irreversible" social and environmental consequences.

"We need food, not electricity," says Mu Panmeesri, a 27-year-old schoolteacher who lives downstream of Sathian Megboon and organizes anti-Xayaburi rallies in his town. "We can't eat electricity."

The Lao PDR, a former French colony that declared independence in 1945, has been governed by one political party for the last 36 years. Usually, if Lao officials want to build something, a road, a bridge, or a massive dam, it would probably be built. But last year's Xayaburi proposal triggered a review process moderated by the Mekong River Commission (MRC), an advisory body formed in 1995 by the four lower Mekong countries—Laos, Thailand,

Vietnam, and Cambodia—to promote sustainable development in the Lower Mekong Basin.

Encouraging sustainable development is no small task in Southeast Asia, where environmental regulations are thin and civil society has little or no voice. Powerful governments often approve energy projects without consulting their citizens or requiring detailed environmental impact assessments. But the four lower Mekong countries agreed in 1995 to initiate a multilateral MRC review if one of them ever proposed a dam on the river's mainstream. So environmental groups and local dam opponents are relying on this untested international body to stop the Xayaburi project.

Xayaburi is the first mainstream dam proposed for the lower Mekong since 1995, and the MRC review that began last fall is the first of its kind. MRC member countries can't legally stop Laos from damming the Mekong, but they can commission feasibility studies and draft statements recommending the best course of action. The MRC review process for the Xayaburi dam was supposed to end in April, but as I write these words in August, analysts aren't sure when the process will end or whether Laos will heed its neighbors' recommendations. Regardless, it's clear that the MRC process has sparked an important public debate about energy, ecology, and food security in Southeast Asia—an achievement in itself for a region plagued by poverty, corruption, and the legacy of war.

Xayaburi appears to be a bellwether of regional environmental diplomacy. Decisions made in the coming months on Xayaburi and other dams will affect millions of people, perhaps for generations. In 50 years, will people who live near the Mekong have enough to water to drink and food to eat? Will Mekong countries establish procedures for sharing their river resources, or will diplomatic scuffles escalate to armed stand-offs? How best can poor countries' demands for energy be balanced with the interests of riverine communities?

These questions invite speculation about whether dams are the best way to meet the region's rising electricity demand, which is predicted to increase each year until 2025. The region has what ICEM calls a "massive potential for hydropower," but sediment buildup in dam reservoirs can hinder medium-and long-term capacity; sedimentation at the Xayaburi dam, according to the MRC, would decrease the dam's output by up to 60 percent in 30 years. Critics charge that the electricity from Mekong dams wouldn't justify the negative impacts the dams would have on watersheds, fisheries, and food production. Rather than dam the Mekong, they say, Mekong countries should promote wind turbines and solar panels.

Laos counters that Mekong dams, in addition to providing energy for the regional grid, would help lift Laotians out of poverty. While many rural development experts say Laos should temporarily or permanently postpone Mekong dams, others suggest the dams, though imperfect, may be a reasonable way for Laos—which was heavily bombed during the Vietnam War and remains one of the world's poorest countries—to achieve its goal of escaping from UN-designated "Least Developed Country" status. Lao officials say they want their country to be the "battery of Southeast Asia."

Dams on the Mekong's mainstream would surely have negative impacts on downstream communities and ecosystems, says Stuart Ling, who directs the Lao office of a Belgian NGO called VECO. "But the Lao government has GDP targets, and these dams that are going ahead were in the system a long time ago because Laos is trying to develop. If Laos doesn't get its revenue from generating hydropower, how will it get its revenue?"

Dams were first proposed for the Mekong in the 1950s, but the Cold War put them on hold. By the time they were re-proposed in the 1990s, writes Philip Hirsch, a Mekong expert at the University of Sydney, resistance from environmental groups made them "unpalatable." Renewed interest in developing hydropower in the Lower Mekong Basin has escalated since 2006 in tandem with rising private investment in power infrastructure. Privately owned hydropower companies are filling a void left by multilateral institutions like the World Bank and the International Monetary Fund, which are less eager than they once were to fund hydropower projects. One reason is that a 2000 report by the World Commission on Dams found dams had displaced 40 to 80 million people worldwide, and that "in too many cases," the associated social and environmental costs were "unacceptable and often unnecessary."

Private developers aren't held to the same environmental regulations that multilateral institutions would be. But the financiers and developers do have to weather criticism from environmental groups—particularly the Berkeley-based NGO International Rivers, which employs a full-time "Mekong Campaigner."

Resistance to Xayaburi raises the question: If the idea of damming the Mekong is unpalatable to so many people, why are developers—and the Lao and Cambodian governments—willing to risk drawing flak?

One explanation is China. The four Mekong dams the Chinese built pump more water downstream in the dry season than the river would otherwise supply. By suppressing the Mekong's flood pulse, they have created an economic incentive for downstream developers, who at the moment have access to a steadier water supply. China has become a hydropower "role model" for downstream neighbors, says Prescott College Professor Ed Grumbine: Southeast Asian countries are thinking, "If China is using the river as a resource and gaining greatly from it, then why can't we do the same thing?"

Indeed, why not? Yes, you can make a case—as developers have and will—that hydropower is a "green" alternative to coal and nuclear power. But it is hard to argue that dams in Laos promote environmental sustainability or benefit the communities they displace. Since the Lao Communist Party seized power in 1975 after a protracted civil war, its primary development strategy has been to sell timber, rubber, mineral, and hydropower rights to foreign bidders.

"The combination of neoliberal economic policies, foreign direct investment, and a nontransparent, nepotistic kind of government [in Laos] is a real toxic mix," says Ian Baird, an environmental expert who has worked there for years. Earnings are not trickling down to everyday Laotians. Vientiane may have fancy hotels and restaurants, but a typical Lao village is a smattering of bamboo huts. According to the United Nations, food insecurity is "widespread" across the country—"alarmingly high" in rural areas—and half of all children under five in Laos are "chronically malnourished."

Baird, a geography professor at the University of Wisconsin-Madison, has seen how hydropower affects Lao communities, and he's not impressed. This spring, as Baird monitored the Xayaburi dam controversy, he was also analyzing a proposal to dam Khone Falls, the only large waterfall on the lower and middle Mekong. He says the project would negatively affect the nutrition of hundreds of thousands of people.

"I'm not an unreasonable person," Baird told me. "If you're building a dam, you should really cost it out in terms of environmental and social impacts and compensate people for the life of the project. But to be honest, I can't tell you about a hydropower project in Laos where they've done a good job."

Environmental activists say a wild Mekong is worth fighting for. The river has the world's most productive inland fishery and is a major source of livelihood for millions of people. The Xayaburi dam would threaten at least 41 of the Mekong's 1,300 fish species and cause $476 million in direct losses through reduced fish harvests—an estimate that doesn't account for fisheries in Vietnam's fertile Mekong River Delta.

Proponents of the dam say it would have "fish ladders" and other fish-friendly technologies, but aquaculture experts say those technologies wouldn't prevent fish extinctions. A study by the World Wildlife Fund and other groups noted that the ladders would be "challenging" even for strong Northern Hemisphere salmon. "If fish can't migrate, they don't breed," explains Stuart Chapman, WWF's Greater Mekong Program manager. "This will lead to a collapse of the fishery."

The dams would also disrupt the production of rice and other crops along Mekong's banks. Thailand and Vietnam are the world's first and second-largest rice exporters, respectively, and farmers in the Mekong River Basin depend on the river to irrigate 6.6 million hectares of farmland. According to ICEM's 2010 study, the dams would destroy about half of the Mekong's riverbank gardens and cost about $25 million per year in lost agricultural land.

How about earthquakes? The concerned residents I met in Thailand worry earthquakes would destroy Mekong dams and cause catastrophic floods worse than the flood that scared them in 2007. After the disastrous March earthquake and tsunami in Japan, two Chinese earthquake experts wrote to the Chinese premier, Wen Jiabao, to warn that building dams in Southwest China on the Mekong and two other rivers (the Salween or Nu and the Yangtze), was ludicrous because the region lies on an active fault line. If an earthquake were to destroy a dam on a river that has other dams, they predicted a torrent of water, mud, and rocks could careen downstream and set off a "chain reaction" of devastation.

And then there is climate change. The Mekong River begins in the Himalayas and flows 4,180 kilometers to the Vietnamese coast. Emerging research suggests that increased glacial melting in mountainous Central Asia will likely increase downstream river flows for a few decades but ultimately cause late-summer water shortages. "Climate change will exacerbate the impact of hydropower dams on the downstream Mekong Delta, but there's a lot we don't know," says Dr. Kien Tran-Mai, the MRC's climate change program officer. Because not enough research has been done on the cumulative impacts of climate change on the Mekong, Dr. Kien told me, the MRC doesn't require developers to plan for climate change in their proposals. The commission is not ignoring climate change, he insists, but as a mere advisory body it cannot require hydropower developers to create proposals based on climate modeling that doesn't yet exist.

Hydropower proponents argue that when climate change alters the Mekong's flow patterns, dam operators could simply increase water flow during dry periods and reduce it when water levels are too high. The argument sounds logical, Dr. Kien says, but it doesn't make him feel better, because dam operators are primarily concerned with "economic benefit and electricity generation."

Sathian Megboon and other concerned citizens in Thailand say they don't believe dam operators will ever look out for their best interests. "We're controlled by China," Somphong Paratphom, a village chief, said in April as he drove me across a Mekong sandbar in his pickup truck. "If the Chinese wanted to kill us, they could just open the dam's gates. And the danger from Xayaburi is twice as high as it was in China, because China is farther away! If they build a dam in Xayaburi, why wouldn't flooding happen again?"

On April 19, as resistance to Xayaburi was building, diplomats from the four lower Mekong countries met at the MRC headquarters in Vientiane to draft their final recommendation. But the leaders couldn't reach a consensus. Instead, they agreed on a follow-up meeting at an unspecified date. Thailand, Cambodia, and Vietnam issued separate statements saying more study was needed on the Xayaburi dam's "transboundary impacts." Vietnam issued the strongest statement, calling for a ten-year moratorium on new Mekong dams. Laos directed the dam's Thai developer, Ch. Karnchang, to fund a new study of the dam's potential environmental impacts.

One thing, Professor Ed Grumbine said, was clear: The decision over whether to build the dam would be based not on science, but politics.

Would Laos really allow a Bangkok developer to build the Xayaburi dam without waiting for the MRC process to formally conclude? Despite a swirl of rumors on the business pages of Thai newspapers suggesting that the dam was still moving forward, it seemed unlikely to some Laos experts that Laos would build the dam over Vietnam's objection. "I think there's a good chance this project is dead," said Baird, the University of Wisconsin professor, noting it was "unprecedented" since 1975 for Vietnam to publicly disagree with Laos, a close ally. "Considering the loss of face, what Lao politician would be willing to put his neck on the line and propose this project again?"

The plot twisted again in late June, when International Rivers released two leaked letters suggesting Laos was indeed moving forward with the project. In the first letter, dated June 8, an official from Laos's Ministry of Energy and Mines informs the Xayaburi Power Company—a subsidiary of Ch. Karnchang—that a consulting firm has reviewed the company's Xayaburi dam proposal and concluded that Laos has "taken all legitimate concerns from member countries into consideration." (A spokesperson for Ch. Karnchang declined to comment for this article.) In the second letter, dated June 9, the chairman of the company's board of directors writes to the governor of the Electricity Generating Authority of Thailand and says the company is ready to execute a power of purchase agreement at the Thai government's earliest convenience.

On August 4, International Rivers reported that a "substantial construction camp" with "at least a few hundred workers" had been established near the dam site. It seemed that the Lao Communist Party was prepared to defy Vietnam, its longtime political ally.

It's hard to get around in Laos. The countryside is mountainous, and the roads are in deplorable shape. To get near the site of the Xayaburi dam, you have to take a rickety school bus from Luang Prabang, the ancient Lao capital, toward Xayaburi City. Thirty minutes into the ride, the bus is traversing mountain passes and barreling past fields of upland sticky rice. When the road turns to dirt, windows shake as the wheels slam into potholes.

I made the journey in late April with a Lao friend. After bumping along for a few hours, the bus crossed the Mekong in a ferry, and a few minutes later it left us in a dusty roadside village about fifteen kilometers north of Xayaburi City. We approached a group of people sitting on the porch of a simple concrete house. The dam site was nearby, they said, but in order to visit we had to get permission from the police.

The police station was across the street, and we approached warily. Inside, two officers sat at a wooden table in a bare room. They didn't look happy to see us,

and they asked for my passport. When I refused, they said we couldn't visit the site.

We returned to the village and paid two men to drive us back to the Mekong on their motorbikes. Then we paid a few Lao villagers to take us downriver in their orange fishing skiff. It would take too long—about five hours—to motor all the way to the dam site, but they could introduce us to fisherman in nearby villages who would be displaced by the dam.

We climbed into the skiff and were soon lurching down the river. The skiff's 5.5 horsepower engine drowned out conversation, but it could barely keep the craft straight in the muddy Mekong's strong current. Against a mountain backdrop, we saw fishermen inspecting nets that were strung across a line of riverside boulders. Our guide recognized the men and pulled the skiff ashore.

What had they heard about the Xayaburi dam? As far as they knew, they said, it was going to be built. The fishermen said a few people representing the Thai dam developer had visited their villages and offered, as compensation for the dam, to build them new homes, plus a hospital and a school, and to give them loans—the equivalent of roughly $250 per family—to help them buy livestock. The men, who support their families on about $500 per year, said they were okay with moving to a new village and planting new gardens, so long as the company followed through on its promises.

Rural development experts tell me that Lao villagers who have been displaced by hydropower dams on Mekong tributaries usually end up worse off even if companies provide compensation. In many cases, villagers not only lose their main source of livelihood but also incur debts as they struggle to survive on their new land, which may be less productive than the parcels they left behind. Eventually, they may move to cities like Bangkok and join the ranks of the urban poor.

"The dam will be good and bad for me," said Harm, a wiry 25-year-old who stood beside a fishing net. "Good because the company will build us a school and we'll get new jobs, bad because we'll have to leave our villages, and we won't be able to fish anymore."

Harm looked at the river. He didn't look angry. "This is all going to become a big pond," he said. "But that's what the government decided, and we have to respect the government."

Mike Ives is a writer based in Hanoi. Before he moved to Vietnam in 2009, he was a staff writer at the Vermont newspaper *Seven Days*.

EXPLORING THE ISSUE

Is Hydropower a Sound Choice for Renewable Energy?

Critical Thinking and Reflection

1. Why may greenhouse gas emissions increase after the land is flooded by a reservoir?
2. In some parts of the world much of the water that rivers carry comes from melting snowpack and glaciers. In a "global warming" world, can a hydroelectric dam on such a river be said to provide "renewable" electricity?
3. How may hydroelectric development affect (for both good and bad) the lives of poor people in developing nations?

Is There Common Ground?

Many of the drawbacks of hydroelectric power have to do with constructing dams. In-stream turbines or hydrokinetic devices, such as those proposed for generating electricity from ocean currents, can also be used in rivers (though they may pose navigational problems). Research this topic (start at the Union of Concerned Scientists' Hydrokinetic Energy page at www.ucsusa.org/clean_energy/our-energy-choices/renewable-energy/how-hydrokinetic-energy-works.html) and answer the following questions:

1. Are there environmental drawbacks?
2. What can turbine operators do if a river dries up?
3. Are they affordable and practical in developing countries?

Create Central

www.mhhe.com/createcentral

Additional Resources

R. Bakis, "The Current Status and Future Opportunities of Hydroelectricity," *Energy Sources: Part B: Economics, Planning, and Policy* (vol. 2, no. 3, 2007)

Kester Gunn and Clym Stock-Williams, "Quantifying the Global Wave Power Resource," *Renewable Energy: An International Journal* (August 2012)

Aviva Imhof and Guy R. Lanza, "Greenwashing Hydropower," *World Watch* (January/February 2010)

Internet References . . .

National Hydropower Association

www.hydro.org/

The Federal Energy Regulatory Commission—Hydropower

www.ferc.gov/industries/hydropower.asp

The National Renewable Energy Laboratory

www.nrel.gov/

Unit 4

UNIT

Food and Population

*T*o many, "sustainability" means arranging things so that the natural world—plants and animals, forests and coral reefs, and fresh water and landscapes—can continue to exist more or less (mostly less) as it did before human beings multiplied, developed technology, and began to cause extinctions, air and water pollution, soil erosion, desertification, climate change, and so on. To others, "sustainability" means arranging things so humankind can continue to survive and thrive, even keeping up its history of growth, technological development, and energy use—as if the environment and its resources were infinite.

The two visions of "sustainability" are logically incompatible. Many fear that we are on a collision course with limited resources, with the National Geographic warning of "The End of Plenty" (June 2009). Can we avoid disaster? Must we reduce the numbers of people on the planet, their use of technology, and their standard of living? If we do, will human well-being be lessened? If we do not, how can we continue to feed everyone? Can ocean fisheries be maintained? Is genetic engineering the answer? Is organic farming the answer? All of these issues provoke considerable debate.

Selected, Edited, and with Issue Framing Material by:
Thomas Easton, *Thomas College*

ISSUE

Do We Have a Population Problem?

YES: David Attenborough, from "This Heaving Planet," *New Statesman* (April 25, 2011)

NO: Sean Lanahan, from "Debunking the Over-Population Myth," *Countryside & Small Stock Journal* (January/February 2013)

Learning Outcomes
After reading this issue, you will be able to:
• Explain the nature of the "population problem."
• Explain the concept of "carrying capacity."
• Explain the potential benefits of stabilizing or reducing population.
• Explain the potential drawbacks of stabilizing or reducing population.

ISSUE SUMMARY

YES: Sir David Attenborough argues that the environmental problems faced by the world are exacerbated by human numbers. Without population reduction, the problems will become ever more difficult—and ultimately impossible—to solve.

NO: Sean Lanahan argues that the world's agricultural system currently produces enough food for at least double the current world population. The "over-population" myth of "unsustainability" is a scare tactic designed to control the lives of individuals and justify dehumanizing acts such as abortion and euthanasia.

In 1798 the British economist Thomas Malthus published his *Essay on the Principle of Population*. In it, he pointed with alarm at the way the human population grew geometrically (a hockey-stick-shaped curve of increase) and at how agricultural productivity grew only arithmetically (a straight-line increase). It was obvious, he said, that the population must inevitably outstrip its food supply and experience famine. Contrary to the conventional wisdom of the time, population growth was not necessarily a good thing. Indeed, it led inexorably to catastrophe. For many years, Malthus was something of a laughing stock. The doom he forecasted kept receding into the future as new lands were opened to agriculture, new agricultural technologies appeared, new ways of preserving food limited the waste of spoilage, and the birth rate dropped in the industrialized nations (the "demographic transition"). The food supply kept ahead of population growth and seemed likely—to most observers—to continue to do so. Malthus's ideas were dismissed as irrelevant fantasies.

Yet overall population kept growing. In Malthus's time, there were about 1 billion human beings on Earth. By 1950—when Warren S. Thompson expressed concern that civilization would be endangered by the rapid growth of Asian and Latin American populations during the next

five decades (see "Population," *Scientific American* (February 1950))—there were a little over 2.5 billion. In 1999, the tally passed 6 billion. It passed 7 billion in 2011. By 2025 it will be over 8 billion. Statistics like these are positively frightening. By 2050 the United Nation (UN) expects the world population to be about 9 billion (see *World Population Prospects: The 2010 Revision Population Database*; http://esa.un.org/unpd/wpp/index.htm; United Nations, 2010). By 2100, it will be 10.1 billion; see Jocelyn Keiser, "10 Billion Plus: Why World Population Projections Were Too Low," *ScienceInsider* (May 4, 2011) (http://scim.ag/_worldpop). While global agricultural production has also increased, it has not kept up with rising demand, and—because of the loss of topsoil to erosion, the exhaustion of aquifers for irrigation water, and the high price of energy for making fertilizer (among other things)—the prospect of improvement seems exceedingly slim to many observers.

Two centuries never saw Malthus's forecasts of doom come to pass. Population continued to grow, and environmentalists pointed with alarm at a great many problems that resulted from human use of the world's resources (air and water pollution, erosion, loss of soil fertility and groundwater, loss of species, and a great deal more). "Cornucopian" economists such as the late Julian Simon insisted that the more people there are on Earth, the more people there are to solve problems and that humans can find ways around all possible resource shortages. See Simon's essay, "Life on Earth Is Getting Better, Not Worse," *The Futurist* (August 1983). See also David Malakoff, "Are More People Necessarily a Problem?" *Science* (July 29, 2011) (a special issue on population). Jonathan A. Foley, "Can We Feed the World and Sustain the Planet?" *Scientific American* (November 2011), argues that if we do the right things—including shifting away from a meat-heavy diet—we can do both.

Was Malthus wrong? Both environmental scientists and many economists now say that if population continues to grow, problems are inevitable. But earlier predictions of a world population of 10–12 billion by 2050 are no longer looking very likely. The UN's population statistics show a slowing of growth, to be followed (after 2100) by an actual decline in population size.

Fred Pearce, *The Coming Population Crash: and Our Planet's Surprising Future* (Beacon, 2010), is optimistic about the effects on human well-being of the coming decline in population. Do we still need to work on controlling population? Historian Matthew Connolly, *Fatal Misconception: The Struggle to Control World Population* (Belknap Press, 2010), argues that the twentieth-century movement to control population was an oppressive movement that failed to deliver on its promises. Now that population growth is slowing, the age of population control is over. Yet there remains the issue of "carrying capacity," defined very simply as the size of the population that the environment can support, or "carry," indefinitely, through both good years and bad. It is not the size of the population that can prosper in good times alone, for such a large population must suffer catastrophically when droughts, floods, or blights arrive or the climate warms or cools. It is a long-term concept, where "long-term" means not decades or generations, nor even centuries, but millennia or more. See Mark Nathan Cohen, "Carrying Capacity," *Free Inquiry* (August/September 2004), and T. C. R. White, "The Role of Food, Weather, and Climate in Limiting the Abundance of Animals," *Biological Reviews* (August 2008).

What is Earth's carrying capacity for human beings? It is surely impossible to set a precise figure on the number of human beings the world can support for the long run. As Joel E. Cohen discusses in *How Many People Can the Earth Support?* (W. W. Norton, 1996), estimates of Earth's carrying capacity range from under a billion to over a trillion. The precise number depends on our choices of diet, standard of living, level of technology, willingness to share with others at home and abroad, and desire for an intact physical, chemical, and biological environment (including wildlife and natural environments), as well as on whether or not our morality permits restraint in reproduction and our political or religious ideology permits educating and empowering women. The key, Cohen stresses, is human choice, and the choices are ones we must make within the next 50 years. Phoebe Hall, "Carrying Capacity," *E—The Environmental Magazine* (March/April 2003), notes that

even countries with large land areas and small populations, such as Australia and Canada, can be overpopulated in terms of resource availability. The critical resource appears to be food supply; see Russell Hopfenberg, "Human Carrying Capacity Is Determined by Food Availability," *Population & Environment* (November 2003). Even if we do not run out of food, rising food prices may provoke widespread civil unrest in the 2012–2013 period; see Marco Lagi, Karla Z. Bertrand and Yaneer Bar-Yam, "The Food Crises and Political Instability in North Africa and the Middle East," http://arxiv.org/abs/1108.2455 (August 2011).

Andrew R. B. Ferguson, in "Perceiving the Population Bomb," *World Watch* (July/August 2001), sets the maximum sustainable human population at about 2 billion. Sandra Postel, in the Worldwatch Institute's *State of the World 1994* (W. W. Norton, 1994), says, "As a result of our population size, consumption patterns, and technology choices, we have surpassed the planet's carrying capacity. This is plainly evident by the extent to which we are damaging and depleting natural capital" (including land and water). The point is reiterated by Robert Kunzig, "By 2045 Global Population Is Projected to Reach Nine Billion. Can the Planet Take the Strain?" *National Geographic* (January 2011) (*National Geographic* ran numerous articles on population-related issues during 2011). Thomas L. Friedman, "The Earth Is Full," *New York Times* (June 7, 2011), thinks a crisis is imminent but we will learn and move on; see also Paul Gilding, *The Great Disruption: Why the Climate Crisis Will Bring On the End of Shopping and the Birth of a New World* (Bloomsbury Press, 2011).

If population growth is now declining and world population will eventually begin to decline, there is clearly hope. But most estimates of carrying capacity put it at well below the current world population size, and it will take a long time for global population to fall far enough to reach such levels. We seem to be moving in the right direction, but it remains an open question whether our numbers will decline far enough soon enough (i.e., before environmental problems become critical). On the other hand, Jeroen Van den Bergh and Piet Rietveld, "Reconsidering the Limits to World Population: Meta-Analysis and Meta-Prediction," *Bioscience* (March 2004), set their best estimate of human global carrying capacity at 7.7 billion, which is distinctly reassuring. However, there is still concern that global population will not stop at that point; see David R. Francis, "'Birth Dearth' Worries Pale in Comparison to Overpopulation," *Christian Science Monitor* (July 14, 2008).

On the other hand, David E. Bloom, "7 Billion and Counting," *Science* (July 29, 2011), notes that "despite alarmist predictions, historical increases in population have not been economically catastrophic. Moreover, changes in population age structure [providing for more workers] have opened the door to increased prosperity." Jonathan A. Foley, "Can We Feed the World & Sustain the Planet?" *Scientific American* (November 2011), thinks that with revisions to the world's agricultural systems, a growing population's demand for food can be met, at least through 2050.

Some people worry that a decline in population will not be good for human welfare. Michael Meyer, "Birth Dearth," *Newsweek* (September 27, 2004), argues that a shrinking population will mean that the economic growth that has meant constantly increasing standards of living must come to an end, government programs (from war to benefits for the poor and elderly) will no longer be affordable, a shrinking number of young people will have to support a growing elderly population, and despite some environmental benefits, quality of life will suffer. China is already feeling some of these effects; see Wang Feng, "China's Population Destiny: The Looming Crisis," *Current History* (September 2010), and Mara Hvistendahl, "Has China Outgrown the One-Child Policy?" *Science* (September 17, 2010). Julia Whitty, "The Last Taboo," *Mother Jones* (May–June 2010), argues that even though the topic of overpopulation has become unpopular, it is clear that we are already using the Earth's resources faster than they can be replenished and the only answer is to slow and eventually reverse population growth. Scott Victor Valentine, "Disarming the Population Bomb," *International Journal of Sustainable Development and World Ecology* (April 2010), calls for "a renewed international focus on managed population reduction as a key enabler of sustainable development." As things stand, the current size and continued growth of population threaten the UN Millennium Development Goals (including alleviating global poverty, improving health, and protecting the environment; see www.un.org/millenniumgoals/); see Willard Cates, Jr., et al., "Family Planning and the Millennium Development Goals," *Science* (September 24, 2010).

In the YES selection, Sir David Attenborough argues that the environmental problems faced by the world are exacerbated by human numbers. Without population reduction, the problems will become ever more difficult— and ultimately impossible—to solve. In the NO selection, Sean Lanahan argues that the world's agricultural system currently produces enough food for at least double the current world population. The "over-population" myth of "unsustainability" is a scare tactic designed to control the lives of individuals and justify dehumanizing acts such as abortion and euthanasia.

YES ⬅

David Attenborough

This Heaving Planet

Fifty years ago, on 29 April 1961, a group of far-sighted people in this country got together to warn the world of an impending disaster. Among them were a distinguished scientist, Sir Julian Huxley; a bird-loving painter, Peter Scott; an advertising executive, Guy Mountford; a powerful and astonishingly effective civil servant, Max Nicholson—and several others.

They were all, in addition to their individual professions, dedicated naturalists, fascinated by the natural world not just in this country but internationally. And they noticed what few others had done—that all over the world, charismatic animals that were once numerous were beginning to disappear.

The Arabian oryx, which once had been widespread all over the Arabian Peninsula, had been reduced to a few hundred. In Spain, there were only about 90 imperial eagles left. The Californian condor was down to about 60. In Hawaii, a goose that once lived in flocks on the lava fields around the great volcanoes had been reduced to 50. And the strange rhinoceros that lived in the dwindling forests of Java—to about 40. These were the most extreme examples. Wherever naturalists looked they found species of animals whose populations were falling rapidly. This planet was in danger of losing a significant number of its inhabitants, both animals and plants.

Something had to be done. And that group determined to do it. They would need scientific advice to discover the causes of these impending disasters and to devise ways of slowing them and, they hoped, of stopping them. They would have to raise awareness and understanding of people everywhere; and, like all such enterprises, they would need money to enable them to take practical action.

They set about raising all three. Since the problem was an international one, they based themselves not in Britain but in the heart of Europe, in Switzerland. They called the organisation that they created the World Wildlife Fund (WWF).

As well as the international committee, separate action groups would be needed in individual countries. A few months after that inaugural meeting in Switzerland, Britain established one—and was the first country to do so.

The methods the WWF used to save these endangered species were several. Some, such as the Hawaiian goose and the oryx, were taken into captivity in zoos, bred up into a significant population and then taken back to their original home and released. Elsewhere—in Africa, for example—great areas of unspoiled country were set aside as national parks, where the animals could be protected from poachers and encroaching human settlement. In the Galápagos Islands and in the home of the mountain gorillas in Rwanda, ways were found of ensuring that local people who also had claims on the land where such animals lived were able to benefit financially by attracting visitors.

Ecotourism was born. The movement as a whole went from strength to strength. Twenty-four countries established their own WWF national appeals. Existing conservation bodies, of which there were a number in many parts of the world but which had been working largely in isolation, acquired new zest and international links. New ones were founded focusing on particular areas or particular species. The world awoke to conservation. Millions—billions—of dollars were raised. And now, 50 years on, conservationists who have worked so hard and with such foresight can justifiably congratulate themselves on having responded magnificently to the challenge.

Yet now, in spite of a great number of individual successes, the problem seems bigger than ever. True, thanks to the vigour and wisdom of conservationists, no major charismatic species has yet disappeared. Many are still trembling on the brink, but they are still hanging on. Today, however, overall there are more problems not fewer, more species at risk of extinction than ever before. Why?

Fifty years ago, when the WWF was founded, there were about three billion people on earth. Now there are almost seven billion—over twice as many—every one of them needing space. Space for their homes, space to grow their food (or to get others to grow it for them), space to build schools, roads and airfields. Where could that come from? A little might be taken from land occupied by other people but most of it could only come from the land which, for millions of years, animals and plants had had to themselves—the natural world.

But the impact of these extra billions of people has spread even beyond the space they physically claimed. The spread of industrialisation has changed the chemical constituents of the atmosphere. The oceans that cover most of the surface of the planet have been polluted and increasingly acidified. The earth is warming. We now realise that the disasters that continue increasingly to afflict the natural world have one element that connects them all—the unprecedented increase in the number of human beings on the planet.

There have been prophets who have warned us of this impending disaster. One of the first was Thomas Malthus. His surname—Malthus—leads some to suppose that he was some continental European philosopher, a German perhaps. But he was not. He was an Englishman, born in Guildford, Surrey, in the middle of the 18th century. His most important book, *An Essay on the Principle of Population*, was published in 1798. In it, he argued that the human population would increase inexorably until it was halted by what he termed "misery and vice." Today, for some reason, that prophecy seems to be largely ignored—or, at any rate, disregarded. It is true that he did not foresee the so-called Green Revolution (from the 1940s to the late 1970s), which greatly increased the amount of food that can be produced in any given area of arable land. And there may be other advances in our food producing skills that we ourselves still cannot foresee. But such advances only delay things. The fundamental truth that Malthus proclaimed remains the truth: there cannot be more people on this earth than can be fed.

Many people would like to deny that this is so. They would like to believe in that oxymoron "sustainable growth."

Kenneth Boulding, President Kennedy's environmental adviser 45 years ago, said something about this: "Anyone who believes in indefinite growth in anything physical, on a physically finite planet, is either mad—or an economist."

The population of the world is now growing by nearly 80 million a year. One and a half million a week. A quarter of a million a day. Ten thousand an hour. In this country [UK] it is projected to grow by 10 million in the next 22 years. That is equivalent to ten more Birminghams.

All these people, in this country and worldwide, rich or poor, need and deserve food, water, energy and space. Will they be able to get it? I don't know. I hope so. But the government's chief scientist and the last president of the Royal Society have both referred to the approaching "perfect storm" of population growth, climate change and peak oil production, leading inexorably to more and more insecurity in the supply of food, water and energy.

Consider food. For animals, hunger is a regular feature of their lives. The stoical desperation of the cheetah cubs whose mother failed in her last few attempts to kill prey for them, and who consequently face starvation, is very touching. But that happens to human beings, too. All of us who have travelled in poor countries have met people for whom hunger is a daily background ache in their lives. There are about a billion such people today—that is four times as many as the entire human population of this planet a mere 2,000 years ago, at the time of Christ.

You may be aware of the government's Foresight project, Global Food and Farming Futures. It shows how hard it is to feed the seven billion of us alive today. It lists the many obstacles that are already making this harder to achieve—soil erosion, salinisation, the depletion of aquifers, over-grazing, the spread of plant diseases as a result

of globalisation, the absurd growing of food crops to turn into biofuels to feed motor cars instead of people—and so on. So it underlines how desperately difficult it is going to be to feed a population that is projected to stabilise "in the range of eight to ten billion people by the year 2050." It recommends the widest possible range of measures across all disciplines to tackle this. And it makes a number of eminently sensible recommendations, including a second green revolution.

But, surprisingly, there are some things that the project report does not say. It doesn't state the obvious fact that it would be much easier to feed eight billion people than ten billion. Nor does it suggest that the measures to achieve such a number—such as family planning and the education and empowerment of women—should be a central part of any programme that aims to secure an adequate food supply for humanity. It doesn't refer to the prescient statement 40 years ago by Norman Borlaug, the Nobel laureate and father of the first green revolution.

Borlaug produced new strains of high-yielding, short-strawed and disease-resistant wheat and in doing so saved many thousands of people in India, Pakistan, Africa and Mexico from starvation. But he warned us that all he had done was to give us a "breathing space" in which to stabilise our numbers. The government's report anticipates that food prices may rise with oil prices, and makes it clear that this will affect poorest people worst and discusses various way to help them. But it doesn't mention what every mother subsisting on the equivalent of a dollar a day already knows—that her children would be better fed if there were four of them around the table instead of ten. These are strange omissions.

How can we ignore the chilling statistics on arable land? In 1960 there was more than one acre of good cropland per person in the world—enough to sustain a reasonable European diet. Today, there is only half an acre each. In China, it is only a quarter of an acre, because of their dramatic problems of soil degradation.

Another impressive government report on biodiversity published this year, *Making Space for Nature in a Changing World,* is rather similar. It discusses all the rising pressures on wildlife in the UK—but it doesn't mention our growing population as being one of them—which is particularly odd when you consider that England is already the most densely populated country in Europe.

Most bizarre of all was a recent report by a royal commission on the environmental impact of demographic change in this country which denied that population size was a problem at all—as though 10 million extra people, more or less, would have no real impact. Of course it is not our only or even our main environmental problem but it is absurd to deny that, as a multiplier of all the others, it is a problem.

I suspect that you could read a score of reports by bodies concerned with global problems—and see that population is one of the drivers that underlies all of them—and yet find no reference to this obvious fact in any of them.

Climate change tops the environmental agenda at present. We all know that every additional person will need to use some carbon energy, if only firewood for cooking, and will therefore create more carbon dioxide—though a rich person will produce vastly more than a poor one. Similarly, we can all see that every extra person is—or will be—an extra victim of climate change—though the poor will undoubtedly suffer more than the rich. Yet not a word of it appeared in the voluminous documents emerging from the Copenhagen and Cancún climate summits.

Why this strange silence? I meet no one who privately disagrees that population growth is a problem. No one—except flat-earthers—can deny that the planet is finite. We can all see it—in that beautiful picture of our earth taken by the Apollo mission. So why does hardly anyone say so publicly? There seems to be some bizarre taboo around the subject.

This taboo doesn't just inhibit politicians and civil servants who attend the big conferences. It even affects the environmental and developmental non-governmental organisations, the people who claim to care most passionately about a sustainable and prosperous future for our children.

Yet their silence implies that their admirable goals can be achieved regardless of how many people there are in the world or the UK, even though they all know that they can't.

I simply don't understand it. It is all getting too serious for such fastidious niceties. It remains an obvious and brutal fact that on a finite planet human population will quite definitely stop at some point. And that can only happen in one of two ways. It can happen sooner, by fewer human births—in a word, by contraception. That is the humane way, the powerful option that allows all of us to deal with the problem, if we collectively choose to do so. The alternative is an increased death rate—the way that all other creatures must suffer, through famine or disease or predation. That, translated into human terms, means famine or disease or war—over oil or water or food or minerals or grazing rights or just living space. There is, alas, no third alternative of indefinite growth.

The sooner we stabilise our numbers, the sooner we stop running up the "down" escalator. Stop population increase—stop the escalator—and we have some chance of reaching the top; that is to say, a decent life for all.

To do that requires several things. First and foremost, it needs a much wider understanding of the problem, and that will not happen while the taboo on discussing it retains such a powerful grip on the minds of so many worthy and intelligent people. Then it needs a change in our culture so that while everyone retains the right to have as many children as they like, they understand that having large families means compounding the problems their children and everyone else's children will face in the future.

It needs action by governments. In my view, all countries should develop a population policy—as many as 70 countries already have them in one form or another—and give it priority. The essential common factor is to make family planning and other reproductive health services freely available to every one, and empower and encourage them to use it—though without any kind of coercion.

According to the Global Footprint Network, there are already more than a hundred countries whose combination of numbers and affluence have already pushed them past the sustainable level. They include almost all developed countries. The UK is one of the worst. There the aim should be to reduce over time both the consumption of natural resources per person and the number of people—while, needless to say, using the best technology to help maintain living standards. It is tragic that the only current population policies in developed countries are, perversely, attempting to increase their birth rates in order to look after the growing number of old people. The notion of ever more old people needing ever more young people, who will in turn grow old and need even more young people, and so on ad infinitum, is an obvious ecological Ponzi scheme.

I am not an economist, nor a sociologist, nor a politician, and it is from their disciplines that answers must come. But I am a naturalist. Being one means that I know something of the factors that keep populations of different species of animals within bounds and what happens when they aren't.

I am aware that every pair of blue tits nesting in my garden is able to lay over 20 eggs a year but, as a result of predation or lack of food, only one or two will, at best, survive. I have watched lions ravage the hundreds of wildebeest fawns that are born each year on the plains of Africa. I have seen how increasing numbers of elephants can devastate their environment until, one year when the rains fail on the already over-grazed land, they die in hundreds.

But we are human beings. Because of our intelligence, and our ever-increasing skills and sophisticated technologies, we can avoid such brutalities. We have medicines that prevent our children from dying of disease. We have developed ways of growing increasing amounts of food. But we have removed the limiters that keep animal populations in check. So now our destiny is in our hands.

There is one glimmer of hope. Wherever women have the vote, wherever they are literate and have the medical facilities to control the number of children they bear, the birth rate falls. All those civilised conditions exist in the southern Indian state of Kerala. In India as a whole, the total fertility rate is 2.8 births per woman. In Kerala, it is 1.7 births per woman. In Thailand last year, it was 1.8 per woman, similar to that in Kerala. But compare that with the mainly Catholic Philippines, where it is 3.3.

Here and there, at last, there are signs of a recognition of the problem. Save the Children mentioned it in its last report. The Royal Society has assembled a working party of scientists across a wide range of disciplines who are examining the problem.

But what can each of us do? Well, there is just one thing that I would ask. Break the taboo, in private and in public—as best you can, as you judge right. Until it is broken there is no hope of the action we need. Wherever and whenever we speak of the environment, we should add a few words to ensure that the population element is not ignored. If you are a member of a relevant NGO, invite them to acknowledge it.

If you belong to a church—and especially if you are a Catholic, because its doctrine on contraception is a major factor in this problem—suggest they consider the ethical issues involved. I see the Anglican bishops in Australia have dared to do so. If you have contacts in government, ask why the growth of our population, which affects every department, is as yet no one's responsibility. Big empty Australia has appointed a sustainable population minister, so why can't small crowded Britain?

The Hawaiian goose, the oryx, and the imperial eagle that sounded the environmental alarm 50 years ago were, you might say, the equivalent of canaries in coal mines—warnings of impending and even wider catastrophe.

Make a list of all the other environmental problems that now afflict us and our poor battered planet—the increase of greenhouse gases and consequential global warming, the acidification of the oceans and the collapse of fish stocks, the loss of rainforest, the spread of deserts, the shortage of arable land, the increase in violent weather, the growth of mega-cities, famine, migration patterns. The list goes on and on. But they all share one underlying cause. Every one of these global problems, social as well as environmental, becomes more difficult—and ultimately impossible—to solve with ever more people.

David Attenborough is a British naturalist and broadcaster who has produced numerous popular wildlife documentaries.

Sean Lanahan

➔ **NO**

Debunking the Over-Population Myth

Many in the world today . . . claim the world is on the brink of unsustainable failure due to our growing population. The entire population of seven billion people could take a nap inside the state of Connecticut. The same amount of people can live inside the state of Texas with the same population density of New York City, leaving the rest of the globe devoid of human life.

But that is not enough for most skeptics. They want to know about the impact of a growing population on food, water, air, waste, forests, oceans, animals, etc. Let's take a look at food. Is there enough? The earth is more than able to support not only seven billion souls, but up to two to four times as much with a little work and little impact to other life.

Most of the raw data that I use to prove my points are from official government sources such as the FAO (Food and Agriculture Organization of the United Nations) and USDA (United States Department of Agriculture).

Calculating Calories

Using the FAO world wide food production reports from 2010 and standard nutritional data, I have calculated the total calories produced by each food category, divided by seven billion people per year (Calories, Per Person, Per Year—C,PP,PY):

- Grains (corn, wheat, rice, etc.)—848,810
- Roots and tubers (potatoes, yams, carrots, etc.) — 113,856
- Oils, pulses & nuts (olives, soybeans, peanuts, etc.)—302,586
- Fruits and vegetables (oranges, tomatoes, lettuce, etc.)—78,091
- Meat and eggs (chicken, beef, pork, etc.)—113,856
- Fish and seafood (salmon, clams, lobster, etc.) — 21,496
- Dairy (milk, cheese, butter, etc.)—61,581
- Raw sugar (sugar cane, sugar beets, etc.)—121,483

Total of all calories produced by God's creation—1,661,460 calories, per person, per year or about 4,552 calories, per person, per day! Multiply by seven billion people and you have an approximate grand total of 11.63 quadrillion calories that were produced in the year 2010.

Considering an average healthy caloric intake of 2,000 calories per person per day (2000 calories × 365 days = 730,000) 730,000 calories are consumed per person, per year.

Generally men consume more than women. In 1971 American men consumed 2,450 calories per day, while women consumed 1,542, a combined average of 1,996 calories per day.

The earth produced 1,661,460 C,PP,PY minus an average consumption of 730,000 C,PP,PY = 855,519 extra calories, per person, per year. More than enough to sustain an additional population of seven billion for a total of 14 billion people!

Land Use

According to the FAO, there are 12.07 billion acres of agricultural land, all land capable of producing food. This includes arable land of 3.41 billion acres capable of producing temporary crops such as grains and vegetables. Permanent crop land of 375.8 million acres producing nuts and fruits in trees or vines. Pasture/meadow land of 8.28 billion acres providing for livestock foraging. Of the land above, only 769.8 million acres is irrigated. Most often irrigated lands can produce twice as much as non-irrigated. Total acreage of agricultural land per person equals 1.725 acres.

All things being equal, 1.725 acres of land produces 1,639,964 calories per year (calories per person, per year, minus calories from fish and seafood). This is well over twice as much land as needed per person with very little that is actually irrigated!

Where Do All the Calories Go?

A good chunk of the extra calories produced each year are consumed by the people of the planet. The average American diet in 1971 was about 2,000 calories a day. According to a survey in 2004, that average consumption has expanded to 2,247, over a 10% increase, and has likely increased even more eight years later. It's not just Americans that are eating more, but most of the world as well. From 1961 to 2003, the available (for consumption) calories per capita has risen from 2,254 to 2,809, over 24% increase in about four decades.

In that same four-decade span, the global population blossomed from three billion in 1961 to 6.3 billion in 2003, over double the population. At the same time, life expectancy rose from an average of 52.2 years in 1961 to 66 years in 2003, over 25% increase across the globe. Not only did the population more than double, but we lived longer and consumed even more calories over a lifetime. The Earth, along with our fruitful domination, creativity, our hard work and ingenuity, and God's blessed provision, has been able to produce more than enough. Because we have been created in God's likeness, we have the ability to do amazing and wonderful things.

Over-Abundance

We've actually created an over-abundance of food. So much so that hundreds of thousands of calories per person, per year (C,PP,PY) are lost or wasted every year.

Roughly 564,532 (34%) calories, per person, per year are lost or wasted through the activities of harvest, post harvest, processing, distribution, and consumption.

The developed world (Europe, North America and Industrial Asia) is very good at preventing losses from harvest to distribution, but is rather wasteful from retail to the dinner plate. The opposite is true for the under-developed world (Africa, Asia and Latin America). Often their largest losses occur from harvest to distribution, but they are very efficient at the consumption level. What is the reason?

The developed world has the equipment, infrastructure, and experience to efficiently move product from harvest to retail with relatively little loss. So much so that now it costs a relatively small portion of personal spending. Thus, what is not valuable to us is often taken for granted, the West has become picky, with little regard to throwing away food that is no longer "appetizing."

For example, for the past 80 years, the citizens of the USA have enjoyed a steadily shrinking food budget. From 1929 to 2010, food expenditures as a percentage of disposable income fell from 23.4% to 9.4%. At the same time, the food demand from a burgeoning population has grown. Food production over the past 80 years has not only kept pace with demand, but surpassed it by a great amount. Though I appreciate Benjamin Franklin's quote "waste not, want not," I can't help but think that in today's over-abundance, the better term is actually "want not, waste much."

The under-developed world unfortunately does not have the equipment, infrastructure, and experience needed to efficiently move product from harvest to retail. Threshing grain by hand, for example, incurs greater losses than if done by machine. However, once the product reaches the store and the dinner plate, the under-developed world becomes very efficient in consumption. Food costs as a percentage of personal income are greater in the developed world, thus they value food more and waste little.

If it were possible for the developed world to consume more efficiently (the same as Sub-Saharan Africa), and the under-developed world were to use modern equipment, infrastructure, and experience (the same as North America and Europe), by my estimation food loss and waste could be reduced from 34% to 19.8%—a savings of about 235,213 C,PP,PY. Keep in mind that if the under-developed world were to use modern farming practices,

not only would they preserve more food, but they would produce more as well.

For example, modern irrigation can increase wheat crop yields per acre from 32.8 bushels per acre to 71.4 bushels per acre, a 117% increase. At 60 lbs. of wheat per bushel, an extra 3,434,628 calories can be produced. This is enough to feed an extra 4.7 people, per acre, per year (at a 2,000 calorie diet per day) just by adding irrigation. Many may wonder why Africa is starving. The answer has nothing to do with the earth's ability to produce what they need. Rather, the cause is a ridiculous geo-political aberration surrounding faulty notions of "sustainability."

Feed People or Cars?

There is also a contingent of food produced and wasted that does not end up in a landfill, but actually makes it into your fuel tank. Vegetable cooking oil is now being recovered on a large scale in the West, chiefly used for bio-fuels, specifically . . . bio-diesel. Some corn, sugar beets, and sugar cane (not included in the food data above), is being grown for bio-fuels such as ethanol. Some agricultural lands that are used to grow fuel could produce food for human consumption instead. More calories would become available to feed more people instead of fueling cars and trucks.

Farming in Israel

From its founding, Israel's farming has been a monumental effort of self-preservation. They have literally had to cultivate hard, rocky soil in an arid climate to produce food for themselves. Over the past 60 years they have become extremely successful.

Wikipedia states: "Modern agriculture developed in the late nineteenth century, when Jews began settling in the land. They purchased land which was mostly semi-arid, although much had been rendered untillable by deforestation, soil erosion, and neglect. They set about clearing rocky fields, constructing terraces, draining swampland, reforesting, counteracting soil erosion, and washing salty land. Since independence in 1948, the total area under cultivation has increased from 408,000 acres (1,650 km2) to 1,070,000 acres (4,300 km2), while the number of agricultural communities has increased from 400 to 725. Agricultural production has expanded 16 times, three times more than population growth.

"Water shortage is a major problem. Rain falls between September and April, with an uneven distribution across the country, from 28 inches (70 cm) in the north to less than two inches (5 cm) in the south. Annual renewable water resources are about 5.6 billion cubic feet (160,000,000 m3), of which 75% . . . is used for agriculture. Most of Israel's fresh water sources have been consequently joined to the National Water Carrier . . . network of pumping stations, reservoirs, canals and pipelines which transfers water from the north to the south.

"The area of irrigated farmland has increased from 74,000 acres (30,000 ha) in 1948 to some 460,000 acres (190,000 ha) today. Israeli agricultural production rose 26% between 1999 and 2009, while the number of farmers dropped from 23,500 to 17,000. Farmers have also grown more with less water, using 12% less water to grow 26% more produce."

The population of Israel has increased from 1.25 million in 1950 to 7.48 million in 2012, a 500% increase. The otherwise "useless" land has been subdued and dominated to the point of not only producing enough food for its rapidly growing population, but also an abundance for export to the tune of 22% of production. Since 1948 they expanded the total area of cultivation from 408,000 acres to 1,070,000 acres (about 150% increase), while increasing production 16-fold. This is from what was once considered desolate wasteland (read Mark Twain) combined with some sweat equity and God's blessing.

New Technology

What about the deserts of the world? Consider the Seawater Greenhouse (seawatergreenhouse.com):

"A single Sahara Forest Project facility with 50 MW of concentrated solar power and 50 hectares of seawater greenhouses would produce 34,000 tons of produce, employ over 800 people, export 155 GWh of electricity and sequester more than 1,500 tons of CO_2 each year."

Let's project this amazing potential. The major hot arid deserts cover 15,577,000 sq.km. If but a meager 1% of that desert space was used to build these types of seawater greenhouses with the above specifications listed in the article:

- 311,540 facilities could be built (50 hectares apiece) 10.592360 billion tons of food could be produced per year (about three times as much as the current grain, tuber, root, fruit and vegetable harvest, equaling 3.580572593 billion tonnes per year) At a paltry 25¢ per pound, the value of that produce could bring $5.29618 trillion dollars in sales.
- 48.2887 trillion kilowatt hours could be generated (at a better than fair price of 10¢ a kilowatt hour, $4.82887 trillion dollars of revenue could be generated). And 249.232 million jobs could be created. These SWGs are not fantasy-land daydreams, many are already in existence and more are slated to come on line.

No Longer Doom and Gloom

At this point, your doom and gloom about the planet's prospects should be lifting slightly, maybe causing a half smile to form across your lips. There is more than enough room, food, water, trees and space for all our garbage. You've heard it said that "necessity is the mother of invention," God, man and the earth have not only met every demand from a burgeoning population but have exceeded them abundantly! It's true that the human family has exploded in growth over the past several decades, but as I have demonstrated, our ability to produce food has grown even more so, creating huge surpluses, bringing down the costs for food for all people.

Consider these calculations. If you were to take every single man, woman and child on the face of the planet (nearly seven billion precious souls to date) and make them lay down on the ground side by side and head to toe, laying out an immense "human carpet," that "carpet" of human beings would not even cover the state of Connecticut.

Here's the math. This allows 18 square feet per individual ($3' \times 6'$). Granted, children would take up much less room, but they are a bit squirmy and don't sit still very well.

- 18 square feet times seven billion people equals 126 billion square feet.
- 27,878,400 square feet are inside one square mile.
- 126 billion square feet divided by 27,878,400 equals 4,519.63 square miles.

Connecticut is the 48th largest state at 5,544 square miles, only beating out Rhode Island and Delaware in size. That's right, the entire population of planet Earth can lay down and take a nap inside the state of Connecticut with some room to spare, leaving the remainder of the globe completely uninhabited.

"But all those people couldn't possibly live like that!" you exclaim. "Come on dude, get real!"

Okay, what if all of those seven billion inhabitants were to live in one big fully functional "mega city," the same size population density of New York City? They would fit inside the foot print of the state of Texas, again leaving the remainder of the globe completely devoid of human life.

Here's the math. The population of New York City is 8,175,133 and covers 303 square miles.

- 8,175,133 divided by 303 equals a population density of 26,980 persons per square mile.
- Seven billion people living at a population density of 26,980 per square mile would require a mega city, which encompasses 259,451 square miles.

The state of Texas sits on 268,580 square miles— more than enough room. But, I doubt them Texans would appreciate all them city slickers moving in on their turf.

The entire land mass on planet Earth (excluding Antarctica, and who would want to live there anyway?) is 52,208,738 square miles. Our fictitious "mega city," housing every single individual on planet Earth, some 259,451 square miles, would only take up 0.5% of the entire land mass. That's right, only half of one percent, leaving 99.5% of terra firma completely uninhabited.

Fortunately for most of us, we don't have to live in such relative "claustrophobic" conditions.

Unfortunately, a population control movement in the world today is seeking to scare the masses with the "over-population" myth of "unsustainability." A scare tac-tic used by them to control the lives of individuals and justify all kinds of dehumanizing acts, primarily abortion and euthanasia.

SEAN LANAHAN is an autodidact who lives in Washington state and also writes articles on financial market technical analysis.

EXPLORING THE ISSUE

Do We Have a Population Problem?

Critical Thinking and Reflection

1. Is it possible to have too many people on Earth?
2. What is wrong with the statement that there is no population problem because all of Earth's human population could fit inside the state of Texas?
3. What does population have to do with sustainability?
4. If the world can feed a population twice as great as today's, how long do we have before we really do have to worry about over-population?

Is There Common Ground?

The essayists for this issue agree that human population continues to grow and that long-term human survival matters. They disagree on whether a large and growing human population can be supported.

1. What factors other than space and food production are important to human welfare?
2. What are the key features of "quality of life"? (One good place to start your research is www.foe.co.uk/community/tools/isew/)
3. An important concept in this context is that of the "ecological footprint" (see www.myfootprint.org/). What is it and how does it influence the Earth's carrying capacity for human beings?

Additional Resources

Joel E. Cohen, *How Many People Can the Earth Support?* (W. W. Norton, 1996)

Jonathan A. Foley, "Can We Feed the World & Sustain the Planet?" *Scientific American* (November 2011)

United Nations, *World Population Prospects: The 2010 Revision Population Database*; http://esa.un.org/unpd/wpp/index.htm (United Nations, 2010)

William Vogt, *Road to Survival* (William Sloane Associates, 1948).

Create Central

www.mhhe.com/createcentral

Internet References . . .

EarthTrends

earthtrends.wri.org/

The Population Council

www.popcouncil.org/

United Nations Population Division

www.un.org/en/development/desa/population/

Selected, Edited, and with Issue Framing Material by:
Thomas Easton, *Thomas College*

ISSUE

Does Commercial Fishing Have a Future?

YES: Carl Safina, from "A Future for U.S. Fisheries," *Issues in Science and Technology* (Summer 2009)

NO: Food and Agriculture Organization of the United Nations, from "World Review of Fisheries and Aquaculture," in *The State of World Fisheries and Aquaculture* (2010)

Learning Outcomes

After reading this issue, you will be able to:

- Explain what overfishing is.
- Explain the factors that drive people to overexploit ecosystem resources such as marine fish.
- Describe possible measures to prevent destruction of oceanic fisheries.
- Describe possible replacements for oceanic fisheries.

ISSUE SUMMARY

Yes: Carl Safina argues that despite an abundance of bad news about the state of the oceans and commercial fisheries, there are some signs that conservation and even restoration of fish stocks to a sustainable state are possible.

NO: The Food and Agriculture Organization of the United Nations argues that the proportion of marine fish stocks that are overexploited has increased tremendously since the 1970s. Despite some progress, there remains "cause for concern." The continuing need for fish as food means there will be continued growth in aquaculture.

Carl Safina called attention to the poor state of the world's fisheries in "Where Have All the Fishes Gone?" *Issues in Science and Technology* (Spring 1994), and "The World's Imperiled Fish," *Scientific American* (November 1995). Expanding population, improved fishing technology, and growing demand had combined to drive down fish stocks around the world. Fishers going further from shore and deploying larger nets kept the catch growing, but the UN's Food and Agriculture Organization had noted that the fisheries situation was already "globally non-sustainable, and major ecological and economic damage [was] already visible."

In 1998, the UN declared the International Year of the Ocean. Kieran Mulvaney, "A Sea of Troubles," *E—The Environmental Magazine* (January–February 1998), reported that, "According to the United Nations Food and Agriculture Organization (FAO), an estimated 70 percent of global fish stocks are 'over-exploited,' 'fully exploited,' 'depleted' or recovering from prior over-exploitation. By 1992, FAO had recorded 16 major fishery species whose global catch had declined by more than 50 percent over the previous three decades-and in half of these, the collapse had begun after 1974." Ocean fishing did not seem sustainable.

Christian Mullon, Pierre Freon, and Philippe Cury, "The Dynamics of Collapse in World Fisheries," *Fish and Fisheries* (June 2005), find that current trends are likely to cause a "global collapse of many more fisheries" than hitherto. See also Carmel Finley, *All the Fish in the Sea: Maximal Sustainable Yield and the Failure of Fisheries Management* (University of Chicago Press, 2011).

As a species becomes depleted, fishers catch more younger fish, and fewer survive to reproduce; see Allen W. L. To and Yvonne Sadovy de Mitcheson, "Shrinking Baseline: The Growth in Juvenile Fisheries, with the Hong Kong Grouper Fishery as a Case Study," *Fish and Fisheries* (December 2009). Daniel Pauly and Reg Watson, "Counting the Last Fish," *Scientific American* (July 2003), note that desirable fish tend to be top predators such as tuna and cod. When the numbers of these fish decline to overfishing, fishers shift their attention to fish lower on the food chain and consumers see a change in what is available at the market. The cod are smaller, and monkfish and other once less desirable fish join them on the crushed ice at the market. And the change is an indicator of trouble in the marine ecosystem.

Karin E. Limburg and John R. Waldman, "Dramatic Declines in North Atlantic Diadromous Fishes," *Bioscience*

(December 2009), report that for 24 fish species studied, "populations have declined dramatically from original baselines." Oran R. Young, "Taking Stock: Management Pitfalls in Fisheries Science," *Environment* (April 2003), notes that despite putting great effort into assessing fish stocks and managing fisheries for sustainable yield, marine fish stocks have continued to decline. This is partly the "result of the inability of managers to resist pressures from interest groups to set total allowable catches too high, even in the face of warnings from scientists about the dangers of triggering stock depletions. The problem also arises, however, from repeated failures on the part of analysts and policy makers to anticipate the collapse of major stocks or to grasp either the current condition or the reproductive dynamics of important stocks." He cautions against putting "blind faith in the validity of scientific assessments" and suggests more use of the precautionary principle despite the risk that this would set allowed catch levels lower than many people would like. According to Marta Coll, et al., "Ecosystem Overfishing in the Ocean," *PLoS One* (www.plosone.org) (2008), "At present, total catch per capita from Large Marine Ecosystems is at least twice the value estimated to ensure fishing at moderate sustainable levels."

In June 2000, the independent Pew Oceans Commission undertook the first national review of ocean policies in more than 30 years. Its report, "America's Living Oceans: Charting a Course for Sea Change" (Pew Oceans Commission, 2003) (www.pewoceans.org/), noted that many commercially fished species are in decline (North Atlantic cod, haddock, and yellowtail flounder reached historic lows in 1989). The reasons include intense fishing pressure to feed demand for seafood, as well as pollution, coastal development, fishing practices such as bottom dragging that destroy habitat, and fragmented ocean policy that makes it difficult to prevent or control the damage. The answers, the Commission suggests, must include such things as the no-fishing zones known as marine protected areas or reserves, which have been shown to restore habitat and fish populations, with clear benefits for commercial fishing. See also Julia Whitty, "The Fate of the Ocean," *Mother Jones* (March/April 2006), Michelle Allsopp, et al., *Oceans in Peril: Protecting Marine Biodiversity* (Worldwatch, 2007), and "The Global Fish Crisis," a special report in *National Geographic* (April 2007). There is also a pressing need to rein in illegal fishing; see Stefan Flothmann, et al., "Closing Loopholes: Getting Illegal Fishing under Control," *Science* (June 4, 2010).

In September 2004, the U.S. Commission on Ocean Policy issued its report, "An Ocean Blueprint for the 21st Century" (www.oceancommission.gov/documents/full_color_rpt/welcome.html), calling for improved management systems "to handle mounting pollution, declining fish populations and coral reefs, and promising new industries such as aquaculture." In many ways it agreed with the Pew report. Carl Safina and Sarah Chasis, "Saving the Oceans," *Issues in Science and Technology* (Fall 2004), discuss the two

reports and say that it is time for Congress to craft a new approach to ocean policy, including fisheries protection, and put scientists in charge of policy. James N. Sanchirico and Susan S. Hanna, "Sink or Swim Time for U.S. Fishery Policy," *Issues in Science and Technology* (Fall 2004), add that an ecosystem approach is essential. According to Rainer Froese and Alexander Proelß, "Rebuilding Fish Stocks No Later than 2015: Will Europe Meet the Deadline?" *Fish & Fisheries* (June 2010), "Maintaining or restoring fish stocks at levels that are capable of producing maximum sustainable yield is a legal obligation under the United Nations Convention on the Law of the Sea (UNCLOS) and has been given the deadline of no later than 2015 in the Johannesburg Plan of Implementation of 2002. . . . But even if fishing were halted in 2010, 22% of the stocks are so depleted that they cannot be rebuilt by 2015. If current trends continue, Europe will miss the 2015 deadline by more than 30 years." They note that just passing laws is not enough. Yimin Ye, et al., "Rebuilding Global Fisheries: The World Summit Goal, Costs and Benefits," *Fish & Fisheries* (June 2013), find that global fisheries are still declining and fishing capacity needs to be cut drastically to let fisheries recover, but vested interests (including fishers) are unwilling to accept the short-term socioeconomic consequences of doing so.

Marine reserves are gaining favor as part of the solution, though they can be difficult to manage; see Christopher Pala, "Giant Marine Reserves Pose Vast Challenges," *Science* (February 8, 2013). So is aquaculture, with researchers struggling to save the bluefin tuna by domesticating it; see Richard Ellis, "The Bluefin in Peril," *Scientific American* (March 2008), and Robert F. Service, "Persevering Researchers Make a Splash with Farm-Bred Tuna," *Science* (June 5, 2009). Conflict does, however, remain over whether such measures are necessary; see Eli Kintisch, "Conservationists and Fishers Face Off over Hawaii's Marine Riches," *Science* (July 20, 2007). Responses to the situation have included government buyouts of fishing fleets and closures of fisheries such as the Canadian cod fishery (see "The Cod-Forsaken Island," *Canada & the World Backgrounder*, December 2010). But the situation has not improved. Janet Raloff, "Big Fishing Yields Small Fish," *Science News* (April 9, 2011), notes that "many alarm bells are ringing."

David Helvarg, in "The Last Fish," *Earth Island Journal* (Spring 2003), concludes that about half of America's commercial seafood species are now overfished. Globally, the figure is still over 70 percent. And the North Atlantic contains only a third as much biomass of commercially valuable fish as it did in the 1950s. He recommends more buying out of excess fishing capacity, limiting the number of people allowed to enter the fishing industry, creating marine reserves, and perhaps most importantly, taking fisheries management out of the hands of people with a vested interest in the status quo.

When Boris Worm, et al., "Impacts of Biodiversity Loss on Ocean Ecosystem Services," *Science* (November 3, 2006), argued that human activities, including overfishing, so threaten marine biodiversity that before the mid-twenty-first century populations of all those ocean fish currently sought will be so reduced that commercial fishing will have ended; critics were vocal (see e.g., Niels Daan, et al., "Apocalypse in World Fisheries? The Reports of Their Death Are Greatly Exaggerated," *ICES Journal of Marine Science* (July 2011). After some debate, according to Eric Stokstad, "Détente in the Fisheries War," *Science* (April 10, 2009), Worm and one major critic, Ray Hilborn of the University of Washington in Seattle teamed up with other researchers and graduate students to study larger data sets in search of a better vision of the future. One recent report from Worm's group (Camilo Mora, et al., "Management Effectiveness of the World's Marine Fisheries," *Public Library of Science Biology* (June 2009) (www.plosbiology.org/article/info%3Adoi%2F10.1371%2Fjournal.pbio.1000131), finds that demand for seafood is rising far beyond what can be met sustainably and only a handful of countries have "a robust scientific basis for management recommendations." Boris Worm, Ray Hilborn, et al., "Rebuilding Global Fisheries," *Science* (July 31, 2009), examine efforts to restore damaged marine ecosystems and fisheries and find that though there is some success, "63% of assessed fish stocks worldwide still require rebuilding, and even lower exploitation rates are needed to reverse the collapse of vulnerable species." Global fisheries remain in danger, as do the marine ecosystems to which the fish belong; see Ellen K. Pikitch, "The Risks of Overfishing," *Science* (October 26, 2012), and Christopher Costello, et al., "Status and Solutions for the World's Unassessed Fisheries," *Science* (October 26, 2012).

In the YES selection, Carl Safina argues that despite an abundance of bad news about the state of the oceans and commercial fisheries, there are some signs that conservation and even restoration of fish stocks to a sustainable state are possible. In the NO selection, the United Nations Food and Agriculture Organization argues that the proportion of marine fish stocks that are overexploited has increased tremendously since the 1970s. Despite some progress, there remains "cause for concern." The continuing need for fish as food means there will be continued growth in aquaculture.

YES ⤶

Carl Safina

A Future for U.S. Fisheries

For the fishing industry in the United States, and for the fishery resources on which the industry depends, there is good news and bad news. Bad news still predominates, as many commercial fishers and their communities have suffered severe financial distress and many fish stocks have declined considerably in numbers. Poor management by the National Marine Fisheries Service (NMFS), which regulates the fishing industry, and some poor choices by many fishers have contributed to the problems. But there are some bright spots, small and scattered, that suggest that improvements are possible.

Starting with the bad news, the federal government's fisheries management remains primitive, simplistic, and, in important cases, ineffectual, despite a fund of knowledge and conceptual tools that could be applied. In many regions—New England and the Pacific Northwest, among others—failed management costs more than the receipts from fisheries. This does not suggest that management should be given up as a lost cause, leaving the industry in a free-for-all, although this strategy might, in fact, be cheaper and not much less effective.

As a key problem, most management efforts today are based primarily on catch quotas that regulate how much fishers can harvest of a particular species in some set period, perhaps a season or a year. The problem is that quotas are set according to estimates of how much of the resource can be taken out of the ocean, rather than on how much should be left in. This may sound like two sides of the same coin, but in practice the emphasis on extraction creates a continual bias on the part of fisheries agencies and unrealistic short-term expectations among fishers. For example, a basic tenet of these approaches is that a virgin fish population should be reduced by about two-thirds to make it more "productive." But this notion is belied in the real world, where it has been proven that larger breeding populations are more productive.

The failure of this approach is readily apparent. The Sustainable Fisheries Act of 1996, reaffirmed by Congress in 2006, states that fish populations may not be fished down below about one-third of their estimated virgin biomass. It also states that in cases where fish stocks already have been pushed below that level, they must be restored (in most cases) to that level within a decade. On paper, this act looked good. (Full disclosure: I drafted the quantitative overfishing and recovery goals and triggers mandated by the act.) Unfortunately, the NMFS wrote implementing regulations interpreting the mandates as meaning that overfishing could continue for some time before rebuilding was required. This too-liberal interpretation blurred the concept and delayed benefits. In its worst cases, it acknowledged that fish populations must be rebuilt in a decade but said that overfishing could continue in the meantime.

Clearly, the nation needs to take a different approach, based solidly on science. As a foundation, regulatory and management agencies must move from basing their actions on "how much can we take?" to concentrating on "how much must we leave?" The goal must be keeping target fish populations and associated living communities functioning, with all components being highly productive and resilient.

The nation must confront another reality as well. So many fisheries are so depleted that the only way to restore them will be to change the basic posture of regulations and management programs to one of recovery. Most fish populations could recover within a decade, even with some commercial fishing. But continuing to bump along at today's depleted levels robs fishing families and communities of income and risks resource collapse.

Ingredients for Success

Moving to a new era of fisheries management will require revising some conventional tools that are functioning below par and adopting an array of new "smart tools." Regulations that set time frames for overfishing and recovery can play a valuable role, if properly interpreted. For example, traditional catch quotas must be based firmly on scientific knowledge about fish stocks, and they must be enforced with an eye toward protecting the resource. Newer tools, adapted to specific environments and needs, would include:

Tradable Catch Shares

In this approach, now being used in some regions in varying degrees, fishery managers allot to fishers specific shares of the total allowable catch and give them the flexibility and the accountability for reaching their shares. Thus, fishers do not own the fish; rather, they own a percentage of the total allowed catch, which may fluctuate from year to year if management agencies adjust it up or down.

In expanding the use of such programs, managers must establish the shares based on the advice of independent scientists who are insulated from industry lobbying. Managers also should allot shares only to working fishers, not to corporations or processors. Of course, finding equitable ways of determining which fishers get catch shares will be critical. Methods of allocating shares may vary from location to location, but the key is ensuring an open process that accounts for fishers' legitimate interests and maintains conservation incentives. In many cases, fewer fishers will be eligible to keep fishing. But those not selected would likely have been forced out of business anyway by the combination of pressure from more successful fishers and reduced fish stocks.

By significantly reducing competition that breeds a race for fish, this approach offers several benefits. For one, it makes for safer fishing. Fishers who own shares know that they have the whole season to fill their quota regardless of what other boats are catching, so they are less likely to feel forced to head out in dangerous weather. In addition, owning a share helps ensure (other factors permitting) that a fisher can earn a decent living, so local, state, or regional politicians will feel less pressure to protect their fishing constituents and push for higher catch quotas. At the same time, marginal operators granted shares would no longer feel trapped, because they would have something to sell if they wished to exit the fishery. By promoting longer-term thinking among fishers and politicians alike, catch-share programs help foster a sense of future investment in which quota holders will benefit from high or recovered fish populations.

The impact of tradable catch shares can be seen in experiences in several regions. In Alaska, where fisheries managers once kept a tight cap on the halibut catch, the fishing season shrank to two days annually because there were so many competing boats. After managers introduced tradable catch shares, the number of boats fell precipitously and the season effectively expanded to whenever the fishers wanted to work toward filling their shares. Safety improved markedly, and the halibut population remained robust. In New England, where the industry resisted tradable shares, the story ended differently. Managers allotted individual fishers a shrinking number of days at sea, which progressively crippled their economic viability, gave them no option to exit the fishery short of foreclosure, and kept fishing pressure so high that the fish stocks never recovered.

Area-Based Fisheries

Although this concept may be relatively new in Western fisheries management, it has underpinned the management of fishing in Pacific islands for millennia. In practice, this approach is most applicable where fish populations spawn in localized areas and do not migrate far from their spawning area. For example, consider the case of clams, which spawn in limited areas and never move far away. In many regions, clamming is regulated on a township-by-township basis. Thus, conserving clams off one port will benefit that port, even if (especially if) the next port eliminates its own clam beds. This model holds promise for greater use with various fish species as well. In New England waters, cod once spawned in many local populations, many of which are now extinct. Overall regional quotas and regional mobility of boats contributed to their extinction. Had managers established local area-based restrictions, these populations might well have been saved, to the benefit of local communities.

In implementing area-based fisheries, managers will need to move deliberately, being mindful of what is scientifically supported and careful not to unduly raise people's expectations. If managers move too hastily, the restrictions may meet a lot of social skepticism and may not work as well as advertised, setting back not only the health of the fish stocks but also the credibility of the managers and scientists who support such actions.

Closed Areas

In recent years, fisheries managers have decided that some stocks are so threatened that the only choice is to close all or part of their habitat to fishing. Such efforts are to be applauded, although they have been too few and too limited in scale to achieve major success. Still, the lessons are instructive, as closures have been found to result in increases in fish populations, in the size of individual fish, and in greater diversity of species.

On Georges Bank in the north Atlantic, for example, success has been mixed, but tantalizing. Managers closed some of the grounds in an effort to protect northern cod, in particular, whose stocks had become severely depleted. So far, cod stocks have not rebounded, for a suite of reasons. But populations of several other important species, notably haddock and sea scallops, have mushroomed. These recovered populations have yielded significant financial benefits to the region, although in the case of sea scallops, fishing interests successfully lobbied to be allowed back into the closed areas, hampering full recovery of the resource.

Mixed Zoning

In many resource-based industries, even competing interests often agree on one thing: They do not want an area closed to them. Yet regarding fishing, conservationists too often have insisted that protected areas be closed to all extraction, and their single-minded pursuit of all-or-nothing solutions has made it easy for commercial interests to unite in demanding that the answer be nothing. A more nuanced approach is needed.

A comprehensive zoning program should designate a mix of areas, including areas that are entirely open to any kind of fishing at any time, areas that are closed to fishers using mobile gear, areas that are closed to fishers using gear that drags along the seafloor, areas that are closed in some seasons, and areas that are fully protected no-take zones. Such integrated zoning would better protect sensitive seafloor habitats and aquatic nursery areas from the kinds of activities that hurt those areas, while allowing harmless activities to proceed. For instance, tuna fishing could be banned in tuna breeding or nursery areas, yet allowed in ocean canyons, even those with deep coral and other important sedentary bottom communities. This type of zoning would also be most likely to gain the support of competing interests, as each party would get something it wants.

Reduction of Incidental Catch

Almost all methods of commercial fishing catch undersized or unmarketable individuals of the target species. Few of these can be returned alive. Fortunately, a number of simple changes in fishing methods and gear, such as the use of nets with larger mesh size, have been developed that can reduce incidental kill by more than 90%, and the government should adopt regulations that require use of these cleaner techniques. In some cases, however, it may be appropriate to require fishers to keep all fish caught—no matter their size, appearance, or even species—in order to reduce the waste that otherwise would result.

Commercial fishers also often catch creatures other than fish, with fatal results. For some creatures, such as sea turtles, capture may endanger their species' very survival. Here, too, advances in fishing technology are helping, but regulators must pay increased attention to finding ways to reduce this problem.

Protection Based on Size

Managers may be able to protect some fish stocks by setting regulations based on graduated fish sizes. This approach, taken almost by default, has led to a spectacular recovery of striped bass along the Atlantic coast. At one time, this population had become deeply depleted, and reproduction rates had fallen precipitously. But one year, environmental conditions arose that favored the survival of eggs and larvae and led to a slight bump in the number of young fish. After much rancor and debate, federal fisheries managers forced states to cooperate in shepherding this class of juveniles to adulthood. They did this primarily by placing a continually increasing limit on the minimum size of fish that fishers could keep. Over the course of more than a decade, the limits protected the fish as they grew and, ultimately, began reproducing. The limits also protected fish hatched in subsequent years, and they, too, grew into adulthood. This simple approach—protecting fish until they have had a chance to reproduce—did more to recover a highly valued, highly sought species than all of the complex calculations, models, and confused politics of previous management efforts.

Subsidy Reform

The federal government provides various segments of the fishing industry with major subsidies that have resulted in a number of adverse consequences. Improperly designed and sized subsidies have propped up bloated and overcapitalized fisheries that have systematically removed too many fish from the seas. Of course, some subsidies will remain necessary. But in most cases, subsidy amounts should be reduced. Also, many subsidies should be redirected to support efforts to develop cleaner technologies and to ease the social pain that fishers and their communities might face in adopting the improved technologies.

Ecologically Integrated Management

Perhaps the worst mistake of traditional fisheries management is that it consider each species in isolation. For example, simply focusing on how much herring fishers can take from the ocean without crashing herring stocks does not address the question of how much herring must be left to avoid crashing the tuna, striped bass, and humpback whales that feed on herring. Management regulations must be revised to reflect such broader food-web considerations.

Sustainable Aquaculture

During the past quarter-century, many nations have turned increasingly to aquaculture to supplement or even replace conventional commercial fishing. Although not at the head of this effort, the United States offers various forms of assistance and incentives to aid the development of the industry. But fish farming is not a panacea. Some operations raise unsustainable monocultures of fish, shrimp, and other aquatic species. Some destroy natural habitats such as marshes that are vital to wild fish. Some transfer pathogens to wild populations. Some pollute natural waters with food, feces, or pesticides necessary to control disease in overcrowded ponds and pens.

As the nation expands fish farming, doing it right should trump doing it fast. Generally, aquaculture will be most successful if it concentrates on raising smaller species and those lower on the food chain. Fish are not cabbages; they do not grow on sunlight. They have to be fed something, and what most fish eat is other fish. Just as the nation's ranchers raise cows and not lions, fish farmers should raise species such as clams, oysters, herring, tilapia, and other vegetarian fish, but not tuna. Farming large carnivores would take more food out of the ocean to feed them than the farming operation would produce. The result would be a loss of food for people, a loss of fish to other fisheries, and a loss to the ocean. Done poorly, aquaculture is as much of a ticking time bomb as were overcapitalized fisheries.

Working Together

Given the magnitude of the problems facing the nation's commercial fishers and fisheries, the various stakeholders must draw together. Although some recent experiences may suggest otherwise, fishers and scientists need each other in order to succeed. Fishers might lack the training to understand the scientific techniques, especially data analysis, that underpin improved management tools, and scientists might lack the experience required to understand the valid concerns and observations of fishers. But without more trust and understanding, adversarial postures that undermine wise management will continue to waste precious time as resources continue to deteriorate and communities and economies suffer. This need not be the case.

Similarly, fishers, fishery managers, and scientists should work together to better inform the public about the conditions and needs of the nation's fishing industry and fish stocks. Consider the example of marine zoning. The less people understand about fishing, the more they insist that closed, no-take marine reserves are the answer. Similarly, the less people understand about conservation, the more they insist that traditional methods of fisheries management, which typically ignore the need for reserves, are adequate tools for protecting fish stocks. As in many other areas, knowledge breeds understanding—and very often solutions.

Carl Safina is the founding President of the Blue Ocean Institute. His latest book is *Voyage of the Turtle* (Henry Holt and Company, 2006).

Food and Agriculture Organization of the United Nations    **NO**

World Review of Fisheries and Aquaculture

Fisheries Resources: Trends in Production, Utilization and Trade

Overview

Capture fisheries and aquaculture supplied the world with about 142 million tonnes of fish in 2008. . . . Of this, 115 million tonnes was used as human food, providing an estimated apparent per capita supply of about 17 kg (live weight equivalent), which is an all-time high. Aquaculture accounted for 46 percent of total food fish supply, a slightly lower proportion than reported in *The State of World Fisheries and Aquaculture 2008* owing to a major downward revision of aquaculture and capture fishery production statistics by China (see below), but representing a continuing increase from 43 percent in 2006. Outside China, per capita supply has remained fairly static in recent years as growth in supply from aquaculture has offset a small decline in capture fishery production and a rising population. In 2008, per capita food fish supply was estimated at 13.7 kg if data for China are excluded. In 2007, fish accounted for 15.7 percent of the global population's intake of animal protein and 6.1 percent of all protein consumed. Globally, fish provides more than 1.5 billion people with almost 20 percent of their average per capita intake of animal protein, and 3.0 billion people with at least 15 percent of such protein. In 2007, the average annual per capita apparent fish supply in developing countries was 15.1 kg, and 14.4 kg in low-income food-deficit countries (LIFDCs). In LIFDCs, which have a relatively low consumption of animal protein, the contribution of fish to total animal protein intake was significant—at 20.1 percent—and is probably higher than that indicated by official statistics in view of the underrecorded contribution of small-scale and subsistence fisheries.

China remains by far the largest fish-producing country, with production of 47.5 million tonnes in 2008 (32.7 and 14.8 million tonnes from aquaculture and capture fisheries, respectively). These figures were derived using a revised statistical methodology adopted by China in 2008 for all aquaculture and capture fishery production statistics and applied to statistics for 2006 onwards. The revision was based on the outcome of China's 2006 National Agricultural Census, which contained questions on fish production for the first time, as well as on results from various pilot sample surveys, most of which were conducted in collaboration with FAO. While revisions varied according to species, area and sector, the overall result was a downward correction of fishery and aquaculture production statistics for 2006 of about 13.5 percent. FAO subsequently estimated revisions for its historical statistics for China for 1997–2005. Notice of the impending revision by China had been given in *The State of World Fisheries and Aquaculture 2008*. Because of the major importance of China in the global context, China is in some cases discussed separately from the rest of the world in this publication.

Global capture fisheries production in 2008 was about 90 million tonnes, with an estimated first-sale value of US$93.9 billion, comprising about 80 million tonnes from marine waters and a record 10 million tonnes from inland waters. World capture fisheries production has been relatively stable in the past decade, with the exception of marked fluctuations driven by catches of anchoveta—a species extremely susceptible to oceanographic conditions determined by the El Niño Southern Oscillation—in the Southeast Pacific. Fluctuations in other species and regions tend to compensate for each other to a large extent. In 2008, China, Peru and Indonesia were the top producing countries. China remained by far the global leader with production of about 15 million tonnes.

Although the revision of China's fishery statistics reduced reported catches by about 2 million tonnes per year in the Northwest Pacific, this area still leads by far the ranking of marine fishing areas, followed by the Southeast Pacific, the Western Central Pacific and the Northeast Atlantic. The same species have dominated marine catches since 2003, with the top ten species accounting for about 30 percent of all marine catches. Catches from inland waters, two-thirds of which were reported as being taken in Asia in 2008, have shown a slowly but steadily rising trend since 1950, owing in part to stock enhancement practices and possibly also to some improvements in reporting, which still remains poor for inland water fisheries (with small-scale and subsistence fisheries substantially underrepresented in the statistics).

Aquaculture continues to be the fastest-growing animal-food-producing sector and to outpace population growth, with per capita supply from aquaculture increasing from 0.7 kg in 1970 to 7.8 kg in 2008, an average annual growth rate of 6.6 percent. It is set to overtake capture fisheries as a source of food fish. While aquaculture production (excluding aquatic plants) was less than 1 million tonnes per year in the early 1950s, production in 2008 was 52.5 million tonnes, with a value of US$98.4 billion.

Aquatic plant production by aquaculture in 2008 was 15.8 million tonnes (live weight equivalent), with a value of US$7.4 billion, representing an average annual growth rate in terms of weight of almost 8 percent since 1970. Thus, if aquatic plants are included, total global aquaculture production in 2008 amounted to 68.3 million tonnes with a first-sale value of US$106 billion. World aquaculture is heavily dominated by the Asia–Pacific region, which accounts for 89 percent of production in terms of quantity and 79 percent in terms of value. This dominance is mainly because of China's enormous production, which accounts for 62 percent of global production in terms of quantity and 51 percent of global value.

Growth rates for aquaculture production are slowing, reflecting the impacts of a wide range of factors, and vary greatly among regions. Latin America and the Caribbean showed the highest average annual growth in the period 1970–2008 (21.1 percent), followed by the Near East (14.1 percent) and Africa (12.6 percent). China's aquaculture production increased at an average annual growth rate of 10.4 percent in the period 1970–2008, but in the new millennium it has declined to 5.4 percent, which is significantly lower than in the 1980s (17.3 percent) and 1990s (12.7 percent). The average annual growth in aquaculture production in Europe and North America since 2000 has also slowed substantially to 1.7 percent and 1.2 percent, respectively. The once-leading countries in aquaculture development such as France, Japan and Spain have shown falling production in the past decade. It is expected that, while world aquaculture production will continue to grow in the coming decade, the rate of increase in most regions will slow. . . .

The proportion of marine fish stocks estimated to be underexploited or moderately exploited declined from 40 percent in the mid-1970s to 15 percent in 2008, whereas the proportion of overexploited, depleted or recovering stocks increased from 10 percent in 1974 to 32 percent in 2008. The proportion of fully exploited stocks has remained relatively stable at about 50 percent since the 1970s. In 2008, 15 percent of the stock groups monitored by FAO were estimated to be underexploited (3 percent) or moderately exploited (12 percent) and able to produce more than their current catches. This is the lowest percentage recorded since the mid-1970s. Slightly more than half of the stocks (53 percent) were estimated to be fully exploited and, therefore, their current catches are at or close to their maximum sustainable productions, with no room for further expansion. The remaining 32 percent were estimated to be either overexploited (28 percent), depleted (3 percent) or recovering from depletion (1 percent) and, thus, yielding less than their maximum potential production owing to excess fishing pressure, with a need for rebuilding plans. This combined percentage is the highest in the time series. The increasing trend in the percentage of overexploited, depleted and recovering stocks and the decreasing trend in underexploited and moderately exploited stocks give cause for concern.

Most of the stocks of the top ten species, which account in total for about 30 percent of the world marine capture fisheries production in terms of quantity, are fully exploited. The two main stocks of anchoveta (*Engraulis ringens*) in the Southeast Pacific and those of Alaska pollock (*Theragra chalcogramma*) in the North Pacific and blue whiting (*Micromesistius poutassou*) in the Atlantic are fully exploited. Several Atlantic herring (*Clupea harengus*) stocks are fully exploited, but some are depleted. Japanese anchovy (*Engraulis japonicus*) in the Northwest Pacific and Chilean jack mackerel (*Trachurus murphyi*) in the Southeast Pacific are considered to be fully exploited. Some limited possibilities for expansion may exist for a few stocks of chub mackerel (*Scomber japonicus*), which are moderately exploited in the Eastern Pacific, while the stock in the Northwest Pacific was estimated to be recovering. In 2008, the largehead hairtail (*Trichiurus lepturus*) was estimated to be overexploited in the main fishing area in the Northwest Pacific. Of the 23 tuna stocks, most are more or less fully exploited (possibly up to 60 percent), some are overexploited or depleted (possibly up to 35 percent) and only a few appear to be underexploited (mainly skipjack). In the long term, because of the substantial demand for tuna and the significant overcapacity of tuna fishing fleets, the status of tuna stocks may deteriorate further if there is no improvement in their management. Concern about the poor status of some bluefin stocks and the difficulties in managing them led to a proposal to the Convention on International Trade in Endangered Species of Wild Fauna and Flora (CITES) in 2010 to ban the international trade of Atlantic bluefin. Although it was hardly in dispute that the stock status of this high-value food fish met the biological criteria for listing on CITES Appendix I, the proposal was ultimately rejected. Many parties that opposed the listing stated that in their view the International Commission for the Conservation of Atlantic Tunas (ICCAT) was the appropriate body for the management of such an important commercially exploited aquatic species. Despite continued reasons for concern in the overall situation, it is encouraging to note that good progress is being made in reducing exploitation rates and restoring overfished fish stocks and marine ecosystems through effective management actions in some areas such as off Australia, on the Newfoundland–Labrador Shelf, the Northeast United States Shelf, the Southern Australian Shelf, and in the California Current ecosystems.

Inland fisheries are a vital component in the livelihoods of people in many parts of the world, in both developing and developed countries. However, irresponsible fishing practices, habitat loss and degradation, water abstraction, drainage of wetlands, dam construction and pollution (including eutrophication) often act together, thus compounding one another's effects. They have caused substantial declines and other changes in inland fishery resources. Although these impacts are not always reflected by a discernable decrease in fishery production (especially when stocking is practised), the fishery may change in composition and value. The poor state of knowledge on inland fishery resources and their ecosystems has led to differing views on the actual status of many resources. One view maintains that the sector is in serious trouble because of the multiple uses of and threats to inland water ecosystems. The other view holds that the sector is in fact growing, that much of the production and growth has gone unreported and that stock enhancement through stocking and other means has played a significant role. Irrespective of these views, the role of inland fisheries in poverty alleviation and food security needs to be better reflected in development and fisheries policies and strategies. The tendency to undervalue inland fisheries in the past has resulted in inadequate representation in national and international agendas. . . .

The Outlook

In spite of the trend of gradually increasing inland catches, it is reported that the abundance of inland water species populations declined by 28 percent between 1970 and 2003. Action is required to secure conservation of aquatic ecosystems and safeguard the resources that form the basis for inland fisheries. A range of factors will directly or indirectly drive the development of the sector. However, there is the possibility to mitigate some negative impacts through technological advances, wealth creation and better management.

Drivers of Inland Fisheries

A General Scenario   For inland fisheries to have a future, there must be fish resources that can be exploited to satisfy people's needs for food, income and/or recreation.

Those now engaged in inland fisheries have fundamentally different reasons to be involved. Commercial, full-time and part-time fishers pursue fisheries because they see the activity as one of their best possibilities to secure a livelihood for themselves and their families. Occasional and subsistence fishers go fishing for additional income or to add fish to their meals, and recreational fishers do so because it is for most of them a leisure-time occupation. However, the sector is highly dynamic with possibilities for people to enter or leave it or increase or decrease their participation in response to developments and available opportunities inside and outside fisheries.

The status of the fisheries resources depends to some extent on the number of fishers and how they are regulated. However, the threats coming from outside the fisheries sector are often more important and can lead to fishers being deprived of their resources and their livelihoods. General social and economic development is a major force influencing the drivers within and outside the fisheries sector, in both a positive and negative manner.

Need for More Food   According to the projections by the United Nations Population Division, the world population will increase from 6.8 billion today to 9 billion by 2050. As stated above, 65–90 percent of the inland capture fish production takes place in the developing and low-income food-deficit countries. The World Bank's forecast for 2020 suggests that 826 million people, or 12.8 percent, of developing country citizens will be living on US$1.25 a day or less and that there will be almost 2 billion poor people living at or below the US$2 a day poverty line. The growing population will need significant increases in food production at affordable prices.

More land (including wetlands) will be used, and some will be used more intensively, as agricultural food production expands during coming decades. This will result in increased use of agrochemicals with serious negative consequences for inland fisheries.

The demand for water for both irrigation and domestic purposes will continue to increase, leading to reduced water availability for fisheries, especially during the dry season. There will be attempts to transfer water between separate basins, with unpredictable consequences for biodiversity. There are also already plans to connect large rivers and transform them into shipping lanes linking distant cities, provinces and countries in areas with poorly developed rail and road infrastructure. There is expected to be increased demand for energy, including hydropower—leading to further damming of rivers.

The need for animal protein, including fish, will increase. Most marine fish stocks are already fully exploited. Notwithstanding increases in aquaculture production, fishing pressure will increase on inland fish stocks, and there will probably be a rise in unsustainable fishing methods, such as the use of explosives and poison, electrofishing and dry pumping of small natural waterbodies. These methods are all capable of killing large amounts of fish indiscriminately.

Aquaculture will continue to grow, and high-value species and products will increasingly come from farms rather than wild stocks. This may reduce capture fishing pressure. In developing countries, improvements in aquaculture technology will allow more fish to be sold more cheaply but, in some markets, cultured species will have problems competing with wild fish because of the need for feed based on fishmeal and fish oil. However, progress is being made on developing feed alternatives derived from locally available animal-waste products or using plant-based proteins instead of animal protein. Where water is available, culture-based and enhanced fisheries will become increasingly important in poor countries with rapidly growing populations because of the lower levels of investment and running costs, but they will require hatcheries to provide the seed. This development will tend to concentrate access to fishing among fewer groups, and the role of fishing as a safety net for the poorest of the poor is likely to be threatened. . . .

The Food and Agriculture Organization (FAO) is a United Nations agency that coordinates efforts to achieve global food security.

EXPLORING THE ISSUE

Does Commercial Fishing Have a Future?

Critical Thinking and Reflection

1. Oceanic fisheries offer an example of a "commons"—a resource "owned" by everyone. Describe at least one problem that follows from this situation.
2. Who should regulate oceanic fisheries?
3. How can one nation's regulators control what happens beyond that nation's own marine boundary (the 200-mile limit)?

Is There Common Ground?

There is a great deal of agreement that global fisheries are in trouble. There is also considerable agreement about the steps that must be taken to make those fisheries sustainable. The problem is the conflict between sustainability and current human needs. If human needs come before the needs of the environment, there may not be much change.

1. Would reducing human needs help achieve sustainability?
2. Is it possible to reduce human needs?
3. How could one possibly reduce human needs without increasing human suffering?

Additional Resources

Carmel Finley, *All the Fish in the Sea: Maximal Sustainable Yield and the Failure of Fisheries Management* (University of Chicago Press, 2011)

National Geographic, "The Global Fish Crisis," a special report in *National Geographic* (April 2007)

Carl Safina, "The World's Imperiled Fish," *Scientific American* (November 1995)

Boris Worm, et al., "Impacts of Biodiversity Loss on Ocean Ecosystem Services," *Science* (November 3, 2006)

Create Central

www.mhhe.com/createcentral

Internet References . . .

American Fisheries Society

www.fisheries.org/afs/index.html

EarthTrends

earthtrends.wri.org/

The Mid-Atlantic Fisheries Management Council

www.mafmc.org/

Selected, Edited, and with Issue Framing Material by:
Thomas Easton, *Thomas College*

ISSUE

Does the World Need High-Tech Agriculture?

YES: N. V. Fedoroff et al., from "Radically Rethinking Agriculture for the 21st Century," *Science* (February 12, 2010)

NO: Satish B. Aher, Bhaveshananda Swami, and B. Sengupta, from "Organic Agriculture: Way towards Sustainable Development," *International Journal of Environmental Sciences* (July 2012)

Learning Outcomes

After reading this issue, you will be able to:

- Explain why genetic modification techniques have not been applied to the majority of food crops.
- Explain why new approaches to agriculture are needed.
- Explain how organic agriculture benefits the environment.
- Discuss whether organic agriculture can supply future world populations with adequate food.
- Describe the benefits of organic farming.

ISSUE SUMMARY

YES: N. V. Fedoroff et al. argue that given the growing human population and demand for food, we need to pursue every available technological means—notably including genetic modification of crops—to increase crop production.

NO: Satish B. Aher, Bhaveshananda Swami, and B. Sengupta argue that organic agriculture (which does not include genetic modification of crops) is good for the soil, can help fight global warming, and has the potential to feed the growing world population, and to do so in a sustainable way.

There was a time when all farming was organic. Fertilizer was compost and manure. Fields were periodically left fallow (unfarmed) to recover soil moisture and nutrients. Crops were rotated to prevent nutrient exhaustion. Pesticides were nonexistent. And farmers were at the mercy of periodic droughts (despite irrigation) and insect infestations.

As population grew, so did the demand for food. In Europe and America, the concomitant demand for fertilizer led in the nineteenth century to a booming trade in guano mined from Caribbean and Pacific islands where deposits of seabird dung could be a hundred and fifty feet thick. When the guano deposits were exhausted, there was an agricultural crisis that was relieved only by the invention of synthetic nitrogen-containing fertilizers early in the twentieth century. See Jimmy Skaggs, *The Great Guano Rush: Entrepreneurs and American Overseas Expansion* (St. Martin's, 1994), G. J. Leigh, *The World's Greatest Fix: A History of Nitrogen and Agriculture* (Oxford, 2004), and Fredrick R. Davis, *Guano and the Opening of the Pacific World: A Global Ecological*

History (Cambridge University Press, 2013). Unfortunately, synthetic fertilizers do not maintain the soil's content of organic matter (humus). This deficit can be amended by tilling in sewage sludge, but the public is not usually very receptive to the idea, partly because of the "yuck factor," but also because sewage sludge may contain human pathogens and chemical contaminants.

Synthetic pesticides, beginning with DDT, came into use in the 1940s. Their history is nicely outlined in Keith S. Delaplane, *Pesticide Usage in the United States: History, Benefits, Risks, and Trends* (University of Georgia Extension Bulletin 1121, 1996) (http://entweb.clemson.edu/pesticid/program/SRPIAP/pestuse.pdf). When they turned out to have problems—target species quickly became resistant, and when the chemicals reached human beings and wildlife on food and in water, they proved to be toxic—some people sought alternatives. On the technological side, these alternatives include genetic modification of crops to produce their own pesticides and resist herbicides (thus permitting more use of weed-killing herbicides). Genetic modification is touted today as promising increased yields while releasing fewer environmental toxins and causing less erosion, both important to sustainability.

Other alternatives to synthetic fertilizers and pesticides (among other things) comprise what is usually meant by "organic farming." Proponents of organic farming have called it holistic, biodynamic, ecological, and natural and claimed a number of advantages for its practice. They say it both preserves the health of the soil and provides healthier food for people. They also argue that it should be used more, even to the point of replacing chemical-based "industrial" agriculture. Because proponents of chemicals hold that fertilizers and pesticides are essential to produce food in the quantities that a world population of over 7 billion people requires, and to hold food prices down to affordable levels, one strand of debate has been over whether organic agriculture can do the job. Jules Pretty, "Can Ecological Agriculture Feed Nine Billion People?" *Monthly Review* (November 2009), puts the emphasis more on achieving sustainability and minimizing damage to the environment and notes that future agricultural systems are likely to include features of both organic and conventional agriculture. No technologies—such as genetic engineering—or practices can be ruled out by fiat.

Catherine Badgley et al., "Organic Agriculture and the Global Food Supply," *Renewable Agriculture & Food Systems* (June 2007), are among the latest to insist that organic methods could produce enough food to sustain a global human population even larger than that of today, and without needing more farmland. Organic agriculture would also decrease the undesirable environmental effects of conventional farming. It is worth noting that Alex Avery, author of *The Truth about Organic Foods* (Henderson, 2006) and director of research and education for the Hudson Institute's Center for Global Food Issues, argues in "'Organic Abundance' Report: Fatally Flawed," Center for Global Food Issues, Hudson Institute (September 2007) that Badgley et al. are guilty of misreporting data, inflating averages by counting high organic yields multiple times, and counting as organic farming clearly nonorganic methods. John J. Miller, "The Organic Myth," *National Review* (February 9, 2004), argues that organic farming is not productive enough to feed today's population, much less larger future populations, it is prone to dangerous biological contamination, and it is not sustainable. "Wishful thinking is at the heart of the organic-food movement." Verena Seufert, Navin Ramankutty, and Jonathan A. Foley, "Comparing the Yields of Organic and Conventional Agriculture," *Nature* (May 10, 2012), find that under certain conditions organic agriculture works well but "overall, organic yields are significantly lower than conventional yields." Tomek de Ponti et al., "The Crop Yield Gap Between Organic and Conventional Agriculture," *Agricultural Systems* (April 2012), did a meta-analysis of 362 yield comparison studies and found that overall organic yields are only 80 percent of conventional agriculture yields.

Is organic farming better for the environment? Soil fertility is in decline in many parts of the world; see

Alfred E. Herteminck, "Assessing Soil Fertility Decline in the Tropics Using Soil Chemical Data," *Advances in Agronomy* (2006). In Africa, the situation is extraordinarily serious. According to the International Center for Soil Fertility and Agricultural Development, "About 75 percent of the farmland in sub-Saharan Africa is plagued by severe degradation, losing basic soil nutrients needed to grow the crops that feed Africa, according to a new report . . . on the precipitous decline in African soil health from 1980 to 2004. Africa's crisis in food production and battle with hunger are largely rooted in this 'soil health crisis'." Proponents of organic farming argue that organic methods are essential to relieving the crisis, but one study of changes in soil fertility, as indicated by crop yield, earthworm numbers, and soil properties, after converting from conventional to organic practices found that different soils responded differently, with some improving and some not; see Anne Kjersti Bakken et al., "Soil Fertility in Three Cropping Systems after Conversion from Conventional to Organic Farming," *Acta Agriculturae Scandinavica: Section B, Soil & Plant Science* (June 2006). Richard Wood et al., "A Comparative Study of Some Environmental Impacts of Conventional and Organic Farming in Australia," *Agricultural Systems* (September 2006), find in a comparison of organic and conventional farms "that direct energy use, energy related emissions, and greenhouse gas emissions are higher for the" former.

Paul Collier, "The Politics of Hunger," *Foreign Affairs* (November–December 2008), says that part of the problem is "the middle- and upper-class love affair with peasant [organic] agriculture." What the world needs is more commercial agriculture and more science, as well as an end to subsidies for biofuels production. According to Catherine M. Cooney, "Sustainable Agriculture Delivers the Crops," *Environmental Science & Technology* (February 15, 2006), "Sustainable agriculture, such as crop rotation, organic farming, and genetically modified seeds, increased crop yields by an average of 79 percent" while also improving the lives of farmers in developing countries. Cong Tu et al., "Responses of Soil Microbial Biomass and N Availability to Transition Strategies from Conventional to Organic Farming Systems," *Agriculture, Ecosystems & Environment* (April, 2006), note that a serious barrier to changing from conventional to organic farming, despite soil improvements, is an initial reduction in yield and increase in pests. David Pimentel et al., "Environmental, Energetic, and Economic Comparisons of Organic and Conventional Farming Systems," *Bioscience* (July 2005), find that organic farming uses less energy, improves soil, and has yields comparable to those of conventional farming but crops probably cannot be grown as often (because of fallowing), which thus reduces long-term yields. However, "because organic foods frequently bring higher prices in the marketplace, the net economic return per [hectare] is often equal to or higher than that of conventionally produced crops." This is one of the factors that prompts Craig J. Pearson to argue in favor of shifting "from conventional open or leaky systems to more closed, regenerative systems" in "Regenerative, Semiclosed Systems: A Priority for Twenty-First-Century Agriculture," *Bioscience* (May 2007).

Better economic return means that organic farming is currently good for the organic farmer. However, the advantage would disappear if the world converted to organic farming. Initial declines in yield mean the conversion would be difficult, but the longer we wait and the more population grows, the more difficult it will be. Whether the conversion is essential depends on the availability of alternative solutions to the problem, and it is worth noting that high energy prices make chemical fertilizers increasingly expensive. Sadly, Stacey Irwin,

"Battle High Fertilizer Costs," *Farm Industry News* (January 2006), does not mention the possibility of using organic methods.

In the following selections, N. V. Fedoroff et al. argue that given the growing human population and demand for food, we need to pursue every available technological means—notably including genetic modification of crops—to increase crop production. Satish B. Aher, Bhaveshananda Swami, and B. Sengupta argue that organic agriculture (which does not include genetic modification of crops) is good for the soil, can help fight global warming, and has the potential to feed the growing world population, and to do so in a sustainable way.

YES ↵

N. V. Fedoroff et al.

Radically Rethinking Agriculture
for the 21st Century

Population experts anticipate the addition of another roughly 3 billion people to the planet's population by the mid-21st century. However, the amount of arable land has not changed appreciably in more than half a century. It is unlikely to increase much in the future because we are losing it to urbanization, salinization, and desertification as fast as or faster than we are adding it.[1] Water scarcity is already a critical concern in parts of the world.[2]

Climate change also has important implications for agriculture. The European heat wave of 2003 killed some 30,000 to 50,000 people.[3] The average temperature that summer was only about 3.5°C above the average for the last century. The 20 to 36% decrease in the yields of grains and fruits that summer drew little attention. But if the climate scientists are right, summers will be that hot on average by midcentury, and by 2090 much of the world will be experiencing summers hotter than the hottest summer now on record.

The yields of our most important food, feed, and fiber crops decline precipitously at temperatures much above 30°C.[4] Among other reasons, this is because photosynthesis has a temperature optimum in the range of 20° to 25°C for our major temperate crops, and plants develop faster as temperature increases, leaving less time to accumulate the carbohydrates, fats, and proteins that constitute the bulk of fruits and grains.[5] Widespread adoption of more effective and sustainable agronomic practices can help buffer crops against warmer and drier environments,[6] but it will be increasingly difficult to maintain, much less increase, yields of our current major crops as temperatures rise and drylands expand.[7]

Climate change will further affect agriculture as the sea level rises, submerging low-lying cropland, and as glaciers melt, causing river systems to experience shorter and more intense seasonal flows, as well as more flooding.[7]

Recent reports on food security emphasize the gains that can be made by bringing existing agronomic and food science technology and know-how to people who do not yet have it,[8,9] as well as by exploring the genetic variability in our existing food crops and developing more ecologically sound farming practices.[10] This requires building local educational, technical, and research capacity, food processing capability, storage capacity, and other aspects of agribusiness, as well as rural transportation and water and communications infrastructure. It also necessitates addressing the many trade, subsidy, intellectual property, and regulatory issues that interfere with trade and inhibit the use of technology.

What people are talking about today, both in the private and public research sectors, is the use and improvement of conventional and molecular breeding, as well as molecular genetic modification (GM), to adapt our existing food crops to increasing temperatures, decreased water availability in some places and flooding in others, rising salinity,[8,9] and changing pathogen and insect threats.[11] Another important goal of such research is increasing crops' nitrogen uptake and use efficiency, because nitrogenous compounds in fertilizers are major contributors to waterway eutrophication and greenhouse gas emissions.

There is a critical need to get beyond popular biases against the use of agricultural biotechnology and develop forward-looking regulatory frameworks based on scientific evidence. In 2008, the most recent year for which statistics are available, GM crops were grown on almost 300 million acres in 25 countries, of which 15 were developing countries.[12] The world has consumed GM crops for 13 years without incident. The first few GM crops that have been grown very widely, including insect-resistant and herbicide-tolerant corn, cotton, canola, and soybeans, have increased agricultural productivity and farmers' incomes. They have also had environmental and health benefits, such as decreased use of pesticides and herbicides and increased use of no-till farming.[13]

Despite the excellent safety and efficacy record of GM crops, regulatory policies remain almost as restrictive as they were when GM crops were first introduced. In the United States, case-by-case review by at least two and sometimes three regulatory agencies (USDA, EPA, and FDA) is still commonly the rule rather than the exception. Perhaps the most detrimental effect of this complex, costly, and time-intensive regulatory apparatus is the virtual exclusion of public-sector researchers from the use of molecular methods to improve crops for farmers. As a result, there are still only a few GM crops, primarily those for which there is a large seed market,[12] and the benefits of biotechnology have not been realized for the vast majority of food crops.

What is needed is a serious reevaluation of the existing regulatory framework in the light of accumulated evidence and experience. An authoritative assessment of existing data on GM crop safety is timely and should encompass protein safety, gene stability, acute toxicity, composition, nutritional value, allergenicity, gene flow, and effects on nontarget organisms. This would establish a foundation for reducing the complexity of the regulatory process without affecting the integrity of the safety assessment. Such an evolution of the regulatory process in the United States would be a welcome precedent globally.

It is also critically important to develop a public facility within the USDA with the mission of conducting the requisite safety testing of GM crops developed in the public sector. This would make it possible for university and other public-sector researchers to use contemporary molecular knowledge and techniques to improve local crops for farmers.

However, it is not at all a foregone conclusion that our current crops can be pushed to perform as well as they do now at much higher temperatures and with much less water and other agricultural inputs. It will take new approaches, new methods, new technology—indeed, perhaps even new crops and new agricultural systems.

Aquaculture is part of the answer. A kilogram of fish can be produced in as little as 50 liters of water,[14] although the total water requirements depend on the feed source. Feed is now commonly derived from wild-caught fish, increasing pressure on marine fisheries. As well, much of the growing aquaculture industry is a source of nutrient pollution of coastal waters, but self-contained and isolated systems are increasingly used to buffer aqua-culture from pathogens and minimize its impact on the environment.[15]

Another part of the answer is in the scale-up of dryland and saline agriculture.[16] Among the research leaders are several centers of the Consultative Group on International Agricultural Research, the International Center for Biosaline Agriculture, and the Jacob Blaustein Institutes for Desert Research of the Ben-Gurion University of the Negev.

Systems that integrate agriculture and aquaculture are rapidly developing in scope and sophistication. A 2001 United Nations Food and Agriculture Organization report[17] describes the development of such systems in many Asian countries. Today, such systems increasingly integrate organisms from multiple trophic levels.[18] An approach particularly well suited for coastal deserts includes inland seawater ponds that support aquaculture, the nutrient efflux from which fertilizes the growth of halophytes, seaweed, salt-tolerant grasses, and mangroves useful for animal feed, human food, and biofuels, and as carbon sinks.[19] Such integrated systems can eliminate today's flow of agricultural nutrients from land to sea. If done on a sufficient scale, inland seawater systems could also compensate for rising sea levels.

The heart of new agricultural paradigms for a hotter and more populous world must be systems that close the loop of nutrient flows from microorganisms and plants to animals and back, powered and irrigated as much as possible by sunlight and seawater. This has the potential to decrease the land, energy, and freshwater demands of agriculture, while at the same time ameliorating the pollution currently associated with agricultural chemicals and animal waste. The design and large-scale implementation of farms based on nontraditional species in arid places will undoubtedly pose new research, engineering, monitoring, and regulatory challenges, with respect to food safety and ecological impacts as well as control of pests and pathogens. But if we are to resume progress toward eliminating hunger, we must scale up and further build on the innovative approaches already under development, and we must do so immediately.

References and Notes

1. *The Land Commodities Global Agriculture & Farmland Investment Report 2009* (Land Commodities Asset Management AG, Baar, Switzerland, 2009; www.landcommodities.com).
2. *Water for Food, Water for Life: A Comprehensive Assessment of Water Management* (International Water Management Institute, Colombo, Sri Lanka, 2007).
3. D. S. Battisti, R. L. Naylor, Historical warnings of future food insecurity with unprecedented seasonal heat. *Science* **323**, 240 (2009). [Abstract/Free Full Text]
4. W. Schlenker, M. J. Roberts, Nonlinear temperature effects indicate severe damages to U.S. crop yields under climate change. *Proc. Natl. Acad. Sci. U.S.A.* **106**, 15594 (2009). [Abstract/Free Full Text]
5. M. M. Qaderi, D. M. Reid, in *Climate Change and Crops,* S. N. Singh, Ed. (Springer-Verlag, Berlin, 2009), pp. 1–9.
6. J. I. L. Morison, N. R. Baker, P. M. Mullineaux, W. J. Davies, Improving water use in crop production. *Philos. Trans. R. Soc. London Ser. B* **363**, 639 (2008). [Abstract/Free Full Text]
7. Intergovernmental Panel on Climate Change, *Climate Change 2007: Impacts, Adaptation and Vulnerability* (Cambridge Univ. Press, Cambridge, 2007; www.ipcc.ch/publications_and_data/publications_ipcc_fourth_assessment_report_wg2_report_impacts_adaptation_and_vulnerability.htm).
8. *Agriculture for Development* (World Bank, Washington, DC, 2008; http://siteresources.worldbank.org/INTWDR2008/Resources/WDR_00_book.pdf).
9. *Reaping the Benefits: Science and the Sustainable Intensification of Global Agriculture* (Royal Society, London, 2009; http://royalsociety.org/Reapingthebenefits).
10. *The Conservation of Global Crop Genetic Resources in the Face of Climate Change* (Summary Statement from a Bellagio Meeting, 2007; http://iis-db.stanford.edu/pubs/22065/Bellagio_final1.pdf).
11. P. J. Gregory, S. N. Johnson, A. C. Newton, J. S. I. Ingram, Integrating pests and pathogens into the climate change/food security debate. *J. Exp. Bot.* **60**, 2827 (2009). [Abstract/Free Full Text]
12. C. James, *Global Status of Commercialized Biotech/GM Crops: 2008* (International Service for the Acquisition of Agribiotech Applications, Ithaca, NY, 2008).

13. G. Brookes, P. Barfoot, *AgBioForum* 11, 21 (2008).
14. S. Rothbard, Y. Peretz, in *Tilapia Farming in the 21st Century*, R. D. Guerrero III, R. Guerrero-del Castillo, Eds. (Philippines Fisheries Associations, Los Baños, Philippines, 2002), pp. 60–65.
15. *The State of World Fisheries and Aquaculture 2008* (United Nations Food and Agriculture Organization, Rome, 2009; www.fao.org/docrep/011/i0250e/i0250e00.HTM).
16. M. A. Lantican, P. L. Pingali, S. Rajaram, Is research on marginal lands catching up? The case of unfavourable wheat growing environments. *Agric. Econ.* 29, 353 (2003). [CrossRef]
17. *Integrated Agriculture-Aquaculture* (United Nations Food and Agriculture Organization, Rome, 2001; www.fao.org / DOCREP/005/Y1187E/y1187e00.htm).
18. T. Chopin *et al.*, in *Encyclopedia of Ecology*, S. E. Jorgensen, B. Fath, Eds. (Elsevier, Amsterdam, 2008), pp. 2463–2475.
19. The Seawater Foundation, www.seawaterfoundation.org.
20. The authors were speakers in a workshop titled "Adapting Agriculture to Climate Change: What Will It Take?" held 14 September 2009 under the auspices of the Office of the Science and Technology Adviser to the Secretary of State. The views expressed here should not be construed as representing those of the U.S. government. N.V.F. is on leave from Pennsylvania State University. C.N.H. is co-chair of Global Seawater, which promotes creation of Integrated Seawater Farms.

N. V. FEDOROFF is a Professor of life sciences at Pennsylvania State University and Science and Technology Adviser to the Secretary of State and to the Administrator of USAID, U.S. Department of State.

D. S. BATTISTI is Professor of Atmospheric Sciences in the Department of Atmospheric Sciences, University of Washington, Seattle, WA.

R. N. BEACHY is with the National Institute of Food and Agriculture, U.S. Department of Agriculture, Washington, DC.

P. J. M. COOPER is with the International Crops Research Institute for the Semi-Arid Tropics, Nairobi, Kenya.

D. A. FISCHHOFF is with Monsanto Company, St. Louis, MO.

C. N. HODGES is with the Seawater Foundation, Tucson, AZ.

V. C. KNAUF is with Arcadia Biosciences, Davis, CA.

D. LOBELL is Associate Professor in the Department of Environmental Earth System Science and Associate Director of the Center on Food Security and the Environment at Stanford University, Stanford, CA.

B. J. MAZUR is with DuPont Agriculture & Nutrition, DuPont Experimental Station, Wilmington, DE.

D. MOLDEN is with the International Water Management Institute, Pelawatte, Battaramulla, Colombo, Sri Lanka.

M. P. REYNOLDS is with the International Maize and Wheat Improvement Center, Carretera Mexico-Veracruz, El Batan, Texcoco, Edo. de México, México.

P. C. RONALD is a Professor in the Department of Plant Pathology, University of California, Davis and the Joint Bioenergy Institute, Emeryville, CA.

M. W. ROSEGRANT is Director of the Environment and Production Technology Division at the International Food Policy Research Institute (IFPRI), Washington, DC.

P. A. SANCHEZ is with the Earth Institute, Columbia University, Palisades, NY.

A. VONSHAK is with the Blaustein Institutes for Desert Research, Ben Gurion University, Sede Boqer Campus, Israel.

J.-K. ZHU is with the Center for Plant Stress Genomics and Technology, King Abdullah University of Science and Technology, Kingdom of Saudi Arabia, and the Department of Botany and Plant Sciences, University of California, Riverside, CA.

Satish B. Aher, Bhaveshananda Swami, and B. Sengupta

Organic Agriculture: Way towards Sustainable Development

Introduction

Agriculture sector is vital for the food and nutritional security of the nation of India. The sector remains the principal source of livelihood for more than 58% of the population though its contribution (14.2%) to the national GDP. Compared to other countries, India faces a greater challenge, since with only 2.3% share in world's total land area; it has to ensure food security of its population which is about 17.5% of world population. This leads to excessive pressure on land and fragmentation of land holdings. On the other side the annual consumption of fertilizers in nutrient terms (N, P & K), has increased from 0.7 lakh [Lakh = 100,000] MT in 1951–52 to 264.86 lakh MT 2009–10, while per hectare consumption of fertilizers, which was less than 1 Kg in 1951–52 has risen to the level of 135.27 Kg (estimated) in 2009–10. Intensive use of inorganic fertilizers and pesticides has been an important tool in the drive for increased crop production. In fact more fertilizers consumption is a good indication of agricultural productivity but depletion of soil fertility is commonly observed in soils. This continuous and massive application of the agrochemicals causes degradation of environment in terms of reduction in soil fertility, water pollution and indirectly significant contribution to the global warming, climate change and ozone layer depletion. According to the National Bureau of Soil Survey and Land Use Planning (NBSSLUP) 21.97 million hectare (mha) of land is degraded in terms of acidity and alkalinity/salinity. Thus, the indiscriminate use of the fertilizer directly affects the soil health in terms of productivity and mineral composition.

Greenhouse gas (GHG) emissions from the agricultural sector account for 10–12% or 5.1–6.1 Gt of the total anthropogenic annual emissions of CO2-equivalents. However, this accounting includes only direct agricultural emissions; emissions due to the production of agricultural inputs such as nitrogen fertilizers, synthetic pesticides and fossil fuels used for agricultural machinery and irrigation are not calculated. Furthermore, land changes in carbon stocks caused by some agricultural practices are not taken into account, e.g., clearing of primary forests. Emissions by deforestation due to land conversion to agriculture, which account for an additional 12% of the global GHG emissions, can be additionally allocated to agriculture. Thus, agriculture production practices emit at least one-quarter of global anthropogenic GHG emissions and, if food handling and processing activities were to be accounted for, the total share of emissions from the agriculture and food sector would be at least one-third of total emissions. Considering the high contribution of agriculture to anthropogenic GHG emissions, the choice of food production practices can be a problem or a solution in addressing global warming. Recent studies have highlighted the substantial contribution of organic agriculture to climate change mitigation and adaptation. The potential of organic agriculture to mitigate climate change is mostly claimed on the basis of assumptions concerning the soil carbon sequestration potential of organic management.

Organic agriculture offers a unique combination of environmentally-sound practices with low external inputs while contributing to food availability. The objective is to describe the potential of the organic agriculture to provide an alternative way for conventional agricultural practices which leads to a sustainable resource utilization and contributes in mitigating global problems like climate change.

Concept of Organic Agriculture

A large number of terms are used as an alternative to organic farming. These are: biological agriculture, ecological agriculture, bio-dynamic, organic-biological agriculture and natural agriculture. According to the National Organic Standards Board of the US Department of Agriculture (USDA) the word 'Organic' has the following official definition: *"An ecological production management system that promotes and enhances biodiversity, biological cycles and soil biological activity. It is based on the minimal use of off-farm inputs and on management practices that restore, maintain and enhance ecological harmony."*

According to the Codex Alimentarius Commission, *"organic agriculture is a holistic production management system that avoids use of synthetic fertilizers, pesticides and genetically modified organisms, minimizes pollution of air, soil and water, and optimizes the health and productivity of interdependent communities of plants, animals and people."* To meet these objectives, organic agriculture farmers need to implement a series of practices that optimize nutrient and energy flows and minimize risk, such as: crop rotations and enhanced crop diversity; different combinations of livestock and plants; symbiotic nitrogen fixation with legumes; application of organic manure; and biological pest control. All these strategies seek to make the best use of local resources.

Organic farming is distinguished from conventional agriculture by exercising particular respect for human values, the environment, nature, and animal welfare, etc. This regard is incorporated in the basic principles of organic farming, as formulated by the International Federation of Organic Agriculture Movements. The main principles for organic farming and food processing include:

1. The production of food of high quality in sufficient quantities,
2. Operation within natural cycles and closed systems as far as possible, drawing upon local resources,
3. The maintenance and long term improvement of the fertility and sustainability of soils,
4. The creation of a harmonious balance between crop production and animal husbandry,
5. The securing of high levels of animal welfare,
6. The fostering of local and regional production and supply chains, and
7. The provision of support for the establishment of an entire production, processing and distribution chain that is both socially and ecologically justifiable.

These basic principles provide organic farming with a platform for ensuring the health of the environment for sustainable development, even though the sustainable development of mankind is not directly specified in the principles.

Organic Agriculture and Sustainable Development

When the World Commission on Environment and Development presented their 1987 report, *Our Common Future* they sought to address the problem of conflicts between environment and development goals by formulating a definition of sustainable development:

Sustainable development is development which meets the needs of the present without compromising the ability of future generations to meet their own needs.

An environmentally sustainable system must maintain a stable resource base, avoiding over-exploitation of renewable resource systems or environmental sink functions, and depleting non-renewable resources only to the extent that investment is made in adequate substitutes. This includes maintenance of biodiversity, atmospheric stability, and other ecosystem functions not ordinarily classed as economic resources. The United Nations report stated: 'All case studies which focused on food production in this research where data have been reported have shown increases in per hectare productivity of food crops, which challenges the popular myth that organic agriculture cannot increase agricultural productivity.'

Comparison of Organic and Conventional Agricultural System

The study carried out in the Central Valley of California showed that tomato yields were quite similar in organic and conventional farms. However, significant differences were found in soil health indicators such as nitrogen mineralization potential and microbial abundance and diversity which were higher in the organic farms. Nitrogen mineralization potential was three times greater in organic compared to conventional fields. The organic fields also had 28% more organic carbon. One of the longest running agricultural trials on record (more than 150 years) [is] the Broadbalk Experiment at the Rothamsted Experimental Station in the United Kingdom. The trials compare a manure based fertilizer farming system (but not certified organic) to a synthetic chemical fertilizer farming system. Wheat yields are shown to be on average slightly higher in the organically fertilized plots (3.45 tones/hectare) than the plots receiving chemical fertilizers (3.40 tones/hectare). More importantly though, soil fertility, measured as soil organic matter and nitrogen levels, increased by 120% in the organic plots, compared with only 20% increase in chemically fertilized plots. Another trial's result from Sustainable Agriculture Farming Systems project (SFAS) at University of California, Davis showed the organic and low-input systems had yields comparable to the conventional systems in all crops which were tested—tomato, safflower, corn and bean, and in some instances yielding higher than conventional systems. Initially tomato yields in the organic system were lower in the first three years, but reached the levels of the conventional tomatoes in the subsequent years and had a higher yield during the last year of the experiment (80 t/ha in the organic compared to 68 t/ha in the conventional). In one such study at South Dakota in midwestern United States shows the higher average yields of soybeans (3.5%) and wheat (4.8%) in the organic compared to conventional farming system. 21 year study compared plots of cropland grown according to both organic and conventional methods at Institute of Organic Agriculture and the Swiss Federal Research Station for Agroecology and Agriculture found that organic yields were less by about 20% but fertilizer, energy and pesticide use were less by 34%, 53% and 97% respectively as compared to conventional. Also organic soils housed a larger and more diverse community of organisms. The study at Iowa State University assessed the agro ecosystem performance of farms which found initially the yield was slightly lower (organic corn & soybean yield averaged 91.8% & 99.6% of conventional respectively) in organic plots but in fourth year organic yield exceeded conventional for both corn and soybean crops. 30 Years Farming System Trial (FST) at Rodale Institute [showed] organic corn yields 31% higher than conventional in years of drought.

These drought yields are remarkable when compared to genetically engineered "drought tolerant" varieties which saw increases of only 6.7% to 13.3% over conventional (non-drought resistant) varieties. Corn and soybean crops in the organic systems tolerated much higher levels of weed competition than their conventional counterparts, while producing equivalent yields. This is especially significant given the rise of herbicide-resistant weeds in conventional systems, and speaks to the increased health and productivity of the organic soil (supporting both weeds and crop yield). The study conducted by ETC Organic Cotton Programme in the district of Karimnagar, Andrha Pradesh India [showed] organic cotton yielded on par at 232 Kg seed cotton/acre vs. conventional cotton at 105 Kg/acre. The pest control expenses was observed about Rs. 220 and Rs. 1624 per acre for organic and in conventional cotton respectively. Study at Washington State University compared yields, economics, soil quality, and other factors resulting from apples grown using organic, conventional, and integrated methods. After combining all of the sustainability indicators, the organic system ranked first in overall sustainability, the integrated second, and the conventional last. A survey conducted by Indian Institute of Soil Science on certified organic farms to evaluate the real benefits and feasibility of organic farming revealed that, on an average, the productivity of crops in organic farming was lower by 9.2% compared to conventional farming. But there was a reduction in the average cost of cultivation in organic farming by 11.7% compared to conventional farming. The average net profit of 22.0% higher in organic farming was observed where 20–40% premium provided. Besides this, overall improvement in soil quality was observed indicating an enhanced soil health and sustainability of crop production in organic farming systems.

Conclusion

A comprehensive review of a number of comparison studies on agricultural yields shows that in all of these studies organic production is equivalent to, and in many cases better than, conventional farming practices. In some, overall lower yield was also reported but the economy [was] still better than the conventional agriculture practices due to the lower external inputs. Besides the yield comparisons, organic practices show higher organic matter in soil, lower energy consumption, lower use of external inputs, better food quality, and also potential to address the global issues like climate change.

Satish B. Aher is affiliated with Environment and Disaster Management, IRDM Faculty Centre, Ramakrishna Mission, Vivekananda University, Narendrapur, Kolkata, West Bengal, India.

Bhaveshananda Swami is affiliated with Environment and Disaster Management, IRDM Faculty Centre, Ramakrishna Mission, Vivekananda University, Narendrapur, Kolkata, West Bengal, India.

B. Sengupta is a former Member Secretary, Central Pollution Control Board, Ministry of Environment and Forests, Government of India.

EXPLORING THE ISSUE

Does the World Need High-Tech Agriculture?

Critical Thinking and Reflection

1. What are the major current threats to global food security?
2. How can the agricultural system be changed to improve global food security?
3. In what sense is organic agriculture more sustainable than conventional agriculture?
4. In what way(s) does organic agriculture help to combat global climate change?

Is There Common Ground?

Both sides in this debate agree that agricultural systems can and need to be improved if growing demand for food is to be satisfied. They differ in precisely what aspects of current agriculture need to be improved.

1. To what extent do both organic and conventional agriculture seek to protect the environment?
2. Discuss how changes in population size affect the answer to the Issue question.
3. Fedoroff et al. briefly discuss the need to "close the loop of nutrient flows." In what way does organic agriculture accomplish this?

Additional Resources

G. J. Leigh, *The World's Greatest Fix: A History of Nitrogen and Agriculture* (Oxford, 2004)

Verena Seufert, Navin Ramankutty, and Jonathan A. Foley, "Comparing the Yields of Organic and Conventional Agriculture," *Nature* (May 10, 2012).

Jimmy Skaggs, *The Great Guano Rush: Entrepreneurs and American Overseas Expansion* (St. Martin's Press, 1994)

Create Central

www.mhhe.com/createcentral

Internet References . . .

The Agriculture Network Information Collaborative

www.agnic.org/

Organic Farming Research Foundation

ofrf.org/

USDA National Organic Program

www.ams.usda.gov/AMSv1.0/nop

Unit 5

UNIT

Hazardous Releases

A great many of today's environmental issues have to do with industrial development, which expanded greatly during the twentieth century. Just since World War II, many thousands of synthetic chemicals—pesticides, plastics, and antibiotics among them—have flooded the environment. We have become dependent on the production and use of energy, particularly in the form of fossil fuels. We have discovered that industrial processes generate huge amounts of waste, much of it toxic. Air and water pollution have become global problems. Most recently, we have begun to develop "synthetic biology" organisms that if released into the environment may prove as bad as any toxic chemicals.

We are now aware that our "hazardous releases" may change the world for generations to come. The following issues by no means exhaust the possible topics for study and debate.

Selected, Edited, and with Issue Framing Material by:
Thomas Easton, *Thomas College*

ISSUE

Should Society Impose a Moratorium on the Use and Release of "Synthetic Biology" Organisms?

YES: Jim Thomas, Eric Hoffman, and Jaydee Hanson, from "Offering Testimony from Civil Society on the Environmental and Societal Implications of Synthetic Biology," *Hearing on Developments in Synthetic Genomics and Implications for Health and Energy* (May 27, 2010)

NO: Gregory E. Kaebnick, from "Testimony to the House Committee on Energy and Commerce," *Hearing on Developments in Synthetic Genomics and Implications for Health and Energy* (May 27, 2010)

Learning Outcomes
After reading this issue, you will be able to: • Explain what "synthetic biology" and its potential benefits are.   • Explain why people are concerned about the impact of "synthetic biology" on the environment.   • Describe measures for controlling the potential risks of "synthetic biology."   • Discuss the difficulty of preventing all the potential risks of a new technology.

ISSUE SUMMARY

YES: Jim Thomas, Eric Hoffman, and Jaydee Hanson, representing the Civil Society on the Environmental and Societal Implications of Synthetic Biology, argue that the risks posed by synthetic biology to human health, the environment, and natural ecosystems are so great that Congress should declare an immediate moratorium on releases to the environment and commercial uses of synthetic organisms and require comprehensive environmental and social impact reviews of all federally funded synthetic biology research.

NO: Gregory E. Kaebnick of the Hastings Center argues that although synthetic biology is surrounded by genuine ethical and moral concerns—including risks to health and environment—which warrant discussion, the potential benefits are too great to call for a general moratorium.

In the past century, biologists have learned an enormous amount about how the cell—the basic functional unit of all living things—works. By the early 1970s, they were beginning to move genes from one organism to another and dream of designing plants and animals (including human beings) with novel combinations of features. By 2002, with Defense Department funding, Jeronimo Cello, Aniko Paul, and Eckard Wimmer were able to construct a live poliovirus from raw laboratory chemicals. This feat was a long ways from constructing a bacterium or animal from raw chemicals, but it was enough to set alarm bells of many kinds ringing. Some people thought this work challenged the divine monopoly on creation. Others feared that if one could construct one virus from scratch, one could construct others, such as the smallpox virus, or even tailor entirely new viruses with which natural immune

systems and medical facilities could not cope. Some even thought that the paper was irresponsible and should not have been published because it pointed the way toward new kinds of terrorism. See Michael J. Selgelid and Lorna Weir, "Reflections on the Synthetic Production of Poliovirus," *Bulletin of the Atomic Scientists* (May/June 2010).

In 2010, the next step was taken. Craig Venter's research group announced that they had successfully synthesized a bacterial chromosome (the set of genes that specifies the function and form of the bacterium) and implanted it in a bacterium of a different species whose chromosome had been removed. The result was the conversion of the recipient bacterium into the synthesized chromosome's species. See Daniel G. Gibson et al., "Creation of a Bacterial Cell Controlled by a Chemically Synthesized Genome," *Science* (July 2, 2010). The report received a great deal of media attention, much of it saying that Venter's group had created a living cell, even though only the chromosome had been synthesized. The chromosome's biochemically complex container—a cell minus its chromosome—had not been synthesized.

The goal of this work is not the creation of life, but rather the ability to exert unprecedented control over what cells do. In testimony before the House Committee on Energy and Commerce Hearing on Developments in Synthetic Genomics and Implications for Health and Energy (May 27, 2010), Venter said "The ability to routinely write the 'software of life' will usher in a new era in science, and with it, new products and applications such as advanced biofuels, clean water technology, food products, and new vaccines and medicines. The field is already having an impact in some of these areas and will continue to do so as long as this powerful new area of science is used wisely." See also Pamela Weintraub, "J. Craig Venter on Biology's Next Leap: Digitally Designed Life-Forms that Could Produce Novel Drugs, Renewable Fuels, and Plentiful Food for Tomorrow's World," *Discover* (January/February 2010); and Michael A. Peters and Priya Venkatesan, "Bioeconomy and Third Industrial Revolution in the Age of Synthetic Life," *Contemporary Readings in Law and Social Justice* (vol. 2, no. 2, 2010). However, the ETC Group, which anticipated a synthetic organism in 2007, condemns the lack of rules governing synthetic biology, calls it "a quintessential Pandora's box moment," and calls for a global moratorium on further work; see "Synthia Is Alive . . . and Breeding: Panacea or Pandora's Box?" ETC Group News Release (May 20, 2010) (www.etcgroup.org/en/node/5142).

Researchers had been working on synthetic biology for a number of years, and well before Craig Venter's group announced their accomplishment, prospects and consequences were already being discussed. Michael Specter, "A Life of Its Own," *New Yorker* (September 28, 2009), describes progress to date and notes "the ultimate goal is to create a synthetic organism made solely from chemical parts and blueprints of DNA." If this sounds rather like manipulating living things the way children manipulate Legos, Drew Endy of MIT and colleagues created in 2005 the BioBricks Foundation to make that metaphor explicit. See also Iina Hellsten and Brigitte Nerlich, "Synthetic Biology: Building the Language for a New Science Brick by Metaphorical Brick," *New Genetics & Society* (December 2011), and Rob Carlson, *Biology Is Technology: The Promise, Peril, and Business of Engineering Life* (Harvard University Press, 2010). David Deamer, "First Life and Next Life," *Technology Review* (May/June 2009), notes that the next step is to create entire cells, not just a single bacterial chromosome. Charles Petit, "Life from Scratch," *Science News* (July 3, 2010), describes the even more ambitious work of Harvard's Jack Szostak, who is trying to understand how life began by constructing a pre-cell just sophisticated enough to take in components, grow, divide, and start evolving. Szostak expects to succeed within a few years. Such efforts, say Steven A. Benner, Zunyi Yang, and Fei Chen, "Synthetic Biology, Tinkering Biology, and Artificial

Biology: What Are We Learning?" *Comptes Rendus Chimie* (April 2011), will drive a better understanding of biology in ways that mere analysis cannot. Allen A. Cheng and Timothy K. Lu, "Synthetic Biology: An Emerging Engineering Discipline," *Annual Review of Biomedical Engineering* (August 2012), see synthetic biology as bringing the engineering mindset to biology, with great potential for human health.

Immediately after the Venter group's announcement of their accomplishment, Vatican representatives declared that synthetic biology was "a potential time bomb, a dangerous double-edged sword for which it is impossible to imagine the consequences" and "Pretending to be God and parroting his power of creation is an enormous risk that can plunge men into barbarity"; see "Vatican Greets First Synthetic Cell with Caution," *America* (June 7–14, 2010). Chuck Colson, "Synthetic Life: The Danger of God-Like Pretensions," *Christian Post* (June 16, 2010), says "God-like control [of risks] isn't only hubris, it's pure fantasy. The only real way to avoid the unthinkable is not to try and play God in the first place. But that would require the kind of humility that Venter and company reject out-of-hand." Nancy Gibbs, "Creation Myths," *Time* (June 28, 2010), says "The path of progress cuts through the four-way intersection of the moral, medical, religious and political—and whichever way you turn, you are likely to run over someone's deeply held beliefs. Venter's bombshell revived the oldest of ethical debates, over whether scientists were playing God or proving he does not exist because someone re-enacted Genesis in suburban Maryland." The "playing God" objection seems likely to grow louder as synthetic biology matures, but it is also likely to fade just as it has done after previous advances such as in vitro fertilization and surrogate mothering. Anna Deplazes-Zemp, "The Conception of Life in Synthetic Biology," *Science & Engineering Ethics* (December 2012), sees a shift toward viewing "life as a toolbox" devoid of higher meaning. Henk van den Belt, "Playing God in Frankenstein's Footsteps: Synthetic Biology and the Meaning of Life," *NanoEthics* (December 2009), notes that "While syntheses of artificial life forms cause some vague uneasiness that life may lose its special meaning, most concerns turn out to be narrowly anthropocentric. As long as synthetic biology creates only new microbial life and does not directly affect human life, it will in all likelihood be considered acceptable." Eleonore Pauwels, "Public Understanding of Synthetic Biology," *BioScience* (February 2013), finds that the general public tends to approve of synthetic biology work aimed at "societal, medical, and sustainability needs."

Maximilian Horner et al., "Synthetic Biology," *Perspectives in Biology & Medicine* (Autumn 2012), see potential benefits for the pharmaceutical industry. Gautam Mukunda et al., "What Rough Beast? Synthetic Biology, Uncertainty, and the Future of Biosecurity," *Politics and the Life Sciences* (September 2009), see synthetic biology as seeking "to create modular biological parts that can be assembled into useful devices, allowing the modification of biological systems with greater reliability, at lower cost, with greater speed, and by a larger pool of people than has been the case with traditional genetic engineering." It is thus a "dual-use" technology, meaning that it has both benign and malign applications. This has clear implications for national security, both offensive and defensive, but they find those implications least alarming in the short term. In the long term, the defensive implications are most important. Because the offensive implications are there, regulation and surveillance of research and development will be necessary in order to forestall terrorists and criminals. Mildred K. Cho and David A. Relman, "Synthetic 'Life,' Ethics, National Security, and Public Discourse," *Science* (July 2, 2010), caution that some concerns about biosecurity and ethics are real but some are imagined; being realistic and avoiding exaggeration are essential if the science is not to become a victim of public mistrust. Meera Lee Sethi and Adam Briggle, "Making Stories Visible: The Task for Bioethics Commissions," *Issues in Science and Technology* (Winter 2011), caution that the stories we tell ourselves about technology (such as "synthetic biology is like computers") may hide issues that warrant deep and careful thought. For the moment, however, the field moves on. The Defense Advanced Research Projects Agency (DARPA) has launched a program called Living Foundries "to create a new manufacturing capability for the United States" using synthetic cells; see Elizabeth Pennisi, "DARPA Offers $30 Million to Jump-Start Cellular Factories," *Science* (July 8, 2011). Some people are already thinking about how synthetic biology will interact with the patent and copyright system, with Jane Calvert, "Ownership and Sharing in Synthetic Biology: A 'Diverse Ecology' of the Open and the Proprietary?" *BioSocieties* (June 2012), noting that we can already see a split between proprietary and open-source work, much as in the computer software industry, but we cannot yet say which approach will win out.

In the YES selection, Jim Thomas, Eric Hoffman, and Jaydee Hanson, representing the Civil Society on the Environmental and Societal Implications of Synthetic Biology, argue that the risks posed by synthetic biology to human health, the environment, and natural ecosystems are so great that Congress should declare an immediate moratorium on releases to the environment and commercial uses of synthetic organisms and require comprehensive environmental and social impact reviews of all federally funded synthetic biology research. In the NO selection, Gregory E. Kaebnick of the Hastings Center argues that although synthetic biology is surrounded by genuine ethical and moral concerns—including risks to health and environment—which warrant discussion, the potential benefits are too great to call for a general moratorium.

YES

Jim Thomas, Eric Hoffman,
and Jaydee Hanson

Offering Testimony from Civil Society on the Environmental and Societal Implications of Synthetic Biology

Last week, the J. Craig Venter Institute announced the creation of the first living organism with a synthetic genome claiming that this technology would be used in applications as diverse as next generation biofuels, vaccine production and the clean up of oil spills. We agree that this is a significant technical feat however; we believe it should be received as a wake-up call to governments around the world that this technology must now be accountably regulated. While attention this week has been on the activities of a team from Synthetic Genomics Inc, the broader field of synthetic biology has in fact quickly and quietly grown into a multi-billion dollar industry with over seventy DNA foundries and dozens of 'pure play' synthetic biology companies entering the marketplace supported by large investments from Fortune 500 energy, forestry, chemical and agribusiness companies. That industry already has at least one product in the marketplace (Du Pont's 'Sorona' bioplastic), and another recently cleared for market entry in 2011 (Amyris Biotechnology's 'No Compromise' biofuel) as well as several dozen near to market applications. We believe the committee should consider the implications of this new industry as a whole in its deliberations not just the technical breakthrough reported last week. Without proper safeguards in place, we risk introducing synthetically constructed living organisms into the environment, intentionally or inadvertently through accident and worker error, that have the potential to destroy ecosystems and threaten human health. We will see the widespread commercial application of techniques with grave dual-use implications. We further risk licensing their use in industrial applications that will unsustainably increase the pressure of human activities on both land and marine ecologies through the increased take of biomass, food resources, water and fertilizer or displacement of wild lands to grow feedstocks for biobased fuel and chemical production.

We call on Congress to:

1. Implement a moratorium on the release of synthetic organisms into the environment and also their use in commercial settings. This moratorium should remain in place until there is an adequate scientific basis on which to justify such activities, and until due consideration of the associated risks for the environment, biodiversity, and human health, and all associated socio-economic repercussions, are fully and transparently considered.
2. As an immediate step, all federally funded synthetic biology research should be subject to a comprehensive environmental and societal impact review carried out with input from civil society, also considering indirect impacts on biodiversity of moving synthetic organisms into commercial use for fuel, chemicals and medicines. This should include the projects that received $305 million from the Department of Energy in 2009 alone.
3. All synthetic biology projects should also be reviewed by the Recombinant DNA Advisory Committee.

On Synthetic Biology for Biofuels—Time for a Reality Check

Much of the purported promise of the emerging Synthetic Biology industry resides in the notion of transforming biomass into next generation biofuels or bio-based chemicals where synthetic organisms work as bio-factories transforming sugars to high value products. On examination much of this promise is unrealistic and unsustainable and if allowed to proceed could hamper ongoing efforts to conserve biological diversity, ensure food security and prevent dangerous climate change. The sobering reality is that a switch to a bio-based industrial economy could exert much more pressure on land, water, soil, fertilizer, forest resources and conservation areas. It may also do little to address greenhouse gas emissions, potentially worsening climate change.

By way of an example, the team associated with Synthetic Genomics Inc who have recently announced the creation of a synthetic cell have specifically claimed that they would use the same technology to develop an algal species that efficiently converts atmospheric carbon dioxide into hydrocarbon fuel, supposedly addressing both the climate crisis and peak oil concerns in one fell swoop. Yet, contrary to the impression put forth by these researchers in the press, algae, synthetic or otherwise, requires much more than just carbon dioxide to grow—it also requires water, nutrients for fertilizer and also sunlight (which therefore means one needs land or open ocean—this can't be done in a vat without also consuming vast quantities of sugar).

In order for Synthetic Genomics or their partners to scale up algal biofuel production to make a dent in the fuel supply, the process would likely exert a massive drain on both water and on fertilizers. Both fresh water and fertilizer (especially phosphate-based fertilizers) are in short supply, both are already prioritized for agricultural food production and both require a large amount of energy either to produce (in the case of fertilizers) or to pump to arid sunlight-rich regions (in the case of water). In a recent life-cycle assessment of algal biofuels published in the journal *Environmental Science and Technology* researchers concluded that algae production consumes more water and energy than other biofuel sources like corn, canola, and switch grass, and also has higher greenhouse gas emissions. "Given what we know about algae production pilot projects over the past 10 to 15 years, we've found that algae's environmental footprint is larger than other terrestrial crops," said Andres Clarens, an assistant professor in U.Virginia's Civil and Environmental Department and lead author on the paper. Moreover scaling-up this technology in the least energy-intensive manner will likely need large open ponds sited in deserts, displacing desert ecosystems. Indeed the federally appointed Invasive Species Advisory Committee has recently warned that non-native algal species employed for such biofuel production could prove ecologically harmful and is currently preparing a fuller report on the matter.

Meanwhile it is not clear that the yield from algal biofuels would go far to meeting our energy needs. MIT inventor Saul Griffiths has recently calculated that even if an algae strain can be made 4 times as efficient as an energy source than it is today it would still be necessary to fill one Olympic-size swimming pool of algae every second for the next twenty five years to offset only half a terawatt of our current energy consumption (which is expected to rise to 16 TW in that time period). That amounts to massive land use change. Emissions from land use change are recognized as one of the biggest contributors to anthropogenic climate change.

Moving Forward—Time for New Regulation

The rapid adoption of synthetic biology is moving the biotechnology industry into the driving seat of industrial production across many previously disparate sectors with downstream consequences for monopoly policy. Meanwhile its application in commercial settings uses a set of new and extreme techniques whose proper oversight and limits has not yet been debated. It also enables many more diverse living organisms to be produced using genetic science at a speed and volume that will challenge and ultimately overwhelm the capacity of existing biosafety regulations. For example, Craig Venter has claimed in press and in his patent applications that when combined with robotic techniques the technology for producing a synthetic cell can be perfected to make millions of new species per day. Neither the US government nor any other country has the capacity to assess such an outpouring of new synthetic species in a timely or detailed manner. The Energy and Commerce Committee urgently needs to suggest provisions for regulating these new organisms and chemicals derived from them under the Toxic Substances Control Act, Climate Change legislation and other legislation under its purview before allowing their release into the environment. It also needs to identify how it intends to ensure that the use of such organisms whether in biorefineries, open ponds or marine settings does not impinge on agriculture, forestry, desert and marine protection, the preservation of conservation lands, rural jobs or livelihoods.

To conclude, Congress must receive this announcement of a significant new lifeform as a warning bell, signifying that the time has come for governments to fully regulate all synthetic biology experiments and products. It is imperative that in the pursuit of scientific experimentation and wealth creation, we do not sacrifice human health, the environment, and natural ecosystems. These technologies could have powerful and unpredictable consequences. These are life forms never seen on the planet before now. Before they are unleashed into the environment and commercial use, we need to understand the consequences, evaluate alternatives properly, and be able to prevent the problems that may arise from them.

JIM THOMAS is Program Manager at ETC Group (Action Group on Erosion, Technology and Concentration).

ERIC HOFFMAN is Genetic Technology Policy Campaigner with Friends of the Earth (www.foe.org/healthy-people /biofuels-synthetic-biology).

JAYDEE HANSON is Policy Director at the International Center for Technology Assessment (www.icta.org).

Gregory E. Kaebnick   ➡ **NO**

Testimony to the House Committee on Energy and Commerce

... **T**he ethical issues raised by synthetic biology are familiar themes in an ongoing conversation this nation has been having about biotechnologies for several decades. . . .

The concerns fall into two general categories. One has to do with whether the creation of synthetic organisms is a good or a bad thing in and of itself, aside from the consequences. These are thought of as intrinsic concerns. Many people had similar intrinsic concerns about reproductive cloning, for example; they just felt it was wrong to do, regardless of benefits. Another has to do with potential consequences—that is, with risks and benefits. The distinction between these categories can be difficult to maintain in practice, but it provides a useful organizational structure.

Intrinsic Concerns

I will start with the more philosophical, maybe more baffling, kind of concern—the intrinsic concerns. They are an appropriate place to start because the work just published by researchers at Synthetic Genomics, Inc., has been billed as advancing our understanding of these issues in addition to making a scientific advance.

This announcement is not the first time we have had a debate about whether biotechnology challenges deeply held views about the status of life and the power that biotechnology and medicine give us over it. There was a similar debate about gene transfer research in the 1970s and 1980s, about cloning and stem cell research in the 1990s, and—particularly in the last decade but also earlier—about various tools for enhancing human beings. They have been addressed by the President's Commission for the Study of Ethical Problems in Medicine and Biomedical and Behavioral Research in 1983, by President Clinton's National Bioethics Advisory Council, and by President Bush's President's Council on Bioethics. These concerns are related to even older concerns in medicine about decisions to withhold or withdraw medical treatment at the end of life.

The fact that we have had this debate before speaks to its importance. I believe the intrinsic concerns deserve respect, and with some kinds of biotechnology I think

they are very important, but for synthetic biology, I do not think they provide a basis for decisions about governance.

Religious or Metaphysical Concerns

The classic concern about synthetic biology is that it puts human beings in a role properly held by God—that scientists who do it are "playing God," as people say. Some may also believe that life is sacred, and that scientists are violating its sacredness. Prince Charles had this in mind in a famous polemic some years ago when he lamented that biotechnology was leading to "the industrialisation of Life."

To object to synthetic biology along these lines is to see a serious moral mistake in it. This kind of objection may be grounded in deeply held beliefs about God's goals in creating the world and the proper role of human beings within God's plan. But these views would belong to particular faiths—not everybody would share them. Moreover, there is a range of opinions even within religious traditions about what human beings may and may not do. Some people celebrate human creativity and science. They may see science as a gift from God that God intends human beings to develop and use.

The announcement that Synthetic Genomics, Inc., has created a synthetic cell appears to some to disprove the view that life is sacred, but I do not agree. Arguably, what has been created is a synthetic genome, not a completely synthetic cell. Even if scientists manage to create a fully synthetic cell, however, people who believe that life is sacred, that it is something more than interacting chemicals, could continue to defend that belief. A similar question arises about the existence of souls in cloned people: If people have souls, then surely they would have souls even if they were created in the laboratory by means of cloning techniques. By the same reasoning, if microbial life is more than a combination of chemicals, then even microbial life created in the laboratory would be more than just chemicals. In general, beliefs about the sacredness of life are not undermined by science. Moreover, even the creation of a truly synthetic cell would still start with existing materials. It would not be the kind of creating with which God is credited, which is creating something from nothing—creation ex nihilo.

From statement before U.S. House of Representatives, May 27, 2010.

Concerns that Synthetic Biology Will Undermine Morally Significant Concepts

A related but different kind of concern is that synthetic biology will simply undermine our shared understanding of important moral concepts. For example, perhaps it will lead us to think that life does not have the specialness we have often found in it, or that we humans are more powerful than we have thought in the past. This kind of concern can be expressed without talking about God's plan.

Synthetic biology need not change our understanding of the value of life, however. The fact that living things are created naturally, rather than by people, would be only one reason for seeing them as valuable, and we could continue to see them as valuable when they are created by people. Further, in its current form, synthetic biology is almost exclusively about engineering single-celled organisms, which may be less troubling to people than engineering more complex organisms. If the work is contained within the laboratory and the factory, then it might not end up broadly changing humans' views of the value of life.

Also, of course, the fact that the work challenges our ideas may not really be a moral problem. It would not be the first time that science has challenged our views of life or our place in the cosmos, and we have weathered these challenges in the past.

Concerns about the Human Relationship to Nature

Another way of saying that there's something intrinsically troubling about synthetic biology, again without necessarily talking about the possibility that people are treading on God's turf, is to see it as a kind of environmentalist concern. Many environmentalists want to do more than make the environment good for humans; they also want to save nature from humans—they want to save endangered species, wildernesses, "wild rivers," old-growth forests, and mountains, canyons, and caves, for example. We should approach the natural world, many feel, with a kind of reverence or gratitude, and some worry that synthetic biology—perhaps along with many other kinds of biotechnology—does not square with this value.

Of course, human beings have been altering nature throughout human history. They have been altering ecosystems, affecting the survival of species, affecting the evolution of species, and even creating new species. Most agricultural crop species, for example, are dramatically different from their ancestral forebears. The issue, then, is where to draw the line. Even people who want to preserve nature accept that there is a balance to be struck between saving trees and harvesting them for wood. There might also be a balance when it comes to biotechnology. The misgiving is that synthetic biology goes too far—it takes human control over nature to the ultimate level, where we are not merely altering existing life forms but creating new forms.

Another environmentalist perspective, however, is that synthetic biology could be developed so that it is beneficial to the environment. Synthetic Genomics, Inc. recently contracted with Exxon Mobil to engineer algae that produce gasoline in ways that not only eliminate some of the usual environmental costs of producing and transporting fuel but simultaneously absorb large amounts of carbon dioxide, thereby offsetting some of the environmental costs of burning fuel (no matter how it is produced). If that could be achieved, many who feel deeply that we should tread more lightly on the natural world might well find synthetic biology attractive. In order to achieve this benefit, however, we must be confident that synthetic organisms will not escape into the environment and cause harms there.

Concerns Involving Consequences

The second category of moral concerns is about consequences—that is, risks and benefits. The promise of synthetic biology includes, for example, better ways of producing medicine, environmentally friendlier ways of producing fuel and other substances, and remediation of past environmental damage. These are not morally trivial considerations. There are also, however, morally serious risks. These, too, fall into three categories.

Concerns about Social Justice

Synthetic biology is sometimes heralded as the start of a new industrial age. Not only will it lead to new products, but it will lead to new modes of production and distribution; instead of pumping oil out of the ground and shipping it around the world, we might be able to produce it from algae in places closer to where it will be used. Inevitably, then, it would have all sorts of large-scale economic and social consequences, some of which could be harmful and unjust. Some commentators hold, for example, that if synthetic biology generates effective ways of producing biofuels from feedstocks such as sugar cane, then farmland in poor countries would be converted from food production to sugar cane production. Another set of concerns arises over the intellectual property rights in synthetic biology. If synthetic biology is the beginning of a new industrial age, and a handful of companies received patents giving them broad control over it, the results could be unjust.

Surely we ought to avoid these consequences. It is my belief that we can do so without avoiding the technology. Also, traditional industrial methods themselves seem to be leading to disastrous long-term social consequences; if so, synthetic biology might provide a way toward better social outcomes.

Concerns about Biosafety

Another concern is about biosafety—about mechanisms for containing and controlling synthetic organisms, both during research and development and in industrial applications. The concern is that organisms will escape, turn out to have properties, at least in their new environment, different from what was intended and predicted, or maybe mutate to acquire them, and then pose a threat to public health, agriculture, or the environment. Alternatively, some of their genes might be transferred to other, wild microbes, producing wild microbes with new properties.

Controlling this risk means controlling the organisms—trying to prevent industrial or laboratory accidents, and then trying to make sure that, when organisms do escape, they are not dangerous. Many synthetic biologists argue that an organism that devotes most of its energy to producing jet fuel or medicine, that is greatly simplified (so that it lacks the genetic complexity and therefore the adaptability of a wild form), and that is designed to work in a controlled, contained environment, will simply be too weak to survive in the wild. For added assurance, perhaps engineering them with failsafe mechanisms will *ensure* that they are incapable of surviving in the wild.

Concerns about Deliberate Misuse

I once heard a well-respected microbiologist say that he was very enthusiastic about synthetic biology, and that the only thing that worries him is the possibility of catastrophe. The kind of thing that worries him is certainly possible. The 1918 flu virus has been recreated in the laboratory. In 2002, a scientist in New York stitched together stretches of nucleotides to produce a string of DNA that was equivalent to RNA polio virus and eventually produced the RNA virus using the DNA string. More recently, the SARS virus was also created in the laboratory. Eventually, it will almost certainly be possible to recreate bacterial pathogens like smallpox. We might also be able to enhance these pathogens. Some work in Australia on mousepox suggests ways of making smallpox more potent, for example. In theory, entirely new pathogens could be created. Pathogens that target crops or livestock are also possible.

Controlling this risk means controlling the people and companies who have access to DNA synthesis or the tools they could use to synthesize DNA themselves. There are some reasons to think that the worst will never actually happen. To be wielded effectively, destructive synthetic organisms would also have to be weaponized; for example, methods must be found to disperse pathogens in forms that will lead to epidemic infection in the target population while sparing one's own population. Arguably, terrorists have better forms of attacking their enemies than with bioweapons, which are still comparatively hard to make and are very hard to control. However, our policy should amount to more than hoping for the best.

Governance

In assessing these risks and establishing oversight over synthetic biology, we do not start from square one. There is an existing framework of laws and regulations, put into action by various agencies and oversight bodies, that will apply to R&D and to different applications. The NIH is extending its guidelines for research on genetic engineering to ensure that they are applicable to research on synthetic biology. These guidelines are enforced by the NIH's Recombinant DNA Advisory Committee and a network of Institutional Biosafety Committees at research institutions receiving federal funding. Many applications would fall under the purview of various federal laws and the agencies that enforce them. For example, a plan to release synthetic organisms into the sea to produce nutrients that would help rebuild ocean food chains would have to pass muster with the EPA. The USDA and FDA also have regulatory authority over applications. The FBI and the NIH's National Science Advisory Board for Biosecurity are formulating policy to regulate the sale of synthetic DNA sequences that might pose a threat to biosecurity.

At the same time, the current regulatory framework may need to be augmented. First, there are questions about whether the existing laws leave gaps. Research conducted by an entirely privately funded laboratory might not be covered by the NIH's Guidelines, for example. Field testing of a synthetic organism—that is, release into the environment as part of basic research—might not be covered by the existing regulations of the EPA or the USDA. Questions about the adequacy of existing regulations are even more pointed when it comes to concerns about biosecurity, particularly if or when powerful benchtop synthesizers are available in every lab.

The other big question is whether the regulatory bodies' ability to do risk assessment of synthetic biology is adequate. Synthetic biology differs from older forms of genetic engineering in that a synthetic organism could combine DNA sequences found originally in many different organisms, or might even contain entirely novel genetic code. The eventual behavior of these organisms in new environments, should they accidentally end up in one, may therefore be hard to predict.

The synthetic biologists' goal of simplicity is crucial. One of the themes of traditional biology is that living things are usually more complex than they first appear. We should not assume at the outset that synthetic organisms will shed the unpredictability inherent to life. Life tends to find a way. As a starting assumption, we should expect that artificial life will try to find a way as well.

Another difficulty in assessing concerns about both biosafety and deliberate misuse is that, if the field evolves so that important and even innovative work could be done in small, private labs, even in homes, then it could be very difficult to monitor and regulate. The threats of biosafety and deliberate misuse would have to be taken yet more seriously.

Concluding Comments

I take seriously concerns that synthetic biology is bad in and of itself, and I believe that they warrant a thorough public airing, but I do not believe that they provide a good basis for restraining the technology, at least if we can be confident that the organisms will not lead to environmental damage. Better yet would be to get out in front of the technology and ensure that it benefits the environment. Possibly, some potential applications of synthetic biology are more troubling than others and should be treated differently.

Ultimately, I think the field should be assessed on its possible outcomes. At the moment, we do not understand the possible outcomes well enough. We need, I believe:

- more study of the emergence, plausibility, and impact of potential risks;
- a strategy for studying the risks that is multidisciplinary, rather than one conducted entirely within the field;
- a strategy that is grounded in good science rather than sheer speculation, yet flexible enough to look for the unexpected; and

- an analysis of whether our current regulatory framework is adequate to deal with these risks and how the framework should be augmented.

Different kinds of applications pose different risks and may call for different responses. Microbes intended for release into the environment, for example, would pose a different set of concerns than microbes designed to be kept in specialized, contained settings. Overall, however, while the risks of synthetic biology are too significant to leave the field alone, its potential benefits are too great to call for a general moratorium.

GREGORY E. KAEBNICK is a Research Scholar at The Hastings Center and Editor of the *Hastings Center Report*.

EXPLORING THE ISSUE

Should Society Impose a Moratorium on the Use and Release of "Synthetic Biology" Organisms?

Critical Thinking and Reflection

1. Do new technologies inevitably pose environmental risks?
2. Does the ability to engineer new forms of life offer at least potential solutions to environmental problems?
3. Should worrisome research be put on hold until all potential risks are analyzed and solutions devised?
4. Should the precautionary principle be applied to "synthetic biology"? In what ways?

Is There Common Ground?

As with many technologies, some people see mostly risks and would, if they could, stop the development of the technology. Others see mostly benefits and think that those benefits are worth putting up with the risks. A more nuanced approach is to determine which risks and benefits seem most likely and then to carefully weigh them against each other. This approach is known as risk-benefit or cost-benefit analysis, and it is used in medicine, engineering, business, and other areas.

1. What seem to be the most likely or worrisome environmental risks associated with "synthetic biology" technology?
2. What seem to be the most likely benefits associated with synthetic biology technology?
3. Do you think the benefits are worth the risks?

Additional Resources

Rob Carlson, *Biology Is Technology: The Promise, Peril, and Business of Engineering Life* (Harvard University Press, 2010)

Michael A. Peters and Priya Venkatesan, "Bioeconomy and Third Industrial Revolution in the Age of Synthetic Life," *Contemporary Readings in Law and Social Justice* (vol. 2, no. 2, 2010)

Michael J. Selgelid and Lorna Weir, "Reflections on the Synthetic Production of Poliovirus," *Bulletin of the Atomic Scientists* (May/June 2010)

Pamela Weintraub, "J. Craig Venter on Biology's Next Leap: Digitally Designed Life-Forms that Could Produce Novel Drugs, Renewable Fuels, and Plentiful Food for Tomorrow's World," *Discover* (January/February 2010)

Create Central

www.mhhe.com/createcentral

Internet References . . .

Friends of the Earth

www.foe.org/projects/food-and-technology/synthetic
-biology

The J. Craig Venter Institute

www.jcvi.org/cms/

The Synthetic Biology Center at MIT

synbio.mit.edu/

Selected, Edited, and with Issue Framing Material by:
Thomas Easton, *Thomas College*

ISSUE

Is Bisphenol A a Potentially Serious Health Threat?

YES: Ted Schettler, from Testimony before Senate Committee on Environment and Public Works Hearing on "EPA's Efforts to Protect Children's Health" (March 17, 2010)

NO: Jon Entine, from "The Troubling Case of Bisphenol A: At What Point Should Science Prevail?" *The American Enterprise* (March 2010)

Learning Outcomes

After reading this issue, you will be able to:

- Explain what endocrine disruptors are.
- Explain how endocrine disruptors may affect normal bodily function.
- Describe how people are exposed to endocrine disruptors.
- Explain what is being done to evaluate and prevent the possible problems posed by endocrine disruptors.
- Describe the scientific evidence that bisphenol A (BPA) is or is not a health hazard.

ISSUE SUMMARY

YES: Ted Schettler argues that a great many synthetic chemicals that have reached the environment interfere with normal hormone function and threaten the reproductive functioning of wildlife and humans. One of these chemicals is bisphenol A, which has been linked to altered brain development, heart disease, and diabetes. The Environmental Protection Agency (EPA) needs to act more quickly to evaluate and regulate these chemicals.

NO: Jon Entine argues that the effects of endocrine disruptors are at most quite modest, and the effects of bisphenol A are inconsequential. What the Environmental Protection Agency (EPA) and Food and Drug Administration (FDA) need to do is to stand by the science and resist political pressures, else "we place the entire system of checks and balances in danger."

Following World War II there was an exponential growth in the industrial use and marketing of synthetic chemicals. These chemicals, now known as "xenobiotics" because they are foreign to living things, were used in numerous products, including solvents, pesticides, refrigerants, coolants, and raw materials for plastics (including bisphenol A or BPA). This resulted in increasing environmental contamination. Many of these chemicals, such as DDT, PCBs, and dioxins, proved to be highly resistant to degradation in the environment; they accumulated in wildlife and were serious contaminants of lakes and estuaries. Carried by winds and ocean currents, these chemicals were soon detected in samples taken from the most remote regions of the planet, far from their points of introduction into the ecosphere.

Until very recently most efforts to assess the potential toxicity of synthetic chemicals to living things, including human beings, focused almost exclusively on their possible role as carcinogens. This was because of legitimate public concern about rising cancer rates and the belief that cancer causation was the most likely outcome of exposure to low levels of synthetic chemicals.

Some environmental scientists urged public health officials to give serious consideration to other possible health effects of xenobiotics. They were generally ignored because of limited funding and the common belief that toxic effects other than cancer required larger exposures than usually resulted from environmental contamination.

In the late 1980s, Theo Colborn, a research scientist for the World Wildlife Fund who was then working on a study of pollution in the Great Lakes, began linking together the results of a growing series of isolated studies. Researchers in the Great Lakes region, as well as in Florida, the West Coast, and Northern Europe, had observed widespread evidence of serious and frequently lethal physiological problems involving abnormal reproductive development, unusual sexual behavior, and neurological problems exhibited by a diverse group of animal species, including fish, reptiles, amphibians, birds, and marine mammals. Through Colborn's insights, communications among these researchers, and further studies, a hypothesis was developed that all of these wildlife problems were manifestations of abnormal estrogenic activity. The causative agents were identified as more than 50 synthetic chemical compounds that have been shown in laboratory studies either to mimic the action or to disrupt the normal function of the powerful estrogenic hormones responsible for female sexual development and many other biological functions. Firouzan Massoomi, "Harm to Fish and Wildlife from Chemicals in the Water Supply," testimony before the House Committee on Natural Resources, Subcommittee on Insular Affairs, Oceans, and Wildlife, hearing on "Overdose: How Drugs and Chemicals in Water and the Environment Are Harming Fish and Wildlife" (June 9, 2009), stresses the problem of pharmaceutical drugs, a great many of which reach the water supply via urine and disposal by flushing down the toilet. Theo Colborn, Dianne Dumanoski, and John Peterson Myers, *Our Stolen Future: Are We Threatening Our Fertility, Intelligence, and Survival?—A Scientific Detective Story* (Dutton, 1996), find the evidence that extensive damage is being done to wildlife by synthetic estrogenic chemicals convincing and say it is likely that humans are experiencing similar health problems. Nancy Langston, *Toxic Bodies: Hormone Disruptors and the Legacy of DES* (Yale University Press, 2010), argues that though the precautionary principle may be difficult to implement, partly because of opposition by industry, its application is warranted in the case of these chemicals.

Concern that human exposure to these ubiquitous environmental contaminants may have serious health repercussions was heightened by a widely publicized European research study which concluded that male sperm counts had decreased by 50 percent over the past several decades (a result that is disputed by other researchers) and that testicular cancer rates have tripled. Some scientists have also proposed a link between breast cancer and estrogen disruptors.

Stephen H. Safe's "Environmental and Dietary Estrogens and Human Health: Is There a Problem?" *Environmental Health Perspectives* (April 1995) is often cited to support the contention that there is no causative link between environmental estrogens and human health problems. He draws a cautious conclusion, calling the link "implausible" and "unproven." Some people actually claim that the battle against environmental estrogens is motivated by environmentalist ideology rather than facts; see Angela Logomasini, "Chemical Warfare: Ideological Environmentalism's Quixotic Campaign Against Synthetic Chemicals," in Ronald Bailey, ed., *Global Warming and Other Eco-Myths: How the Environmental Movement Uses False Science to Scare Us to Death* (Prima, 2002). Some caution is

certainly warranted, for the complex and variable manner by which different compounds with estrogenic properties may affect organisms makes projections from animal effects to human effects risky.

Sheldon Krimsky, in "Hormone Disruptors: A Clue to Understanding the Environmental Cause of Disease," *Environment* (June 2001), summarizes the evidence that many chemicals released to the environment affect—both singly and in combination or synergistically—the endocrine systems of animals and humans and may threaten human health with cancers, reproductive anomalies, and neurological effects. He cautions that the regulatory machinery is likely to move very slowly, adding that we cannot wait for scientific certainty about the hazards before we act. See also M. Gochfeld, "Why Epidemiology of Endocrine Disruptors Warrants the Precautionary Principle," *Pure & Applied Chemistry* (December 1, 2003); Rebecca Renner, "Human Estrogens Linked to Endocrine Disruption," *Environmental Science and Technology* (January 1, 1998); and Ted Schettler et al., *Generations at Risk: Reproductive Health and the Environment* (MIT Press, 1999). Julia R. Barrett, "Phthalates and Baby Boys," *Environmental Health Perspectives* (August 2005), reports effects of phthalate plastic-softeners on the development of the human male reproductive system. Laboratory work has found that at least one phthalate reduces the number of sperm-generating cells in fetal testis tissue; see Romain Lambrot et al., "Phthalates Impair Germ Cell Development in the Human Fetal Testis in Vitro Without Change in Testosterone Production," *Environmental Health Perspectives* (January 2009). When researchers added synthetic estrogen to a Canadian lake, they found that the reproduction of fathead minnows was so severely affected that the population nearly died out; see Karen A. Kidd et al., "Collapse of a Fish Population After Exposure to a Synthetic Estrogen," *Proceedings of the National Academy of Sciences of the United States of America* (May 22, 2007). In 1999 the National Research Council published *Hormonally Active Agents in the Environment* (National Academy Press), in which the Council's Committee on Hormonally Active Agents in the Environment reports on its evaluation of the scientific evidence pertaining to endocrine disruptors. The National Environmental Health Association has called for more research and product testing; see Ginger L. Gist, "National Environmental Health Association Position on Endocrine Disruptors," *Journal of Environmental Health* (January–February 1998).

Elisabete Silva, Nissanka Rajapakse, and Andreas Kortenkamp, in "Something from 'Nothing'—Eight Weak Estrogenic Chemicals Combined at Concentrations Below NOECS Produce Significant Mixture Effects," *Environmental Science and Technology* (April 2002), find synergistic effects of exactly the kind dismissed by critics. Also, after reviewing the evidence, the U.S. National Toxicity Program found that low-dose effects had been demonstrated in animals; see Ronald Melnick et al., "Summary of the National Toxicology Program's Report of the Endocrine

Disruptors Low-Dose Peer-Review," *Environmental Health Perspectives* (April 2002). It is perhaps not encouraging that Monika Plotan et al., "Endocrine Disruptor Activity in Bottled Mineral and Flavoured Water," *Food Chemistry* (February 2013), find low levels of endocrine disruptors in 78 percent of bottled water samples and note that source, long-term effects, and synergistic effects remain to be determined. Some researchers say there is reason to think endocrine disruptors are linked to the epidemic of obesity in the United States and to a decline in the proportion of male births in the United States and Japan; Julia R. Barrett, "Shift in the Sexes," *Environmental Health Perspectives* (June 2007). Retha R. Newbold et al., "Effects of Endocrine Disruptors on Obesity," *International Journal of Andrology* (April 2008), review the literature on the obesity link and conclude that the evidence is strong enough to warrant expanding "the focus on obesity [and consequent diabetes and heart disease] from intervention and treatment to include prevention and avoidance of these chemical modifiers."

The view that environmental hormone mimics or disruptors have potentially serious effects continues to gain support. Evidence continues to accumulate, and changes are beginning to happen. The U.S. Congress passed legislation requiring that all pesticides be screened for estrogenic activity and the Environmental Protection Agency (EPA) develop procedures for detecting environmental estrogenic contaminants in drinking water supplies; see the EPA's Endocrine Disruptor Screening Program website at www.epa.gov/scipoly/oscpendo/index.htm. In April 2009, the EPA released its first list of chemicals for initial screening; in November 2010, a second list was released (it was revised in June 2013; see www.epa.gov/endo/). Medical plastics, softened by phthalates, are being reformulated; see "Medical Devices to Get Phthalate Substitution Rule," *European Environment & Packaging Law Weekly* (June 1, 2007). Arnold Schechter et al., "Bisphenol A (BPA) in U.S. Food," *Environmental Science and Technology* (December 15, 2010), found BPA (an endocrine disruptor) in numerous commercially available foods. The authors note that BPA "is a chemical used for lining metal cans and in polycarbonate plastics, such as baby bottles. In rodents, BPA is associated with early sexual maturation, altered behavior, and effects on prostate and mammary glands. In humans, BPA is associated with cardiovascular disease, diabetes, and male sexual dysfunction in exposed workers. Food is a major exposure source." Plastic bottles, which contain BPA, are already being replaced by metal bottles; see Nancy Macdonald, "Plastic Bottles Get the Eco-Boot," *Maclean's* (October 15, 2007). Janet Raloff, "Concerns over Plastics Chemical Continue to Grow," *Science News* (July 18, 2009), reports that the House Committee on Energy and Commerce has asked the Food and Drug Administration (FDA) to "reconsider the Bush Administration's position that BPA is safe at current estimated exposure levels." Europe and many U.S. states have already banned many uses of BPA. On the other hand, Denise Grady, "In Feast of Data on BPA Plastic, No Final Answer," *New York Times* (September 7, 2010), notes that despite the laboratory data, evidence of harm to humans is unclear.

In the YES selection, Ted Schettler argues that a great many synthetic chemicals that have reached the environment interfere with normal hormone function and threaten the reproductive functioning of wildlife and humans. One of these chemicals is bisphenol A, which has been linked to altered brain development, heart disease, and diabetes. The EPA needs to act more quickly to evaluate and regulate these chemicals. In the NO selection, Jon Entine argues that the effects of endocrine disruptors are at most quite modest, and the effects of bisphenol A are inconsequential. What the EPA and FDA need to do is to stand by the science and resist political pressures, else "we place the entire system of checks and balances in danger."

YES ↵

Ted Schettler

Testimony before Senate Committee on Environment and Public Works Hearing on "EPA's Efforts to Protect Children's Health"

... The Vulnerability of Developing Children

From an extremely large body of scientific work we know that, compared to adults, developing children are uniquely susceptible to hazardous environmental exposures. Windows of vulnerability during *in utero* development, infancy, and childhood increase risks of some adverse health outcomes resulting from exposures, often with lifelong consequences. Among the better known examples, lead exposures that have minimal or no discernable impacts in adults can permanently alter brain development and function in a child. Similarly, fetal alcohol exposures can have lifelong impacts in children, while the same exposure in adults has only mild, transient effects.

Many of the reasons for this vulnerability are well understood and others are being worked out at the molecular, cellular, and tissue levels. During fetal, infant, and child development, cells rapidly divide, tissues and organs are formed, and signaling mechanisms, hormone levels, feedback loops, and their set points are established. Exposures to hazardous chemicals as well as other environmental influences may perturb these events through various mechanisms with long-term consequences.

It is also important to recognize the substantial and growing evidence showing that environmental exposures during development can increase the risk of chronic, degenerative diseases much later in life. For example, lifelong cumulative exposures to lead, including developmental exposures, increase the risk of cognitive decline and Parkinson's disease in people decades later. Animal studies show that early life exposure to certain pesticides seem to prime the brain, making it more susceptible to further exposures in adulthood, resulting in neurodegeneration in areas responsible for Parkinson's disease. Indeed, epidemiologic studies show an increased risk of Parkinson's disease in agricultural communities where pesticides are heavily used. Thus, while protecting children, we are also lowering the risk of various diseases and disabilities much later in life.

Endocrine Disruptors

One area of concern that I would like to highlight is the potential for some pesticides, metals, and various other industrial chemicals to disrupt the function of hormones and other chemical messengers that are vital to normal human development and function. These chemicals are known as endocrine disruptors.

An endocrine disruptor is "an exogenous agent or mixture of agents that interferes with or alters the synthesis, secretion, transport, metabolism, binding action, or elimination of hormones that are present in the body and are responsible for homeostasis, growth, neurological signaling, reproduction and developmental processes." Endocrine disruptors interfere with the body's key signaling pathways and can cause harm, especially during fetal and early life development.

Endocrine disruptors gained increased public and scientific attention during the 1990s, although the capacity for certain industrial chemicals to mimic or otherwise interfere with hormone function was known at least as long ago as the 1930s. For example, in 1938, scientists showed that bisphenol A, a chemical used to make many consumer products today, has estrogen-like properties, although its molecular structure is quite different from naturally-occurring estrogen. The use of this chemical is now so widespread that, according to the Centers for Disease Control and Prevention, 93% of all Americans have residues of bisphenol A in their urine. Recent studies link bisphenol A levels to altered brain development, heart disease, and diabetes.

In the 1950s, 1960s, and early 1970s the potent synthetic estrogen, diethylstilbestrol was purposely given to many pregnant women with the unfounded promise that it would help to prevent miscarriages and promote healthier pregnancies. Tragically, fetal exposure to DES resulted in abnormalities of reproductive tract development in females and males and a sharply increased risk of reproductive tract cancers in women decades later. Thus,

U.S. House of Representatives Committee on Environment and Public Works, March 17, 2010.

we learned through uncontrolled human experimentation that certain chemicals could profoundly alter development with consequences that were often not apparent at birth and might only become manifest decades later.

During the 1980s and 1990s exposures of wildlife to industrial chemicals and their health effects were increasingly reported in the scientific literature. Reproduction and development of birds, amphibians, reptiles, and mammals have been affected by exposure to endocrine disrupting chemicals. Fish in numerous rivers, including the Potomac, have disrupted sexual development—specifically feminized male fish. When this finding was first noted in England in the 1990's, it was considered unusual. It is now recognized as a widespread, pervasive phenomenon.

Based on findings in wildlife and laboratory animal studies, many scientists are concerned that in humans, the increasing incidence of cancer of the testis, prostate, and breast, birth defects of the male reproductive tract, lower sperm counts, behavioral disorders, diabetes, and a wide range of other abnormalities may result, at least in part, from exposures to endocrine disrupting chemicals.

A recent report shedding new light on a puzzling observation that has baffled scientists for years is illustrative. Many studies find a higher incidence of testicular cancer and male reproductive tract abnormalities in Danish men than in nearby Finland. Finnish boys have larger testes and higher sperm counts than Danish boys. Reasons for these differences have been unclear. Recently, scientists analyzed the breast milk of 68 women from the two countries for 121 different chemicals and found significantly higher levels in the milk of the Danish women. The chemicals tested for included flame retardants, pesticides, phthalates, polychlorinated biphenyls, dioxins, and furans. These chemicals are commonly identified in biomonitoring studies around the world, including in the US. Their concentration in breast milk is a good indicator of fetal exposures during pregnancy. Clearly, this kind of study cannot definitively establish a causal relationship between the different levels of these industrial chemicals in mothers in Denmark and Finland and the patterns of male reproductive tract abnormalities in the two countries. But a causal relationship is entirely plausible, based on what we know about the effects of many of these chemicals in laboratory animal studies. Current environmental exposures also include hundreds if not thousands of chemicals that were not tested for in this study that may also be part of the problem.

Because of growing concern about endocrine disrupting chemicals, in 1996 the EPA created the Endocrine Disruptor Screening and Testing Advisory Committee (EDSTAC) in response to a Congressional mandate in the Food Quality Protection Act and authorization in the Safe Drinking Water Act Amendments of 1996.

These laws specified that EPA:

"... develop a screening program, using appropriate validated test systems and other scientifically relevant information, to determine whether certain substances may have an effect in humans that is similar to an effect produced by a naturally occurring estrogen, or other such endocrine effect as the Administrator may designate."

The laws required EPA to develop a screening program by August 1998, to implement the program by August 1999, and to report on the program's progress by August 2000. Unfortunately, EPA is now about a decade behind.

I served on the EDSTAC. The committee included representatives from industry, government, environmental and public health groups, and academia. We were charged with developing consensus-based recommendations for a screening program that would provide EPA the necessary information to make regulatory decisions about endocrine effects of chemicals.

The committee delivered a final report by the statutory deadline of August 1998. It included a groundbreaking priority setting, screening and testing approach that encompasses the universe of chemicals in use today, evaluates a range of human health and ecological effects, and recommends a feasible, health-protective, approach. The committee:

- recognized that problems with endocrine disruption go beyond estrogen, and also called for screening of chemicals for interference with male androgens and thyroid hormone.
- recommended the use of new technologies to rapidly pre-screen numerous chemicals to see if they interact with hormone receptors *in vitro* (in the "test-tube"). The committee recommended that this technology be used to rapidly evaluate the ten thousand most widely-used chemicals within one year.
- recommended a computer-based tracking system allowing information about health effects and exposure to be collected in one place to facilitate prioritization. That database didn't exist then, and it doesn't exist today.
- urged EPA to accept nominations from the public of chemicals or *chemical mixtures* for expedited testing. This would allow workers, or impacted communities to press for more information about chemicals to which they are exposed.

Unfortunately, EPA missed deadline after deadline and became bogged down in an endless set of validation exercises that remain unfinished. Many of the recommendations were discarded. Finally, a decade late, EPA implemented an extremely scaled down version of the program when it issued the first test orders in October 2009. Only 67 chemicals are on the list for this first round of screening—mostly pesticides, including a number of chemicals that are already well-known endocrine disruptors. Meanwhile tens of thousands of chemicals in consumer products, food, water, and air have not been tested for endocrine disrupting properties.

In 2009 the Endocrine Society evaluated the science on endocrine disruptors and concluded:

> "The evidence for adverse reproductive outcomes (infertility, cancers, malformations) from exposure to endocrine disrupting chemicals is strong, and there is mounting evidence for effects on other endocrine systems, including thyroid, neuroendocrine, obesity and metabolism, and insulin and glucose homeostasis."

The Endocrine Society is the premier professional organization devoted to research on hormones and the clinical practice of endocrinology. It is comprised of over 14,000 research scientists and physicians from over 100 countries. This statement has since been endorsed by the American Medical Association, which is joining the Endocrine Society in calling for decreased public exposure to endocrine disrupting chemicals. The American Chemical Society just issued a similar statement with additional recommendations for: "More rapid advancement of the congressionally-mandated effort by the EPA, called the Endocrine Disruptor Screening Program (EDSP)."

As a result of EPA's failure to implement a strong endocrine disruptor screening program, the *Endocrine Disruption Prevention Act* was introduced in Congress in 2009. This act would authorize a new research program at the National Institute of Environmental Health Sciences (NIEHS) to identify endocrine disrupting chemicals, using the most current science. It would establish an independent panel of scientists to oversee research and develop a prioritized list of chemicals for investigation. If the panel determined that a chemical presented endocrine-disrupting concerns, it would compel the federal agencies with established regulatory authority to report to Congress and propose next steps within six months. NIEHS has the capacity to carry out such a research program if provided with appropriate resources. But EPA remains the regulatory authority responsible for protecting children from environmental threats.

I have focused here on endocrine disrupting chemicals, but my concerns about human exposures to industrial chemicals are not limited to those with endocrine disrupting properties. Well-known flaws in the Toxic Substances Control Act (TSCA) have allowed tens of thousands of untested industrial chemicals to stay on the market and new ones brought to market with limited or no toxicity information. These include chemicals to which workers and people in the general population, including pregnant women and children, are regularly exposed.

The EPA Office of the Inspector General's (OIG) report, released in February 2010, and previous GAO reports clearly describe these problems." Not only are basic safety data lacking, but whatever limited information is submitted to the agency is frequently accompanied by requests to protect it from public disclosure. The OIG report concludes that the agency's process is "predisposed to protect industry information rather than to provide public access to health and safety studies." Physicians and other health care professionals do not have access to the data they need in order to appropriately advise patients, and workers and communities remain ignorant of the potential hazards of the chemicals to which they may be exposed.

Meaningful TSCA reform is essential in order to protect developing children and people of all ages from the impacts of exposure to hazardous chemicals in consumer products, food, water, and air.

The Impacts of Industrial Chemicals, Including Pesticides, on Brain Development and Function

Another area of concern to bring to your attention is the failure of EPA to require adequate evaluation of the impacts of industrial chemicals, including pesticides, on brain development and function in children. Ample scientific evidence confirms the unique susceptibility of the developing brain to chemical exposures that can disrupt one or more of a number of biologic processes that must proceed in an orderly fashion as brain architecture and chemistry are established throughout pregnancy, infancy, and childhood.

Lead, mercury, polychlorinated biphenyls (PCBs), arsenic, ethyl alcohol, and toluene are recognized causes of neurodevelopmental disorders. A large body of experimental and human epidemiologic evidence shows diverse, long-lasting impacts of these substances on the ability of children to learn, remember, pay attention, and behave appropriately. The effects can occur after relatively low-level exposures that have no discernable effects in adults.

Policies that reduce exposures to these substances have been successful. For example, the removal of lead from gasoline resulted in a sharp decline in average blood levels in children throughout the US. Even so, the economic consequences of lower IQ resulting from lead levels in children in the US today are conservatively estimated to be in excess of $40 billion annually. That figure does not take into account costs to society incurred by responding to special educational needs and disruptive or criminal behavior.

Unfortunately, these well-studied substances are the exception. The large majority of industrial chemicals have never been evaluated for their potential impact on the developing brain of children. This is true even for those chemicals known to be toxic to the nervous system more generally. . . .

Recommendations

EPA should:

1. Move more quickly to implement the Endocrine Disruptor Screening Program for chemicals in consumer products, air, food, and water, using current, up-to-date scientific methods. Evaluation should include commonly encountered

mixtures as identified in environmental media (air, water, food) and biomonitoring studies. If NIEHS becomes the institution in which the endocrine disrupting properties of chemicals are evaluated, EPA must then promptly respond to the findings with health protective interventions. . . .

Congress should pass comprehensive chemical regulatory policy reform. Effective reform should:

- **Immediately Initiate Action on the Worst Chemicals:** Persistent, bioaccumulative toxicants (PBTs) are uniquely hazardous. Any such chemical to which people could be exposed should be phased out of commerce.
- **Require Basic Information for All Chemicals:** Manufacturers should be required to provide basic information on the health hazards associated with their chemicals, how they are used, and the ways that the public or workers could be exposed.
- **Protect the Most Vulnerable:** Chemicals should be assessed against a health standard that explicitly requires protection of the most vulnerable subpopulations. That population is likely to include children, but it could also be workers, pregnant women, or another vulnerable group.
- **Use the Best Science and Methods:** The National Academy of Sciences' recommendations for reframing risk assessment at the EPA should be adopted. Regulators should expand development and use of information gleaned from "biomonitoring" for setting priorities.
- **Hold Industry Responsible for Demonstrating Chemical Safety:** Chemical manufacturers should be responsible for evaluating and demonstrating the safety of their products.
- **Ensure Environmental Justice:** Effective reform should contribute substantially to reducing the disproportionate burden of toxic chemical exposure placed on people of color, low-income people, and indigenous communities.
- **Enhance Government Coordination:** The EPA should work effectively with other agencies such as the Food and Drug Administration that have jurisdiction over some chemical exposures. The ability of the states to enact stricter chemical policies should be maintained and state/federal cooperation on chemical safety encouraged.
- **Promote Safer Alternatives:** There should be national support for basic and applied research into green chemistry and engineering, and policies should favor chemicals and products that are benign over those that are hazardous.
- **Ensure the Right to Know:** The public, workers, and the marketplace should have full access to chemical safety data and information about the way in which government safety decisions are made.

Congress should also adopt legislation establishing the Endocrine Disruption Prevention Program so that 1) environmental chemicals can be screened for endocrine disrupting properties using the most current science in a timely manner and 2) regulatory agencies are obligated to take action to protect public health based on the best available science.

TED SCHETTLER is Science Director of the Science and Environmental Health Network. He has served on the U.S. Environmental Protection Agency's Endocrine Disruptor Screening and Testing Advisory Committee (1996–1998) and the Endocrine Disruptor Methods Validation Subcommittee (2001–2003). He has also served on the National Academy of Sciences' committee on defining concerns associated with products of animal biotechnology.

Jon Entine   ➡ **NO**

The Troubling Case of Bisphenol A: At What Point Should Science Prevail?

On January 15, 2010, the U.S. Food and Drug Administration (FDA) issued a long-awaited update of its policy regarding bisphenol A (BPA)—an industrial chemical used to add strength and flexibility to many plastic products—finding it safe as currently used. The FDA review was undertaken after intense campaigning by advocacy groups and the media to ban or severely restrict BPA use, which continues even in the wake of the FDA decision. The campaigners' focus has now expanded to include other regulatory bodies, as well as states and localities. If they are successful, they will jeopardize the system for making regulatory decisions based on sound science.

BPA is one of the most ubiquitous chemicals in the world. Plastics made from it have been in use since the early 1950s. Approximately 6 billion pounds are produced globally each year by fifteen different corporations. When used as a building block in plastic, BPA makes it stronger—hard enough to replace steel and transparent enough to substitute for glass. BPA can withstand high heat and has high electrical resistance. It is found in electronics, DVDs, car dashboards, water bottles, eyeglass lenses, and microwavable plastic containers and is a key ingredient in epoxy resins used to make dental applications. At present, viable alternatives for many of its uses—such as in the plastic coating of metal can liners, where it does not affect taste, helps prevent bacterial contamination, and extends shelf life at a relatively low cost—do not exist.[1]

BPA is also one of the world's most studied chemicals. In 1982, the National Cancer Institute and the National Toxicology Program (NTP) cleared it as a potential carcinogen,[2] and a massive review by the Environmental Protection Agency (EPA) endorsed its safety in 1988.[3] Twenty years later, the FDA reviewed the studies to date and declared it safe at estimated levels of human exposure.[4] Over the past forty years, BPA has undergone more than 4,500 evaluations and counting. It has been declared safe based on peer-reviewed scientific evidence by every major government agency in every major industrial country in the world.

Last summer, after members of the Obama administration's incoming, politically appointed executive team at the FDA assumed their new positions, the agency announced it was reviewing the 2008 decision that had endorsed the safety of BPA—a decision heavily criticized by activists as an example of the Bush administration's alleged anti-science bias.

BPA has been under constant attack by some environmental advocacy groups and journalists campaigning to ban the chemical outright or restrict its use, particularly in products handled by infants and children. The point organization for much of this criticism is the Washington-based Environmental Working Group (EWG), which has been actively campaigning for a ban since 2007. EWG is most noted for its work in demonizing phthalates, which are relatively innocuous plastic softeners. EWG has not done any original research and does not have any scientists with targeted expertise in chemical additives to plastics. It regularly seeds the Internet with sensational and often-misleading interpretations of serious studies. For example, in November 2009, as the environmental community anxiously awaited the FDA's decision regarding BPA, EWG posted a report on the Huffington Post with the headline, "BPA Wrecks Sex, Fouls Food—and Probably Worse."[5]

The public campaign conducted by EWG and other activist organizations has led to thousands of news reports by mainstream news organizations. The *Milwaukee Sentinel* alone has published no fewer than fifty stories—for which it has won a bushel of journalism awards—excoriating the government for not restricting the use of BPA. It consistently frames the issue using what can only be characterized as scare tactics. In its "Watchdog Report," the *Sentinel* warned that BPA could cause "cancers of the breast, brain and testicles; lowered sperm counts, early puberty and other reproductive system defects; diabetes; attention deficit disorder, asthma and autism"[6]—none of which has been authenticated by scientific studies.

Over the past year, a feedback loop has developed among news organizations, environmental groups, and consumer organizations promoting the view that BPA is unsafe. In its December 2009 issue, *Consumer Reports* repeated unfounded allegations that "BPA has been linked to a wide array of health effects including reproductive abnormalities, heightened risk of breast and prostate cancers, diabetes, and heart disease" in humans. It urged the FDA to revise its "inadequate and out of date" standards.[7] The report inspired panic-inducing stories at ABC News, the *Los Angeles Times*, Fox News, and the *New York Times*. In a separate commentary, *New York Times* columnist Nicholas Kristoff compared the danger of BPA to dangers he has faced as a reporter, such as "threats from warlords, bandits and tarantulas."[8] News accounts rarely noted that

Entine, Jon. From *The American Enterprise*, March 2010, pp. 1–7. Copyright © 2010 by American Enterprise Institute. Reprinted by permission.

no regulatory body in the world had reached conclusions even remotely similar to the scare headlines.

Considering the change in ideological complexion at the FDA, ban proponents were taken aback on January 15, 2010, when the agency upheld its prior finding that BPA is safe. It declared that BPA posed "negligible" or "minimal" concern for most adults and "is not proven to harm children or adults." Although activist groups have charged that the FDA's evaluation protocol is inadequate or outdated, according to the report, "Studies employing standardized toxicity tests used globally for regulatory decision making thus far have supported the safety of current low levels of human exposure to BPA."[9] When asked directly if adults or children faced any real health dangers, Joshua Sharfstein, M.D., the FDA's principal deputy commissioner, minced no words: "If we thought it was unsafe, we would be taking strong regulatory action."[10]

Although the FDA report repeatedly stated that the FDA study had not found any scientific basis to justify restrictions on products containing BPA—even those regularly handled by infants or pregnant women—the report still voiced "some concern," adding, "we need to know more." The FDA said it would continue to fund studies, and it expects results within twenty-four months. (Last fall, the National Institute of Environmental Health Sciences awarded $30 million in grants, $14 million of which was funded through the stimulus package, to study BPA.)[11] While reaffirming there were no dangers, the FDA report recommended ways to limit exposure to BPA.

If the FDA had taken stronger action, it would have come as a shock to regulators worldwide. In the past few years, BPA has undergone comprehensive reviews by ten other regulatory bodies, including ones in Asia, Australia, New Zealand, and Canada.[12] In what is considered the most comprehensive and definitive review ever, a 2006 European Union (EU) review certified that BPA is safe.[13] This is particularly striking because the EU evaluates chemicals using the "precautionary principle," a controversial notion that environmental policy should be based on the suspected risk of causing harm rather than on evidence that a substance actually causes harm. Even under such broad standards, the EU has consistently found that BPA is safe for use in products handled by adults and infants. (For a list of regulatory BPA studies, surveys, and reviews, see www.aei.org/outlook/100946.)

What is the back story here? How does it happen that a substance consistently deemed safe by reviewing bodies and scientific studies remains in the cross hairs of campaigning journalists and environmentalists? What does this controversy suggest about how scientific decisions are made in a highly charged political environment?

The Low-Dose Hypothesis

Researchers generally agree BPA is neither mutagenic nor likely to be a carcinogen.[14] They disagree, however, about whether the chemical presents danger to children or infants. The controversy over BPA rests with the way science evaluates risk in chemicals. Activists cite a small but growing body of laboratory research using a novel theory that suggests that low doses of chemical exposure might have more impact than high doses. These studies—many of which have not been replicated—suggest that BPA may cause metabolic disorders in rodents.[15] On the basis of these concerns, and invoking the precautionary principle, some scientists have urged that the use of BPA be curtailed.

The notion that lower levels of exposure may cause problems while higher doses do not is referred to as the new paradigm. It grew out of studies pioneered most famously by the zoologist Theo Colborn. While examining fish in the Great Lakes during the 1980s, Colborn found evidence that prenatal exposure to some chemicals impacted development in newborn fish. The findings elicited concern and warranted serious follow up to evaluate the potential effect of these chemicals on humans. She convened a conference of like-minded scientists and activists in Wingspread, Wisconsin, in 1991, which culminated in the Wingspread Consensus Statement. As a group, these scientists and activists believed many chemicals—not just BPA—might not harm people at high doses but may be harmful at extremely low doses undetected by traditional tests. They promoted this thesis in the sensational bestselling book *Our Stolen Future*. This is the beginning of what became known as the low-dose hypothesis.[16]

Activists also zeroed in on evidence that particles of BPA leached from the plastic and showed a laboratory response on estrogen-responsive cancer cells.[17] Many natural substances also subtly alter the way the hormones in our endocrine system work, including clover, many fruits, and soy. The Wingspread activists coined the term "endocrine disruptor" to describe this modest effect, rebranding what until that point had been known, more innocuously, as an endocrine mediator. The phrase was clearly designed to be a slogan, not unlike the Right calling the estate tax a "death tax" or campaigners on abortion issues branding themselves "prochoice" or "prolife." At first, journalists and other scientists carefully put the term in quotation marks to signify that this was a highly charged political term, but over time the quotation marks disappeared.

Certainly, BPA can affect the endocrine system, as can many substances, including foods such as tofu. Reproductive and neurodevelopmental effects of BPA at low doses in animals, including environmental doses potentially relevant to humans, have been the subject of ongoing scientific reviews and study. In its January announcement, the FDA said that this "novel" idea does "describe BPA effects" in laboratory animals, but most scientists and regulators remain doubtful that low doses of certain chemicals, natural or synthetic, significantly disrupt the human metabolic and reproductive systems. Only a tiny fraction of studies on BPA indicate serious toxic or hormonal effects on rodents, and then only when consumed at levels at least five hundred thousand times greater than humans consume.[18] Some studies have shown estrogenic effects and reproductive impacts on male rodents, while others have not. Doubts also have been raised about the meaningfulness of these studies because of what scientists call nonrepeatability.

Also in question is to what degree the results of studies conducted on rodents can be applied to humans. Animal tests are considered an important step in evaluating chemical toxicity, but by themselves they are of "limited usefulness to human health," as noted in an article in the *British Medical Journal*.[19] An axiom of laboratory research is that chemicals tested on animals rarely have identical effects on humans at comparable dosages, and sometimes have no discernible effect because of inherent flaws in the studies and significant differences between the species in biochemistry, immune systems, and other anatomical systems.

More problematic, the novel low-dose endocrine-disruptor hypothesis is based almost entirely on administering BPA to rats by injection. Regulatory agencies do not put much stock in tests in which a substance is introduced to subjects in a different way than humans would come into contact with it. The European Food Safety Authority, which uses the precautionary principle in its deliberations, as well as the FDA, the NTP, Health Canada, and, in fact, every regulatory body that has systematically assessed the risks of BPA either rejects studies of injected BPA outright or gives preference to studies in which BPA is ingested instead.

Because humans are exposed to BPA almost exclusively through food,[20] why would scientists seeking to evaluate whether it is dangerous design tests in which it is injected rather than ingested? For one, it is much easier to do those tests. But there is another, more problematic, explanation: orally ingested BPA quickly becomes harmless, while the toxic qualities of injected BPA are preserved, so studies testing the orally injected chemical are more likely to show effects.

Peer-reviewed studies consistently show that BPA taken orally is rapidly detoxified, first in the gastrointestinal tract and then in the liver, by enzymes that transform BPA into a water-soluble chemical known as BPA-glucuronide, which repeated studies have shown is harmless. This sugar substance has a half-life of six hours and is easily excreted in urine.[21] Even when used in dental sealants, BPA exits the system in fewer than twenty-four hours.[22] While some studies in which rodents were injected with BPA have shown some effects, studies in which rats receive the chemical orally have shown little effect, if any.[23] In other words, repeated tests over decades have shown that BPA that makes its way into human bodies is quickly neutralized and made innocuous.

Perhaps the most common, and seemingly damning, allegation against BPA, and one that shows up repeatedly in media reports, is that the Centers for Disease Control and Prevention (CDC) has found the chemical in the urine of 95 percent of adults and 93 percent of children over six years old.[24] That sounds frightening. But is it?

How journalists and social scientists frame this often-stated fact is a good barometer of whether they are genuinely wrestling with complex science or whether they are mouthpieces, intentionally or not, for a predetermined, chemophobic perspective.

Consider, for example, a news report on the FDA's January announcement in the *New York Times*. Denise Grady dispassionately writes that a "study of more than 2,000 people found that more than 90 percent of them had BPA in their urine."[25] But such findings are little more than artifacts of state-of-the-art biomonitoring techniques that find potentially toxic chemicals of all kinds—natural and synthetic—omnipresent in the human body. Because available technology can detect even vanishingly low concentrations, scientists are likely to find virtually anything for which they look. To put this in perspective, CDC tests have also found dietary estrogens (also called phytoestrogens)—known hormone disruptors that occur naturally in a vast array of products such as nuts, seeds, soy, tofu, wheat, berries, bourbon, and beer—in the urine of more than 90 percent of people and at levels one hundred times higher than traces of BPA. Repeated studies have shown that neither BPA nor dietary estrogens bioaccumulate. "In animal and human studies, bisphenol A is well absorbed orally," the CDC reports. "Finding a measurable amount of bisphenol A in the urine does not mean that the levels of bisphenol A cause an adverse health effect."[26]

Of course, none of these qualifications appeared in news reports, not even in the *New York Times*, which has the resources to know better. Instead, they recirculated talking points from environmental activists. Without this contextual information, an innocent reader could easily be forgiven for concluding that the FDA overlooked damning facts in deciding not to ban BPA.

Scientific Split

The question of whether BPA is harmful revolves around the murky issue of toxicity. As Paracelsus, the father of toxicology, observed, "All things are poison and nothing is without poison, only the dose permits something not to be poisonous." The world is awash in chemicals, natural and human-made. Approximately one hundred thousand synthetic chemicals are approved for consumer products and industrial processes, and very few are considered harmful at levels found in humans. Many natural substances in our foods are toxic, including some essential and beneficial vitamins and minerals, when consumed in large quantities. The only significant science-based question is whether a particular substance is harmful at the trace level at which it is metabolized in the human body.

The debate over BPA has been riddled with distortions over dose. Advocacy groups quickly jumped on a study from China that suggested that BPA could cause male sexual dysfunction, as the EWG, the *Los Angeles Times,* and other organizations warned in outrageous headlines. But the study that prompted this news blizzard focused on Chinese workers who handled the chemical in bulk, not on normal levels of exposure to BPA in foods.[27] The NTP has reported "negligible concern" that men exposed at nonoccupational capacities—in other words, men who are exposed to BPA from using plastic containers or consuming canned foods—would experience reproductive effects.[28]

The scientific community appears to be divided into two conflicting camps when it comes to assessing BPA's risks. Regulatory authorities and scientists who rely on long-established study protocols are on one side, and they have concluded, almost unanimously, that BPA presents no serious harm. They represent the majority by far, but their views are often downplayed or even ridiculed by advocacy groups and a small, but media-savvy, faction of scientists who embrace precautionary notions.

These sharp differences appear every time a new study comes out finding BPA is safe. In 2001, the NTP released an independent peer-reviewed study of the evidence for and against the novel hypothesis. In its conclusion, the report says, "The Subpanel is not persuaded that a low dose effect of BPA has been conclusively established as a general or reproducible finding," although it did recommend further review.[29] Numerous studies followed including by the Harvard Center for Risk Analysis.[30] All of them raised doubts about the validity of the low-dose hypothesis and the reproducibility of findings based on tests performed on animals injected with BPA. Nevertheless, after each of these studies, the authors were attacked. Frederick S. vom Saal, an endocrinologist at the University of Missouri who has emerged as the most vocal critic of BPA, argued that these reports all failed to take into account the "latest knowledge" in endocrinology, developmental biology, and estrogen-receptor research.[31] . . .

To respond to the growing consensus of BPA's comparative safety, in 2006, vom Saal coordinated a conference that brought together thirty-eight scientists who advocated the low-dose endocrine-disruptor theory. Considering the lack of dissenting viewpoints, their summary conclusion, known as the Chapel Hill Consensus Statement, was hardly surprising. It found BPA associated with "organizational changes in the prostate, breast, testis, mammary glands, body size, brain structure and chemistry, and behavior of laboratory animals."[32] Using inflammatory language uncharacteristic of science, vom Saal summed up their conclusion: "The science is clear and the findings are not just scary, they are horrific. When you feed a baby out of a clear, hard plastic bottle, it's like giving the baby a birth control pill."[33]

The consensus viewpoint has been systematically rejected by the EPA and government regulators around the world. In 2008, the NTP released yet another extensive and expensive peer-reviewed analysis of BPA that again found no reason for serious concern about its effects on human reproduction or development in adults or children.[34] The NTP used its well-known code term "some concern" to characterize the possible effects of BPA

on fetuses, which means it recommended further study but did not consider the chemical harmful or worthy of restrictions or health warnings. It made it clear that it reached that qualified conclusion because the rodent studies, which suggested some problems, were not "experimentally consistent"—many showed no problems, and tests could not be replicated. Yet news organizations consistently have misrepresented the NTP's 2008 findings. In the most recent gross example, *Fast Company* began a wildly sensational exposé of BPA by baldly and inaccurately claiming the NTP had concluded: "BPA is dangerous to human health."[35]

In an attempt to address this persistent controversy, the EPA recently funded two multigenerational studies. Both studies failed to support the low-dose hypothesis. The most recent analysis, which appeared in November in *Toxicological Sciences*, a leading scientific journal, was particularly definitive. Carried out at the EPA's National Health and Environmental Effects Research Laboratory based at North Carolina State University in Raleigh, it was specifically designed to cover a wide range of BPA doses. L. Earl Gray Jr. and his colleagues concluded that BPA is an extremely weak estrogen not worthy of being called an "endocrine disruptor." BPA was found to be so weak that even at levels of exposure four thousand times higher than the maximum exposure of humans in the general population, there were no discernible effects.[36]

This comprehensive EPA study—which should have ended any reasonable debate over the low-dose, endocrine-disruptor, precautionary principle–fed hypothesis—immediately set off a new round of contretemps. *Consumer Reports*, other nonscientific media such as *Fast Company*, and activist blogs accused researchers of using a strain of rat that is extremely insensitive to estrogen. This is a technical catfight over study design, but it is worth addressing, as it highlights the beleaguered position of ban proponents trying to explain why other scientists have not been able to consistently replicate the few studies showing that BPA has no or only modest estrogenic effects. The "you used insensitive rats" argument originated with vom Saal, who has advanced it repeatedly,[37] although the EPA's senior reproductive toxicologist, who leads the agency's team that investigates endocrine disruption, has dismissed it as absurd.[38]

Vom Saal and twenty-three of his colleagues responded to the L. Earl Gray Jr. article with a blistering letter rehashing the same discredited argument.[39] Gray and his team returned the volley, dismantling the critique. They address in particular why the appropriate rat strains were used, citing similar studies by Health Canada and the NTP, among numerous organizations. Its key statement: while vom Saal dismisses studies by Gray and others as "flawed," governmental regulatory agencies have consistently relied on them while dismissing as "inadequate," "not replicable," and "extremely limited" the hodgepodge of low-dose studies. Their conclusion: "BPA did not display any estrogenicity."[40]

Going Forward

Unable to prevail on the science, ban proponents have begun shifting their focus from the laboratory to politics. In December, Charles Schumer and Kirsten Gillibrand, both Democratic senators from New York, proposed the BPA-Free Kids Act, intended to outlaw the use of BPA in food container linings for infants and toddlers.[41] Activists are also targeting state and local governments, employing the same tactics that failed to persuade the FDA but that have worked wonders in manipulating media coverage and fanning public anxiety. Minnesota, Connecticut, Washington, Wisconsin, and three counties in New York have passed bans on products or beverage containers for children, basing it on the precautionary principle and citing fears of harm, and more jurisdictions are considering restrictions.

That fear strategy prevailed in Canada in 2008. Activists mounted a massive campaign designed to frighten parents and pressure the media. The effort was chronicled gleefully in the book *Slow Death by Rubber Duck*.[42] Public concerns led to an intense investigation by Health Canada. When Mark Richardson, the chief scientist and head of the investigation, said the evidence showed that the dangers of BPA were "so low as to be totally inconsequential" and compared its estrogenic effects to tofu, activists and the media, led by the *Globe and Mail* of Toronto, mounted an attack on his credibility that led to his reassignment.[43] Months later, when the report was finally issued, Health Canada firmly rejected claims that BPA was unsafe. "The current research tells us the general public need not be concerned," the report declared after reviewing hundreds of studies. "Bisphenol A does not pose a risk to the general population, including adults, teenagers and children."[44]

Nonetheless, the precautionary principle is embodied in Canadian (and EU, but not yet in U.S.) law. Considering the hysteria generated, and even absent convincing scientific evidence, Health Canada was compelled to ban BPA in baby products. Citing public anxiety, Health Canada banned BPA for use in products for infants and children in April 2008. "Even though scientific information may be inconclusive," it wrote, "decisions have to be made to meet society's expectations that risks be addressed and living standards maintained." It was the first (and still only) restriction on BPA at a national level. Activists now regularly cite the Canadian ban, arrived at through fear rather than based on scientific evidence, as "proof" that regulatory bodies are now finding BPA harmful. . . .

Considering the immense efforts to demonize BPA, regulators are under intense political pressure to fund studies and tests, perhaps indefinitely. In a world of finite research dollars, policymakers must ask whether spending hundreds of millions of dollars on yet more research into BPA imposes an unnecessarily high cost on science and society. Richard Sharpe, head of the Centre for Reproductive Biology at the Medical Research Institute in Edinburgh and an internationally recognized pioneer on the effects on endocrine disruption by chemicals in the environment, suggests in a recent article that governments are squandering research dollars on what has evolved from a once-legitimate scientific inquiry into an obsession. Writing in *Toxicological Sciences*, before the FDA's recent reaffirmation of BPA's safety, Sharpe writes:

> Fundamental, repetitive work on bisphenol A has sucked in tens, probably hundreds of millions of dollars from government bodies and industry which, at a time when research money is thin on the ground, looks increasingly like an investment with a nil return. . . . If this opinion piece does nothing else, I hope that it will remind us all of the central importance to be attached to the repeatability of experiments and how we should react when a study proves to be unrepeatable.[45]

Scientific "facts" are constantly being reevaluated. That is the essence of the research ethic that depends on subjecting empirical data to challenge and review. An explanation accepted as truth is discarded when more robust data suggest a more empirically solid alternative. Such creative destruction embodied in the process of scientific hypothesizing and empirical studies provides a useful framework for understanding what the science community should do about BPA.

Now that the current FDA has weighed in on this issue and yet another international oversight agency has reaffirmed the chemical's safety, we might legitimately echo Sharpe's statement and ask when enough is enough. To navigate the complexities of modern society, policymakers need standards and established systems—objective science—to guide them in weighing the benefits and potential hazards of chemicals, drugs, and any other concerns. But the moment we abandon standards for fashion or due to political pressure—no matter how superficially attractive that may seem—we place the entire system of checks and balances in danger.

References

1. Lyndsey Layton, "Alternatives to BPA Containers Not Easy for U.S. Foodmakers to Find," *Washington Post*, February 23, 2010.
2. National Toxicology Program (NTP), *NTP Technical Report on the Carcinogenesis Bioassay of Bisphenol A (CAS No. 80-0507) in F344 Rats and B6C3F1 Mice (Feed Study)*, no. 82–1771 (Bethesda, MD: Public Health Service, National Institutes of Health [NIH], 1982).
3. Integrated Risk Information System, "Bisphenol A. (CASRN 80-05-7)" (Washington, DC: Environmental Protection Agency [EPA], 1988), available at www.epa.gov/ncea/iris/subst/0356.htm (accessed March 17, 2010).
4. U.S. Food and Drug Administration (FDA), *Draft Assessment of Bisphenol A for Use in Food Contact Applications*, 110th Cong., 2d sess. (Washington, DC, August 2008), available at www.fda.gov

/ohrms/dockets/AC/08/briefing/2008-0038b1_01_02 _FDA BPA Draft Assessment.pdf (accessed March 17, 2010).

5. Elaine Shannon, "BPA Wrecks Sex, Fouls Food—and Probably Worse," Huffington Post, November 12, 2009, available at www .huffingtonpost.com/elaine-shannon/bpa-wrecks-sex -fouls-food_b_354268.html (accessed March 16, 2010).

6. "Chemical Fallout," *Watchdog Report,* JS Online, 2010, available at www.jsonline.com/watchdog /34405049.html (accessed March 19, 2010).

7. "Concern over Canned Foods: Our Tests Find Wide Range of Bisphenol A in Soups, Juice, and More," *Consumer Reports,* December 2009, available at www.consumerreports.org/cro/magazine-archive /december-2009/food/bpa/overview/bisphenol-a-ov .htm (accessed March 19, 2010).

8. Nicholas D. Kristoff, "Chemicals in Our Food, and Bodies," *New York Times,* November 8, 2009.

9. FDA, *Update on Bisphenol A (BPA) for Use in Food Contact Applications,* 111th Cong., 2d sess. (Washington, DC, January 2010), 1, available at www .fda.gov/downloads/NewsEvents/PublicHealthFocus /UCM197778.pdf (accessed March 19, 2010).

10. Medline Plus of the U.S. National Library of Medicine and NIH, "Health Agencies Express Concern over BPA," HealthDay, January 15, 2010, available at www.nlm.nih.gov/medlineplus/news /fullstory_94170.html (accessed March 19, 2010).

11. National Institute of Environmental Health Sciences (NIEHS)–NIH, "NIEHS Awards Recovery Act Funds to Address Bisphenol A Research Gaps," news release, October 28, 2009, available at www .niehs.nih.gov/news/releases/2009/bisphenol-research .cfm (accessed March 19, 2010).

12. Trevor Butterworth, "New Independent Study by EPA Refutes BPA Risk," Statistical Assessment Service, October 30, 2009, available at http://stats .org/stories/2009/breaking_news_bpa_oct30_09.html (accessed March 19, 2010).

13. European Food Safety Authority (EFSA), "Opinion of the Scientific Panel on Food Additives, Flavourings, Processing Aids and Materials in Contact with Food (AFC) Related to 2,2-Bis (4-Hydroxyphenyl)Propane," *EFSA Journal* 428 (2006): 1–75, available at www.efsa.europa.eu/en /scdocs/doc/s428.pdf (accessed March 19, 2010).

14. Lois A. Haighton et al., "An Evaluation of the Possible Carcino-genicity of Bisphenol A to Humans," *Regulatory Toxicology and Pharmacology* 35, no. 2 (April 2002): 238–54, available at http://dx.doi.org/ doi:10.1006/rtph.2001.1525 (accessed March 19, 2010).

15. Catherine A. Richter et al., "*In Vivo* Effects of Bisphenol A in Laboratory Rodent Studies," *Reproductive Toxicology* 24, no. 2 (August–September 2007): 199–224, available at http://dx.doi.org/doi:10.1016 /j.reprotox.2007.06.004 (accessed March 16, 2010).

16. Theo Colborn, Dianne Dumanoski, and John Peter Meyers, *Our Stolen Future: Are We Threatening Our Fertility, Intelligence and Survival? A Scientific Detective Story* (New York: Penguin Books, 1996).

17. A. V. Krishnan et al., "Bisphenol-A: An Estrogenic Substance Is Released from Polycarbonate Flasks during Autoclaving," *Endocrinology* 132 (1993): 2279–86.

18. Wolfgang Dekant and Wolfgang Völkel, "Human Exposure to Bisphenol A by Biomonitoring: Methods, Results and Assessment of Environmental Exposures," *Toxicology and Applied Pharmacology* 228, no. 1 (April 2008): 114–34.

19. "Animal Studies 'of Limited Use': Tests of Drugs on Animals Are Not Reliable in All Cases, a Study Warns," BBC News, December 15, 2006, available at http://news.bbc.co.uk/2/hi/health/6179687.stm (accessed March 21, 2010).

20. Nancy K. Wilson et al., "An Observational Study of the Potential Exposures of Preschool Children to Pentachlorophenol, Bisphenol-A, and Nonylphenol at Home and Daycare," *Environmental Research* 103, no.1 (2007): 9–20.

21. For one of many studies addressing this point, see Wolfgang Völkel et al., "Metabolism and Kinetics of Bisphenol A in Humans at Low Doses Following Oral Administration," *Chemical Research in Toxicology* 15, no. 10 (2002): 1281–87.

22. Renée Joskow et al., "Exposure to Bisphenol A from Bis-glycidyl Dimethacrylate–based Dental Sealants," *Journal of the American Dental Association* 137, no. 3 (2006): 353–62.

23. For one of many studies, see Kembra L. Howdeshell et al., "Gestational and Lactational Exposure to Ethinyl Estradiol, but Not Bisphenol A, Decreases Androgen-Dependent Reproductive Organ Weights and Epididymal Sperm Abundance in the Male Long Evans Hooded Rat," *Toxicological Sciences* 102, no. 2 (2008): 371–82.

24. Antonia M. Calafat et al., "Exposure of the U.S. Population to Bisphenol A and 4-Tertiary-Octylphenol: 2003–2004," *Environmental Health Perspectives* 116 (2007): 39–44.

25. Denise Grady, "U.S. Concerned about the Risks from a Plastic," *New York Times,* January 16, 2010.

26. Centers for Disease Control and Prevention, "Bisphenol A: Cas no. 80-05-7," *National Report on Human Exposure to Environmental Chemicals,* February 11, 2010, available at www.cdc.gov/exposurereport/data _tables/BisphenolA_ChemicalInformation.html (accessed March 21, 2010).

27. Kaiser Permanente Division of Research, "Workplace BPA Exposure Increases Risk of Male Sexual Dysfunction: First Human Study to Measure Effects of BPA on Male Reproductive System," news release, November 11, 2009, available at www.eurekalert.org/pub_releases/2009-11/kp -wbe110309.php (accessed March 21, 2010).

28. NTP and U.S. Department of Health and Human Services (HHS) Center for the Evaluation of Risks to Human Reproduction (CERHR), *NTP-CERHR Monograph on the Potential Human Reproductive and Developmental Effects of Bisphenol A,* 110th Cong. 2d sess. (Washington, DC, September 2008), available at http://cerhr.niehs.nih.gov/chemicals/bisphenol /bisphenol.pdf (accessed March 21, 2010).

29. NTP, HHS, and the NIEHS, *National Toxicology Program's Report of the Endocrine Disruptors Low-Dose Peer Review* (Research Triangle Park, NC: NTP, August 2001), available at http://ntp.niehs.nih.gov /ntp/htdocs/liason/LowDosePeerFinalRpt.pdf (accessed March 21, 2010).

30. George M. Gray et al., "Weight of the Evidence Evaluation of Low-Dose Reproductive and Developmental Effects of Bisphenol A," *Human and Ecological Risk Assessment* 10, no. 5 (2004): 875–921.

31. Frederick S. vom Saal and Claude Hughes, "An Extensive New Literature Concerning Low-Dose Effects of Bisphenol A Shows the Need for a New Risk Assessment," *Environmental Health Perspectives* 113, no. 8 (2005): 926–33.

32. Frederick S. vom Saal et al., "Chapel Hill Bisphenol A Expert Panel Consensus Statement: Integration of Mechanisms, Effects in Animals and Potential to Impact Human Health at Current Levels of Exposure," *Reproductive Toxicology* 24, no. 2 (2007): 131–38.

33. University of Missouri College of Arts and Sciences, "Evidence Mounts against Chemical Used Widely in Everyday Plastic Products," April 5, 2005, available at http://rcp.missouri.edu/articles /vomsaal.html (accessed March 21, 2010).

34. NTP, HHS, and NIEHS, *National Toxicology Program's Report of the Endocrine Disruptors Low-Dose Peer Review.*

35. David Case, "The Real Story behind Bisphenol A," *Fast Company,* February 1, 2009, available at www .fastcompany.com/magazine/132/the-real-story-on -bpa.html (accessed March 21, 2010).

36. Bryce C. Ryan, Andrew K. Hotchkiss, Kevin M. Crofton, and L. Earl Gray Jr., "In Utero and Lactational Exposure to Bisphenol A, in Contrast to Ethinyl Estradiol, Does Not Alter Sexually Dimorphic Behavior, Puberty, Fertility, and Anatomy of Female LE Rats," *Toxicological Sciences* 114, no. 1 (2010): 133–48.

37. John Peterson Myers et al., "Why Public Health Agencies Cannot Depend on Good Laboratory Practices as a Criterion for Selecting Data: The Case of Bisphenol A," *Environmental Health Perspectives* 117, no. 3 (March 2009): 309–15, available at http://toxsci.oxfordjournals.org/cgi/external _ref?access_num=000263933600014&link_type=ISI (accessed March 21, 2010).

38. Trevor Butterworth, "Top US EPA Scientist Rejects Consumer Reports' BPA Claim," Statistical Assessment Service, November 10, 2009, available at www.stats.org/stories/2009/top_epa_scientist _rejects_consumer_reports_bpa_claim_nov10_09 .html (accessed March 21, 2010).

39. Frederick S. vom Saal, letter to the editor, *Toxicological Sciences,* February 17, 2010.

40. L. Earl Gray Jr., Bryce C. Ryan, Andrew K. Hotchkiss, and Kevin M. Crofton, letter to the editor, *Toxicological Sciences,* March 5, 2010.

41. *BPA-Free Kids Act of 2009,* S. 753, 111th Cong., 1st sess. (March 31, 2009), available at www .govtrack.us/congress/ bill.xpd?bill=s111-753 (accessed March 21, 2010).

42. Rick Smith, Bruce Lourie, and Sarah Dopp, *Slow Death by Rubber Duck: How the Toxic Chemistry of Everyday Life Affects Our Health* (Knopf Canada: Toronto, 2009).

43. Martin Mittelstaedt, "Scientist's Endorsement of Bisphenol A under Review," *Globe and Mail* (Toronto), June 20, 2007.

44. "Chemical Substances: Bisphenol A, Fact Sheet," Government of Canada, available at www .chemicalsubstanceschimiques.gc.ca/fact-fait /bisphenol-a-eng.php (accessed March 21, 2010).

45. Richard M. Sharpe, "Is It Time to End Concerns over the Estrogenic Effects of Bisphenol A?" *Toxicological Sciences* 114, no. 1 (2010): 1–4.

Jon Entine, a former Emmy-winning Producer for NBC News and ABC News, is a Senior Fellow, Center for Risk & Health Communication, George Mason University. His latest books include *No Crime But Prejudice: Fischer Homes, the Immigration Fiasco, and Extra-Judicial Prosecution* (TFG Books, May 2009), and *Abraham's Children: Race, Identity, and the DNA of the Chosen People* (Grand Central Publishing, 2007).

EXPLORING THE ISSUE

Is Bisphenol A a Potentially Serious Health Threat?

Critical Thinking and Reflection

1. The possible health effects of exposure to synthetic chemicals include cancer, reproductive malfunctions, and neurological problems. Which effects are most worthy of public concern?
2. Should we refrain from evaluating and regulating possibly toxic chemicals because the result may be more expensive food?
3. Most of the thousands of chemicals in use by industry have never been thoroughly tested for toxicity. Should they be?
4. Should the precautionary principle be applied to endocrine disruptors? In what ways?

Is There Common Ground?

Few people disagree that human health is worth protecting. They do, however, disagree on what constitutes acceptable evidence that human health is threatened. The endocrine disruptors debate began when signs were detected that synthetic chemicals were affecting wildlife. Effects on humans remain debatable.

1. Research the history of concerns over DDT (and other pesticides). Did wildlife effects play a role here?
2. What is the justification for animal testing of drugs (and other consumer products)? (Start here: www.fda.gov/Cosmetics/ProductandIngredientSafety/ProductTesting/ucm072268.htm)
3. In what ways is looking at effects of chemicals on wildlife similar to testing those effects on lab animals? In what ways are they different?

Create Central

www.mhhe.com/createcentral

Additional Resources

Theo Colborn, Dianne Dumanoski, and John P. Myers, *Our Stolen Future: Are We Threatening Our Fertility, Intelligence, and Survival?—A Scientific Detective Story* (Dutton, 1996)

Nancy Langston, *Toxic Bodies: Hormone Disruptors and the Legacy of DES* (Yale University Press, 2010)

Stephen H. Safe, "Environmental and Dietary Estrogens and Human Health: Is There a Problem?" *Environmental Health Perspectives* (April 1995)

Internet References . . .

e.hormone

e.hormone.tulane.edu

National Institute of Environmental Health Sciences

www.niehs.nih.gov/health/topics/agents/endocrine/

The Natural Resources Defense Council

www.nrdc.org/

The Science and Environmental Health Network

www.sehn.org/

The Silent Spring Institute

www.silentspring.org/

Selected, Edited, and with Issue Framing Material by:
Thomas Easton, *Thomas College*

ISSUE

Should Agricultural Animal Wastes Be Exempt from the Requirements of Superfund Legislation?

YES: Walter Bradley, from "Testimony before the House Committee on Energy and Commerce, Subcommittee on Environment and the Economy, regarding H.R. 2997, the Superfund Common Sense Act of 2011" (June 27, 2012)

NO: Ed Hopkins, from "Testimony before the House Committee on Energy and Commerce, Subcommittee on Environment and the Economy, regarding H.R. 2997, the Superfund Common Sense Act of 2011" (June 27, 2012)

Learning Outcomes

After reading this issue, you will be able to:

- Explain the need for the Comprehensive Environmental Response, Compensation, and Liability Act of 1980 (CERCLA), commonly called "Superfund."
- Explain how Superfund cleanups have been funded in the past.
- Explain the purpose of "polluter pays" taxes.
- Discuss whether Superfund regulations should be applied only to industrial emitters of toxic chemicals.

ISSUE SUMMARY

YES: Walter Bradley, of the Dairy Farmers of America, Inc., argues that Superfund legislation is and has been targeted at industrial chemicals, not animal manure. Despite agricultural exemptions in existing law, there is a need to make such exemptions more explicit and thereby protect the ability of dairy farmers to make a living.

NO: Ed Hopkins of the Sierra Club argues that large animal feeding operations (or factory farms) emit such large quantities of toxic chemicals such as hydrogen sulfide and ammonia, as well as pollutants that damage water supplies and ecosystems, that they should not be exempted from Superfund legislation.

The potentially disastrous consequences of ignoring hazardous wastes and disposing of them improperly burst upon the consciousness of the American public in the late 1970s. The problem was dramatized by the evacuation of residents of Niagara Falls, New York, whose health was being threatened by chemicals leaking from the abandoned Love Canal, which was used for many years as an industrial waste dump. Awakened to the dangers posed by chemical dumping, numerous communities bordering on industrial manufacturing areas across the country began to discover and report local sites where chemicals had been disposed of in open lagoons or were leaking from disintegrating steel drums. Such esoteric chemical names as dioxins and PCBs have become part of the common lexicon, and numerous local citizens' groups have been

mobilized to prevent human exposure to these and other toxins.

The expansion of the industrial use of synthetic chemicals following World War II resulted in the need to dispose of vast quantities of wastes laden with organic and inorganic chemical toxins. For the most part, industry adopted a casual attitude toward this problem and, in the absence of regulatory restraint, chose the least expensive means of disposal available. Little attention was paid to the ultimate fate of chemicals that could seep into surface water or groundwater. Scientists have estimated that less than 10 percent of the waste was disposed of in an environmentally sound manner.

The magnitude of the problem is mind-boggling: Over 275 million tons of hazardous waste is produced in the United States each year; as many as 10,000 dump sites may pose a serious threat to public health, according to the federal Office of Technology Assessment; other government estimates indicate that more than 350,000 waste sites may ultimately require corrective action at a cost that could easily exceed $500 billion.

Congressional response to the hazardous waste threat is embodied in two complex legislative initiatives. The Resource Conservation and Recovery Act (RCRA) of 1976 mandated action by the Environmental Protection Agency (EPA) to create "cradle to grave" oversight of newly generated waste. The Comprehensive Environmental Response, Compensation, and Liability Act of 1980 (CERCLA), commonly called "Superfund," gave the EPA broad authority to clean up existing hazardous waste sites. The implementation of this legislation has been severely criticized by environmental organizations, citizens' groups, and members of Congress who have accused the EPA of foot-dragging and a variety of politically motivated improprieties. Less than 20 percent of the original $1.6 billion Superfund allocation was actually spent on waste cleanup. In 2009, the U.S. Inspector General issued a report accusing the EPA of mismanagement; see "Inspector General's Report Faults EPA for Management of 819 Superfund Accounts," *BNA's Environmental Compliance Bulletin* (April 6, 2009) (for EPA commentary, visit www .epa.gov/oig/reports/2009/20090318-09-P-0119_glance.pdf).

Amendments designed to close RCRA loopholes were enacted in 1984, and the Superfund Amendments and Reauthorization Act (SARA) added $8.6 billion to a strengthened cleanup effort in 1986 and an additional $5.1 billion in 1990. At the same time, Congress passed the Emergency Planning and Community Right-to-Know Act (EPCRA) to enhance public awareness of potential problems and facilitate dealing with emergencies. Superfund cleanups, when those responsible for contaminated sites could not pay or could not be found (there are many such "orphan" sites), were to have been funded by taxes on industry (e.g., the Crude Oil Tax, the Chemical Feedstock Tax, the Toxic Chemicals Importation Tax, and the Corporate Environmental Income Tax, also known as "polluter pays" fees or taxes).

The Competitive Enterprise Institute objects that taxes such as the superfund taxes are an assault on consumer pocketbooks, as is CERCLA's "joint and several liability" clause, which can make minor contributors to toxics sites liable for large cleanup costs even when they acted according to all laws and regulations in force at the time. However, in May 2009, the U.S. Supreme Court ruled that some contributors may not be liable for cleanup costs; see "US Supreme Court Ruling Limits Superfund Liability," *Oil Spill Intelligence Report* (May 7, 2009). While acknowledging some improvement, both environmentalists and industrial policy analysts remain very critical about the way that both RCRA and Superfund/SARA are being implemented. See, e.g., Sheila Kaplan and Marilyn Berlin Snell, "Sapping the Superfund's Strength," *Nation* (May 3, 2010). Efforts to reauthorize and modify both of these hazardous

waste laws have been stalled in Congress since the early 1990s. But the work went on. The Superfund program continued to identify hazardous waste sites that warranted cleanup and to clean up sites; see www.epa.gov/superfund/ for the latest news. In 2004, EPA even declared that the infamous Love Canal site was finally safe.

The "polluter pays" taxes expired in 1995, and Congress has so far refused to reauthorize them, but the EPA has said they are necessary (Jay Landers, "EPA Calls on Congress to Reinstate Superfund Taxes," *Civil Engineering*, August 2010), and Representative Earl Blumenauer (D-OR) and Senator Frank R. Lautenberg (D-NJ) have proposed bills that would reinstate them. Support has come from the Obama administration, whose 2013 budget proposal (released in February 2012) included their reinstatement, the American Society of Civil Engineers, and others. See also Stephen Lester and Anne Rabe, *Superfund: In the Eye of the Storm* (Center for Health, Environment & Justice, June 2010). Opposition has come from the chemical industry (see David J. Hanson, "Waste Site Cleanup Tax Opposed," *Chemical & Engineering News*, April 19, 2010), which has been hiring lobbyists to fight the bills (see Kevin Bogardus, "Superfund Tax Push Spurs Rush for New Lobbyists," *The Hill*, June 23, 2010). On the other hand, the American Society of Civil Engineers favors the taxes, partly because they relieve some pressure on the federal budget; see "ASCE Endorses Bill to Enact Superfund Taxes, Take CERCLA Spending Off-Budget," *This Week in Washington* (April 15, 2011). At the Senate Committee on Environment & Public Works, Subcommittee on Superfund, Toxics and Environmental Health, hearing on "Oversight of the Environmental Protection Agency's Superfund Program," held June 22, 2010, Senator James M. Inhofe (R-OK) testified that cleanups are necessary but the EPA needs to improve efficiency (in part by not wasting money on public-relations campaigns). Since responsible parties already pay for most cleanups, "polluter-pays" taxes are an unnecessary burden on small businesses.

The "unnecessary burden" complaint is at the heart of many attacks on federal regulations such as CERCLA, RCRA, and EPCRA (the Superfund legislation) and on attempts to broaden the coverage of such regulations. Agricultural operations are a case in point, for the original impetus for the Superfund legislation was clearly contamination of the environment by industrial chemicals. The agricultural industry was barely in the picture. But in the last few decades, the practice of raising cows, pigs, and chickens in enormous "factory farms" has grown and so has the amount of animal waste produced on these farms, as well as the chemicals (some of which are covered by Superfund regulations in more conventionally industrial settings) contributed by these wastes to air and water. Should these wastes and chemicals be regulated under the name of Superfund? In the YES selection, Walter Bradley of the Dairy Farmers of America, Inc., argues that Superfund legislation is and has been targeted at industrial chemicals, not animal manure. Despite agricultural exemptions in existing law, there is a need to make such exemptions more explicit and thereby protect the ability of dairy farmers to make a living. In the NO selection, Ed Hopkins of the Sierra Club argues that large animal feeding operations (or factory farms) emit such large quantities of toxic chemicals such as hydrogen sulfide and ammonia, as well as pollutants that damage water supplies and ecosystems, that they should not be exempted from Superfund legislation.

YES ↵

Walter Bradley

Testimony before the House Committee on Energy and Commerce, Subcommittee on Environment and the Economy, regarding H.R. 2997, the Superfund Common Sense Act of 2011

... I am testifying today on behalf of the nearly 15,000 farmer-owners of the Dairy Farmers of America (DFA), who raise their herds and their families in 48 states across the nation, in strong support of HR 2997. H.R. 2977 will once and for all affirm that the very same livestock manure used to fertilize our nation's organic crops is not a hazardous or toxic substance under Superfund.

DFA's members are diverse in size, thought and farm management. They are all similar, however, in that they are proud stewards of the land who understand quality feed and quality milk comes from land that is respected and well cared for. Additionally, DFA's members and the manure generated from their herds are already regulated by state and federal law.

I have the pleasure of specifically serving the dairy industry of New Mexico. New Mexico's average dairy herd is about 2,500 cows, considerably larger than the average dairy nationally, and the state is the eight largest producer of the nation's milk. While farms are large in our state, they are owned and run by families who live and work on their farms. Most support several family members and are multi-generational.

The last few years have posed extraordinary challenges for producers across the country, but especially in my region, where a majority of feed is imported from other regions. In 2009 and part of 2010, our producers dealt with very volatile milk prices and input costs, a supply demand imbalance, and other world factors, which drove down price and operating margins. Many lost a generation's worth of equity while others left the business. In New Mexico, most took on more debt.

This year, we are facing another low margin cycle. Some in our area are still paying for feed and other bills acquired during 2009. Since January of this year, we have lost 21 dairies in the area and 5 in New Mexico specifically with several on the cusp of exiting the business. Prices, feed costs, supply demand fluctuations and weather, are all things farmers cannot always control. The uncertainty that these and other factors bring to the industry is startling.

One thing we can deliver—that we should be delivering to our dairymen—is regulatory surety. I ask this committee to do just that. Bring certainty to the nation's dairy farmers on the treatment of manure, a resource and potential source of income, under the Comprehensive Environmental Response, Compensation and Liability Act (CERCLA) and Emergency Planning and Community Right to Know Act (EPCRA) regulation.

Specifically, make clear the intent of Congress to not regulate manure under these laws. We are not seeking an exemption from the federal Clean Water Act (CWA) or the Clean Air Act (CAA) or similar state laws including any federal or state worker protection laws. We are merely seeking clarification under CERCLA and EPCRA that animal manure does not necessitate an emergency response nor does it create a Superfund site.

Congress created the Superfund to deal specifically with the problem of cleaning up toxic waste sites, including hazardous materials such as petrochemicals, inorganic raw materials and petroleum oil used to make hazardous products and waste. At the same time, EPCRA was adopted to require the reporting of releases of hazardous chemicals and to enable emergency responses from governmental authorities when necessary. The composition and use of animal manure by farmers does not meet the threshold of being considered a hazardous waste.

It should also be pointed out that both the CERCLA and EPCRA include exemptions for animal operations. For example the definition of a "hazardous chemical" under CERCLA excludes "any substance to the extent it is used in agriculture operations." At the same time, EPCRA specifically exempts "any substance to the extent that it is used in routine agricultural operations or is a fertilizer held for sale by a retailer to the ultimate customer from the definition of being a hazardous chemical."

U.S. House of Representatives Committee on Energy and Commerce, June 27, 2012.

CERCLA was passed in the wake of Love Canal for the purpose of dealing with the "legacy of hazardous substances and wastes which pose a serious threat to human health and the environment." The law states that it was intended "to clean the worst abandoned hazardous waste sites in the country." The legislative history contains a litany of references to "synthetic," "manmade" chemicals, "chemical contamination," and the results of "modern chemical technology" as the problems CERCLA intended to address. It contains no reference to an intention to clean up manure or urea, or their byproducts, from cattle or any other animal agricultural operations.

Without this clarity, the courts are left to redefine the regulation. Animal manure has been safely used as a fertilizer and soil amendment all over the world for centuries. In recent years, however, we have seen litigation challenge the use of animal manure as a fertilizer by claiming contamination and damage to natural resources.

The issue of CERCLA/EPCRA's applicability to the livestock industry has been discussed in Congress several times in the last decade. I believe congressional intent is clear. When the law was passed Congress did not intend for manure to be regulated as a hazardous substance. Moreover, recent history demonstrates that Congress understands the value of manure to producers and has encouraged its creative use. Laws have been passed and initiatives undertaken to encourage rural America to participate in the renewable energy field through the development of on-farm energy production. Whether it is producing biogas, electricity or biodiesel derived from manure. Congress has acknowledged manure's value by funding research, passing tax credits and mandates for its use. How can we possibly ask dairy producers to invest millions of dollars in technologies to support the nation's energy needs without addressing the threat that manure might be classified as a hazardous substance.

Besides being used for bio-energy production, manure is frequently spread on fields for fertilizer. When waterways are near, farmers often use a buffer ship as they are sensitive to keeping water sources clean. The buffer strip produces no income or feed, but protects the environment. This simple, long standing and environmentally respectful practice is threatened by the insecurity surrounding manure's possible regulation under CERCLA/EPCRA. Conversely, I find it interesting that petroleum-based fertilizers, the alternative to this naturally occurring fertilizer, are exempt from such laws.

As a fertilizer, manure is excellent as it contains nitrogen, phosphorus, potassium and other nutrients. Manure not only supplies many nutrients for crop production, including micronutrients, but it is also a valuable source of organic matter. Increasing soil organic matter improves soil structure or tilth, increases the water-holding capacity of coarse-textured sandy soils, improves drainage in fine-textured clay soils, provides a source of slow release nutrients, reduces wind and water erosion, and promotes growth of earthworms and other beneficial soil organisms. Additionally, its use reduces an operations dependence on man-made chemical fertilizers, which have become very expensive.

If Congress does not act, if the courts are allowed to determine the specifics of this law, how will this law be applied to dairy and other agriculture producers? Will producers need a special permit to dispose of manure? What about the phosphates used by people on their lawns, golf clubs or other community green areas? Would they be classified a Superfund site? On this issue, the science and common sense are in agreement. The phosphates in manure are not now, nor have they ever been, equivalent to the harmful chemicals that CERCLA has been addressing for the last 32 years.

In 2005, the United States Environmental Protection Agency (EPA) entered into a Consent Agreement with animal agriculture to address emissions of air pollutants from animal feeding operations that may be subject to requirements of the CAA, the hazardous substance release notification provisions of CERCLA and the emergency notification provisions of EPCRA. In order to secure a substantive sampling, covering different types of operations in different geographic regions, the impacted industries paid for their portion of the study themselves. The data is currently being analyzed for the purpose of proposing "threshold requirements" for reporting an emissions from animal operations of various sizes that could be put into effect next year. It is clear that CWA and the CAA already provide sufficient authority to address the needs and challenges associated with animal agriculture. EPA is currently moving under the provisions of the CAA to deal with air emissions from ammonia, hydrogen sulfide and other substances. Using CERCLA and EPCRA authority to enforce water and air quality standards on animal agriculture is another example of "regulatory overkill" at its worst.

I will also note that legal action can already be brought against animal operations that are not complying with the CWA to require cleanup. At the same time. Total Maximum Daily Load requirements have been instituted in watersheds requiring farmers to comply with nutrient runoff.

Lastly, animal agricultural operations are subject to a vast array of federal, state and local environmental laws and authority to deal with every conceivable environmental problem presented by them. They include the CAA, CWA, the Resource Conservation and Recovery Act, the Toxic Substances Control Act, FIFRA, soil conservation, dust and odor mitigation controls, as well as nuisance laws, which have been applied broadly throughout the country to provide environmental protection from every conceivable aspect of animal agricultural operations. In New Mexico, we have the Ground Water Protection Bureau, the Surface Water Bureau, the Air Qualify Bureau and a set of New Mexico Dairy Rules for permitting of all dairies in addition to all the federal rules and regulations. There has been no indication that environmental laws such as these are inadequate.

The statute is clear in my view. However, that has not been enough to prevent litigation over applying the Superfund Laws to manure from animal agriculture, and decisions that they apply. I hope Congress addresses this issue and makes clear their original intent that manure from animal agriculture is exempt as a CERCLA hazardous substance. . . .

***

WALTER BRADLEY works in government and business relations at the Dairy Farmers of America, Inc. In the past he served as a New Mexico State Senator and Lieutenant Governor.

Ed Hopkins

NO

Testimony before the House Committee on Energy and Commerce, Subcommittee on Environment and the Economy, regarding H.R. 2997, the Superfund Common Sense Act of 2011

Summary

Production of livestock and poultry has concentrated greatly in the last several decades, and this has concentrated large amounts of animal waste. While manure is not a hazardous waste, improper handling can release phosphorus into water supplies, and its degradation can release large amounts of ammonia and hydrogen sulfide into the air. These are hazardous substances under the Comprehensive Environmental Response Compensation and Liability Act (CERCLA) and the Emergency Planning and Community Right to Know Act (EPCRA). Some large livestock and poultry operations release more of these chemicals than major industrial facilities.

Phosphorus can contaminate communities' drinking water supplies, burdening ratepayers and taxpayers with cleanup costs. At least 29 states have reported damage to lakes, rivers and streams from large animal feeding operations. Air releases of ammonia and hydrogen sulfide—manure degradation products—trigger respiratory problems, eye and nose irritation, and in some extreme cases, death. Many university reports, peer-reviewed literature and government-sponsored studies have documented the adverse public health and environmental effects of animal feeding operations.

The Sierra Club opposes H.R. 2997, which would create a special exemption for livestock manure from CERCLA and EPCRA. If Congress creates this exemption, communities whose waterways or drinking water has been damaged by hazardous substances witi lose a vital tool for recovering cleanup costs. The cost of cleaning up water damaged by excess phosphorus and other hazardous substances in manure would fall on communities and ratepayers rather than those responsible for causing the contamination. Communities would remain in the dark about toxic chemicals these facilities release into the air. Poultry and livestock operators will also lose a powerful incentive to manage their waste responsibly. . . .

Modern Livestock Operations Concentrate Large Amounts of Animal Waste

Most animal feeding operations do not resemble the livestock farms of years past. Instead, many are industrialized operations that confine thousands of animals at a single location, often generating the waste equivalent of a small city. Unlike traditional livestock farms where the animals grazed on pastureland, these facilities confine thousands, or even millions, of the animals in closed buildings for most of their lives, where they are fed a regimented diet in a closely controlled indoor environment.

A General Accountability Office study estimated that between 1982 and 2002, the number of large farms that raise animals increased by 234 percent and that almost half of all animals were raised on large farms. These large operations produce about 300 million tons of manure annually or three times more waste than humans generate each year in the United States. Depending on the type and number of animals, the GAO estimated that individual farms can produce from over 2,800 tons to more than 1.6 million tons of manure each year. The GAO estimated that one large hog farm could generate 1.6 million tons of manure a year, which is one and one-half times the amount of sanitary waste generated by the 1.5 million residents of Philadelphia in a year. An important difference: the City of Philadelphia treats its wastewater, the large hog operation does not.

Livestock Waste Threatens Public Health and the Environment

With so much manure concentrated in small areas come threats to public health, water and air. According to the most recent National Water Quality Inventory, 29 states specifically identified animal feeding operations as contributing to water quality impairments. Waste pits full of manure fail, inundating rivers and killing fish. In 1995, approximately 25 million gallons of manure were discharged from a single hog operation in North Carolina. Similarly, discharges of thousands of gallons of animal waste have been reported in Iowa, Illinois, Minnesota, Missouri, Ohio and New York.

Waste applied to fields in large quantities can run off into lakes and rivers. The nutrient-rich runoff alters the chemical composition of receiving waters, and triggers a surge in algae and other aquatic vegetative growth. This vegetative growth can choke out fish and other marine life, and lead to increased treatment requirements for drinking water supplies. According to the EPA, "over-enrichment of waters by nutrients (nitrogen and phosphorus) is the biggest overall source of impairment of the nation's rivers and streams, lakes and reservoirs, and estuaries."

When large farms are clustered in a region for easy access to processing facilities, the GAO reported that: "According to agricultural experts and government officials that we spoke to, such clustering of operations raises concerns that the amount of manure produced could result in the overapplication of manure to croplands in these areas and the release of excessive levels of some pollutants that could potentially damage water quality." According to a U.S. Department of Agriculture report, the numbers of counties with excess manure nitrogen increased by 103 percent, from 36 counties in 1982 to 73 counties in 1997. Similarly, the number of counties with excess manure phosphorus increased by 57 percent, from 102 counties in 1982 to 160 counties in 1997.

This contamination poses serious risks to human health. Manure-related microbes in water can cause severe gastrointestinal disease, complications and even death. In May 2000 in Walkerton, Ontario, an estimated 2,321 people became ill and seven died after drinking water from a municipal well contaminated with *E. coli* and *Camplyobacter* from runoff resulting from manure spread onto fields by a nearby livestock operation. Manure can also carry arsenic and other toxic metal compounds, as well as antibiotics, into water contributing to antibiotic resistance. Finally, pollution from animal confinements can cause nitrate contamination of drinking water supplies, which can result in significant human health problems including methemoglobinemia in infants ("blue baby syndrome"), spontaneous abortions and increased incidence of stomach and esophageal cancers.

Air emissions also cause significant health problems in workers and in nearby residents. Livestock and poultry operations emit significant amounts of particulate matter (fecal matter, feed materials, skin cells, bioaerosols, etc.), ammonia, hydrogen sulfide, sulfur dioxide, volatile organic compounds, and other harmful contaminants into the air. Many adverse human health effects associated with air pollution from these operations, including respiratory diseases (asthma, hypersensitivity pneumonitis, industrial bronchitis), cardiovascular events (sudden

death associated with particulate air pollution), and neuropsychiatric conditions (due to odor as well as delayed effects of toxic inhalations). Other problems include increased headaches, sore throats, excessive coughing, diarrhea, burning eyes, and reduced quality of life for nearby residents. This air pollution is especially problematic, because neighboring communities are exposed on a near constant basis.

Ammonia is a human toxin that EPA lists alongside arsenic, cyanide, and benzene as a hazardous substance under CERCLA. 40. C.F.R. § 302.4. The livestock sector produces roughly 73% of all ammonia emissions nationwide. Human exposure to ammonia triggers respiratory problems, causes nasal and eye irritation, and in extreme circumstances, is fatal. Ammonia also contributes to the development of fine particulate matter. Fine particulate matter causes significant health problems, including aggravated asthma, difficult or painful breathing, chronic bronchitis, decreased lung function, and premature death. Fine particulate matter has been linked to increased hospital emissions and emergency room visits for people with heart and lung disease, and decreased work and school attendance.

Animal feeding operations expose downwind neighbors to elevated ammonia levels, as well as other pollutants. For example, the Missouri Department of Health and Senior Services documented ambient ammonia levels downwind of a swine operation ranging from 153 to 875 ppb. The EPA submitted comments on the Missouri study, comparing the ambient ammonia levels to recommended exposure limits and noted that "the conclusion could be drawn that a *public health hazard* did exist at the time the . . . data was acquired." Some of the largest facilities produce staggering quantities of ammonia gas—comparable to pollution from the nation's largest manufacturing plants. For example, Threemile Canyon Farms in Boardman, Oregon, reported that its 52,300 dairy cow operation emits 15,500 pounds of ammonia per day, more than 5,675,000 pounds per year. That is 75,000 pounds more than the nation's number one manufacturing source of ammonia air pollution (CF Industries of Donaldson, Louisiana) reported releasing that year. Buckeye Egg Farm's facility in Croton, Ohio reported ammonia emissions of over 4,300 pounds per day—43 times the reporting threshold under CERCLA and EPCRA.

In addition to ammonia, EPA also lists hydrogen sulfide as a hazardous pollutant under CERCLA. High-level exposures of hydrogen sulfide, an asphyxiate, can cause loss of consciousness, coma and death. At least 19 workers have died from sudden hydrogen sulfide exposure during liquid manure agitation. Epidemiological studies of communities exposed to hydrogen sulfide reported symptoms such as asthma, chronic bronchitis, shortness of breath, eye irritation, nausea, headaches and loss of sleep.

The GAO study found that "Since 2002, at least 68 government-sponsored or peer-reviewed studies have been completed on air and water pollutants from animal feeding operations. Of these 68 studies, 15 have directly linked pollutants from animal waste generated by these operations to specific health or environmental impacts, 7 have found no impacts, and 12 have made indirect linkages between these pollutants and health and environmental impacts. In addition, 34 of the studies have focused on measuring the amount of certain pollutants emitted by animal feeding operations that are known to cause human health or environmental impacts at certain concentrations."

These risks to public health led the American Public Health Association to call for a moratorium on new concentrated animal feeding operations "until scientific data on the attendant risks to public health have been collected and uncertainties resolved."

EPCRA and CERCLA Requirements

CERCLA has two main policy objectives. First, Congress intended to give the federal government the necessary tools for a prompt and effective response to problems of national magnitude resulting from hazardous waste disposal. Second, Congress intended that the polluters bear the costs and responsibility for remedying the harmful conditions that they created.

Specifically, section 103 of CERCLA provides that any person in charge of a facility from which a hazardous substance has been released in a reportable quantity (RQ) must immediately notify the National Response Center ("NRC"). For example, releases of ammonia and hydrogen sulfide that exceed 100 pounds per day must be reported under section 103.42. Section 103(f)(2) of CERCLA further provides for relaxed reporting requirements for substances that are classified as a continuous release. If a reported release demands a response, the government may act, pursuant to section 104, to respond to that release. And if the government acts, it may recoup the costs of the recovery action under CERCLA section 107.

In addition to the reporting requirements under CERCLA, owners and operators of facilities must also provide immediate notice of the release of an extremely hazardous substance under EPCRA. Section 304(a) requires an owner or operator of a facility to report the release of an extremely hazardous substance to designated state and local officials, if "such release requires notification of section 103(a) of the Comprehensive Environmental Response, Compensation, and Liability Act of 1980." The EPCRA emergency reporting requirements, therefore, track the CERCLA requirements and ensure that federal, state and local authorities are notified of potentially dangerous chemical releases.

The right-to know provisions of CERCLA and EPCRA not only empower government but also citizens. Information about chemical releases enables citizens to hold companies and local governments accountable in terms of how toxic chemicals are managed. Transparency also often spurs companies to focus on their chemical management

practices since they are being measured and made public. In addition, the data serve as a rough indicator of environmental progress over time.

CERCLA/EPCRA Fill Important Gaps in Permitting Statutes

CERCLA and EPCRA require the reporting of only nonfederally permitted releases. Therefore, if a facility's emissions are authorized by a permit under another federal statute, they do not have to report these emissions. Releases that are federally permitted are exempt not only from CERCLA and EPCRA notification requirements but from CERCLA liability as well. Although EPA and the States have permitted some feeding operations under other federal statutes, CERCLA is still necessary to fill critical gaps. Although the Clean Water Act has required large livestock operations to obtain permits for almost 40 years, noncompliance has been widespread. As EPA indicates in a proposed information collection rule, only about 8,000 concentrated animal feeding operations out of a universe of about 20,000 facilities—about 40 percent—have obtained Clean Water Act NPDES permits.

Even if a facility were to have a federal permit, the permit would not necessarily address all of the releases of hazardous chemicals. A Clean Water Act permit, for example, would not address releases of hazardous chemicals to the air and, conversely, a Clean Air Act permit would not address releases of hazardous chemicals to water. Furthermore, not all statutes regulate the same chemicals. For example, the Clean Air Act does not regulate ammonia or hydrogen sulfide as hazardous air pollutants. Although CERCLA's list of hazardous substances were first identified under other statutes, including the Clean Water Act, the Clean Air Act and the Resource Conservation and Recovery Act, CERCLA authorizes the Administrator of EPA to add to this list "substances [like ammonia and hydrogen sulfide] which, when released to the environment may present a substantial danger to public health or welfare or the environment. . . ."

Thus, EPCRA and CERCLA are necessary complements to federal permitting statutes to address hazardous pollutants that would not otherwise be regulated. They do not duplicate other federal laws.

Animal Production Operations Should Not Be Exempted from EPCRA/CERCLA

The poultry and livestock industry argues that Congress never intended to apply CERCLA and EPCRA requirements to animal agriculture. However, they cite no authority for this claim. If Congress had intended such a result, it could have excluded animal production facilities, like hog or poultry facilities, from the reporting requirements of CERCLA. Instead, Congress only chose to exempt "the normal application of fertilizer" from the CERCLA definition of release, and provided an exemption under EPCRA for reporting releases when the regulated substance "is used in routine agricultural operations or is a fertilizer held for sale by a retailer to the ultimate consumer."

Both of these exemptions were considered by a federal district court in Kentucky which held that neither of the exemptions should apply to Tyson's poultry production operations. Tyson did not qualify for the routine agricultural use exemption, because it did not store ammonia in the chicken houses for agricultural use, nor did it use the ammonia in an agricultural operation. Rather, it used exhaust fans and vents to release the ammonia to the environment so that it would not kill the chickens. Tyson did not qualify for the normal application of fertilizer exemption, because they were not applying ammonia to farm fields as fertilizer when they vented it into the atmosphere.

A federal court in Texas also considered the normal application of fertilizer exemption. The court ruled that the exemption does not apply if Plaintiffs prove that the Defendants improperly stored and maintained large amounts of waste on their property, causing hazardous releases of phosphorus and other pollutants to nearby sources of drinking water. Industry representatives also argue that the CERCLA exclusion for "naturally occurring substances" should apply to livestock operations. Section 104(a)(3)(A) of CERCLA prohibits the President [through EPA] from ordering a remedial or response action "in response to a release or threat of release . . . of a naturally occurring substance in its unaltered form, or altered solely through naturally occurring processes or phenomena, from a location where it is naturally found. . ." Industry argues that CERCLA should not apply to farming operations because "[s]ubstances, such as orthophosphate, ammonia and hydrogen sulfide, occur naturally in the environment in the same forms as they occur as byproducts of biological processes on farming operations." However, releases of hazardous substances from agribusinesses would not qualify for the exemption, because they occur as a result of activities associated with milk or meat production. For example, as discussed below, in both of the response actions taken to date, the governments' actions were not based on releases of naturally occurring phosphorus or orthophosphate undisturbed by human activity. Rather, the governments sought to remove hazardous substances that were added to the environment and disposed of by the operations during the improper storage and handling of waste.

CERCLA/EPCRA Cases against Agribusinesses, Not Family Farms

There have only been a handful of cases filed against poultry and livestock operations for violations of CERCLA and EPCRA. In most of the cases, the defendants have been large corporate agribusinesses, not family farmers, and the releases of hazardous chemicals have been significant. Courts have consistently held that CERCLA and EPCRA reporting requirements apply to agricultural operations if releases of regulated hazardous substances meet regulatory thresholds.

Premium Standard Farms—In November 2001, the United States and Citizens Legal Environmental Action Network, Inc. settled a case against Premium Standard Farms, Inc. (PSF), the nation's second largest pork producer and Continental Grain Company. PSF's and Continental's operations in Missouri consist of more than 1,000 hog barns, 163 animal waste lagoons and 1.25 million hogs, primarily located on 21 large-scale farms in five counties. The settlement resolved numerous claims of violations under the CWA, CAA, CERCLA and EPCRA.

PSF exposed downwind neighbors to elevated ammonia levels, as well as other pollutants. Measurements taken pursuant to the settlement agreement reveal that PSF releases 3 million pounds of ammonia annually from the cluster of barns and lagoons at its Somerset facility. At the time, these emissions made PSF the fifth largest industrial emitter of ammonia in the United States. This data does not include the ammonia gases released when liquid manure is sprayed on the company's nearby fields.

Seaboard Corporation—On January 7, 2003, the Sierra Club reached partial settlement of a lawsuit against the Seaboard Corporation, concerning pollution at one of the largest hog factories in North America. The settlement resolved all claims, except for Sierra Club's CERCLA and EPCRA claims. CERCLA requires a person to report releases of a hazardous substance from a "facility." In an effort to avoid regulation, Seaboard argued that each pit and building should be counted separately. An appellate court found Seaboard's arguments "unconvincing." The Court held that the entire 25,000-head hog operation was a single "facility" and that Seaboard must report the combined emissions from all its waste pits and confinement buildings. Seaboard estimates that the total average daily emissions of ammonia from its Dorman Sow Facility is 192 pounds per day, almost double the 100 pound per day reporting threshold under CERCLA.

Tyson Foods, Inc.—On January 26, 2005, the Sierra Club entered into a settlement agreement with Tyson Foods. Tyson is the number one poultry producer in the nation, and each of its four facilities that were involved in the case could confine approximately 600,000 chickens at one time. Under the decree, Tyson agreed to study and report on emissions from its chicken operations and mitigate ammonia emissions that have been plaguing rural residents for years. The settlement came in the wake of a court decision in 2003, when a federal judge ruled that the term "facility" should be interpreted broadly, including facilities operated together for a single purpose at one site, and that the whole farm site is the proper regulated entity for purposes of the CERCLA and EPCRA reporting requirements.

City of Tulsa—The City of Tulsa filed suit against some of the largest poultry producers in the nation including Tyson, Simmons and Cargill. The City alleged that the Defendants' growers polluted Lakes Eucha and Spavinaw, from which Tulsa draws its water supply, by applying excess litter to land application areas. As of September 1, 2002, just one of the Defendant's growers produced approximately 40,715,200 birds and an estimated 39,859 tons of litter in the affected watershed. The City's complaint included claims for cost recovery and contribution under CERCLA. A federal court ruled that phosphorus contained in the poultry litter in the form of phosphate is a hazardous substance under CERCLA.

The City of Tulsa continues to experience water quality problems as a result of pollution from animal feeding operations. The following comment, submitted in response to the EPA's proposed information collection request for large animal feeding operations, supports the need for more regulation of these operations, not less: "The City of Tulsa used a significant amount of financial resources in an attempt to coordinate with the poultry industry on ways to promote environmental stewardship, improve nutrient management practices and stakeholder communication, but with no success. Success only came from subsequent court order directives."

City of Waco—In 2004, the City of Waco filed suit against fourteen commercial dairies for failure to properly manage and dispose of waste. The complaint alleges that hazardous pollution from these dairies contaminated Lake Waco, which is the sole source of drinking water for the City of Waco and a significant source of drinking water for surrounding communities. The City's complaint includes claims for cost recovery and contribution costs under CERCLA. The Court denied the dairies' Motion to Dismiss and held, among other things, that the type of phosphorus that was released by the dairies was a hazardous substance under CERCLA. The Court also held that the normal application of fertilizer exemption would not apply if Plaintiffs could prove that the releases of hazardous substances were caused by the dairies' improper handling of animal waste. The City subsequently settled its case. To address taste and odor problems caused by excessive phosphorus in its water supply, the City is spending more than $54 million in upgrades to its drinking water treatment system. The City also opposed legislation similar to H.R. 2977 in a previous session of Congress.

State of Oklahoma—On June 18, 2005, the Oklahoma Attorney General's Office filed a lawsuit against some of the nation's largest producers of chickens, turkeys and eggs for water pollution in the Illinois River watershed caused by the improper dumping and storage of poultry waste. The watershed contains elevated levels of a number of pollutants found in poultry waste. For example, the phosphorus from the poultry waste dumped into the Illinois River watershed is equivalent to the waste that would be generated by 10.7 million people, a population greater than the states of Arkansas, Kansas and Oklahoma combined. The watershed also serves as the source of drinking water for 22 public water supplies in eastern Oklahoma. The Attorney General's complaint alleges violations of state and federal nuisance laws, trespass, as well as other violations of state environmental regulations. The State also seeks to recover the costs that it has had to incur, and will incur, to respond to the pollution. These costs include "the costs of monitoring, assessing and evaluating water quality, wildlife and biota in the [Illinois River Watershed]." The State also seeks to recover Natural Resource Damages for the injury to, destruction of, and loss of natural resources. This case remains unresolved.

Citizens Cannot Recover Natural Resources Damages or Penalties Under the Response Sections of CERCLA

Industry representatives have incorrectly asserted that citizen suits threaten to impose natural resource damage liability under CERCLA. In fact, natural resource damages may only be recovered by a designated federal, state or tribal trustee. Industry has also raised alarms about high penalties from citizen suits and cases brought by municipal and state governments. Again, there is no rational basis for this assertion. Tyson and Seaboard did not pay a single penny in their cases brought by Sierra Club for failure to report their hazardous air emissions under CERCLA and EPCRA. Furthermore, penalties are unavailable under CERCLA for removal or remedial actions, regardless of whether they are initiated by government or by a private party. Finally, citizens are even limited in their cost recovery actions. A private party must prove as part of its prima facie case that the cleanup activities for which it incurred response costs were consistent with the National Contingency Plan.

Exempting Agribusinesses from EPCRA/CERCLA Requirements Would Prevent EPA from Gathering Critical Data

By exempting reporting requirements for poultry and livestock waste emissions, the EPA would be prevented from even knowing the scope or consequences of this pollution. Ignoring this problem will not make it go away; virtually every study that has been done on this subject emphasizes the need for more information. The National Academy of Sciences (NAS) issued a report in 2003 in which it expressed concern over air pollution from animal feeding operations and criticized EPA and USDA for not devoting the necessary technical or financial resources to estimate air emissions and to develop mitigation technologies. The 2008 GAO report underscored the need for more information: "Although EPA is aware of the potential impacts of air and water pollutants from animal feeding operations, it lacks data on the number of animal feeding operations and the amount of discharges actually occurring. Without such data, according to EPA officials, the agency is unable to assess the extent to which these pollutants are harming human health and the environment."

Failing to require reporting may impede responses to acute health threats. Emissions of hydrogen sulfide from the Excel Dairy in Marshall County, Minnesota, illustrate the importance of retaining the reporting requirements so that health officials can respond to emergencies. According to a September, 2008, Exposure Investigation by the Agency for Toxic Substances and Disease Registry, high and persistent emissions of hydrogen sulfide from the dairy prompted a finding of a "public health hazard associated with community exposures." With concentrations of hydrogen sulfide exceeding the measuring capability of the Minnesota Pollution Control Agency's monitoring

equipment, ATSDR recommended immediate action to reduce emissions from the dairy, more sophisticated air monitoring, and restricted access to the dairy property to reduce exposures.

In conclusion, because of the demonstrable public health and environmental threats that animal feeding operations pose, CERCLA and EPCRA provide critical safeguards complementing other statutes. Sierra Club strongly opposes legislation like H.R. 2997 that would create special exemptions for hazardous substances released from poultry and livestock operations. This bill serves only to shift cleanup costs to taxpayers and ratepayers and keep regulatory agencies and the public in the dark about exposures to chemicals that these facilities release. . . .

ED HOPKINS is the Director of the Sierra Club's Environmental Quality Program.

EXPLORING THE ISSUE

Should Agricultural Animal Wastes Be Exempt from the Requirements of Superfund Legislation?

Critical Thinking and Reflection

1. In the past many industrial chemical wastes were disposed of according to the "best practice" of the time (as defined by laws and regulations). Later it was realized that "best practice" was not very good. Should companies be held liable for public health effects in such cases?
2. In the past many industrial chemical wastes were disposed of in the cheapest possible way, without regard to potential public health effects. Should companies be held liable for public health effects in such cases?
3. Should the way the public (and regulators) regards emissions of toxic chemicals depend on the scale of the operation responsible for the emissions (compare the old-fashioned family farm with modern factory farms)?
4. Should the impact of potential regulations on the profitability of regulated businesses be a factor in deciding whether to implement the regulations?

Is There Common Ground?

An interesting point hidden in the readings is the question of scale. Walter Bradley speaks of dairy operations as if they were all traditional "family farms." Ed Hopkins stresses that a great many modern dairy (and pig and chicken) farms are much, much bigger than that and the chemical emissions they generate are much, much larger than anything produced by a family farm. Both agree that there is a need to control pollution, but Walter Bradley thinks existing laws are entirely adequate for the purpose. Ed Hopkins argues that we need the reporting provisions of the Superfund regulations to determine just how bad the problem is.

1. Is it reasonable to say that a substance is a toxic waste of great concern when produced in large amounts, but not when produced in small amounts? Where should one draw the line between "small" and "large"?
2. Are there other ways besides the reporting requirements of the Emergency Planning and Community Right-to-Know Act (EPCRA) to determine the extent of toxic emissions from animal-raising operations?

3. The rhetoric of American political debate involves a number of very traditional themes. Among them are family values and the dependence of America's strength and identity on the family farmer. Examine news reports for a week and see how often these themes appear (the count will be greater in an election year). Did you notice other rhetorical themes as well?

Create Central

www.mhhe.com/createcentral

Additional Resources

Sheila Kaplan and Marilyn B. Snell, "Sapping the Superfund's Strength," *Nation* (May 3, 2010)

Stephen Lester and Anne Rabe, *Superfund: In the Eye of the Storm* (Center for Health, Environment & Justice, June 2010)

Iwonna Rummel-Bulska, "The Basel Convention: A Global Approach for the Management of Hazardous Wastes," *Environmental Policy and Law* (vol. 24, no. 1, 1994)

Internet References . . .

Natural Resource Conservation Service

http://go.usa.gov/KoB

Superfund

www.epa.gov/superfund/

The Center for Health, Environment & Justice (CHEJ)

chej.org/

Selected, Edited, and with Issue Framing Material by:
Thomas Easton, *Thomas College*

Should the United States Reprocess Spent Nuclear Fuel?

YES: Kate J. Dennis et al., from "The Case for Reprocessing," *Bulletin of the Atomic Scientists* (November/December 2009)

NO: David M. Romps et al., from "The Case Against Nuclear Reprocessing," *Bulletin of the Atomic Scientists* (November/December 2009)

Learning Outcomes

After reading this issue, you will be able to:

- Explain why fear of nuclear proliferation has kept the United States from developing nuclear fuel processing facilities.
- Explain how nuclear fuel reprocessing can help deal with the nuclear waste problem.
- Explain how nuclear fuel reprocessing supports expended use of nuclear power.
- Explain why it will be difficult to find a publicly acceptable site for a nuclear fuel reprocessing plant.

ISSUE SUMMARY

YES: Kate J. Dennis, Jason Rugolo, Lee T. Murray, and Justin Parrella argue that nuclear fuel reprocessing extracts more energy from nuclear fuel and reduces the amount of nuclear waste to be disposed of. "If the United States truly wants to proceed with nuclear energy as a viable, low-carbon emitting source of energy, it should pursue reprocessing in combination with the development of fast reactors. Once such a decision is made, the debate should turn to how best to develop cheaper and safer reprocessing options, rather than denying its general benefit."

NO: David M. Romps, Christopher D. Holmes, Kurt Z. House, Benjamin G. Lee, and Mark T. Winkler argue that reprocessing is both dangerous and unnecessary. "It is in the best interests of the United States—from the perspective of waste management, national security, economics, and environmental protection—to maintain its *de facto* moratorium on reprocessing and encourage other countries to follow suit."

As nuclear reactors operate, the nuclei of uranium-235 atoms split, releasing neutrons and nuclei of smaller atoms called fission products, which are themselves radioactive. Some of the neutrons are absorbed by uranium-238, which then becomes plutonium. The fission product atoms eventually accumulate to the point where the reactor fuel no longer releases as much energy as it used to. It is said to be "spent." At this point, the spent fuel is removed from the reactor and replaced with fresh fuel.

Once removed from the reactor, the spent fuel poses a problem. Currently it is regarded as high-level nuclear waste which must be stored on the site of the reactor, initially in a swimming pool-sized tank and later, once the radioactivity levels have subsided a bit, in "dry casks." Until 2009, there was a plan to dispose of the casks permanently in a subterranean repository being built at Yucca Mountain, Nevada, but the Obama administration has ended funding for Yucca Mountain (see Dan Charles, "A Lifetime of Work Gone to Waste?" *Science*, March 20, 2009). Robert Alvarez, "Improving Spent-Fuel Storage at Nuclear Reactors," *Issues in Science and Technology* (Winter 2012), argues that on-site dry-cask storage is both safe and affordable, especially if user fees collected to pay for permanent storage, as at the now-defunct Yucca Mountain, are used to pay the bills.

It is worth noting that spent fuel still contains useful components. Not all the uranium-235 has been burned up, and the plutonium created as fuel is burned can itself be used as fuel. When spent fuel is treated as waste, these components of the waste are discarded. Early in the Nuclear Age, it was seen that if these components could be recovered, the amount of waste to be disposed of could be reduced. The fuel supply could also be extended, and in fact, since plutonium is made from otherwise useless uranium-238, new fuel could be created. Reactors designed to maximize plutonium creation, known as "breeder" reactors because they "breed" fuel, were built and are still in use as power plants in Europe. In the United States, breeder reactors have been built and operated only by the Department of Defense, for plutonium extracted from spent fuel is required for making nuclear bombs. They have not seen civilian use in part because of fear that bomb-grade material could fall into the wrong hands.

The separation and recycling of unused fuel from spent fuel is known as reprocessing. In the United States, a reprocessing plant operated in West Valley, NY, from 1966 to 1972 (see "Plutonium Recovery from Spent Fuel Reprocessing by Nuclear Fuel Services at West Valley, New York, from 1966 to 1972" (DOE, 1996; www.osti.gov /opennet/forms.jsp?formurl=document/purecov/nfsrepo.html). After the Nuclear Nonproliferation Treaty went into force in 1970, it became U.S. policy not to reprocess spent nuclear fuel and thereby to limit the availability of bomb-grade material. As a consequence, spent fuel was not recycled, it was regarded as high-level waste to be disposed of, and the waste continued to accumulate.

Despite the termination of the Yucca Mountain nuclear waste disposal site, the nuclear waste disposal problem is real and it must be dealt with. If it is not, we may face the same kinds of problems created by the former Soviet Union, which disposed of some nuclear waste simply by dumping it at sea. For a summary of the nuclear waste problem and the disposal controversy, see Michael E. Long, "Half Life: The Lethal Legacy of America's Nuclear Waste," *National Geographic* (July 2002). The need for care in nuclear waste disposal is underlined by Tom Carpenter and Clare Gilbert, "Don't Breathe the Air," *Bulletin of the Atomic Scientists* (May/June 2004); they describe the Hanford Site in Hanford, Washington, where wastes from nuclear weapons production were stored in underground tanks. Leaks from the tanks have contaminated groundwater, and an extensive cleanup program is under way. But cleanup workers are being exposed to both radioactive materials and toxic chemicals, and they are falling ill. And in June 2004, the U.S. Senate voted to ease cleanup requirements. Per F. Peterson, William E. Kastenberg, and Michael Corradini, "Nuclear Waste and the Distant Future," *Issues in Science and Technology* (Summer 2006), argue that the risks of waste disposal have been sensibly addressed and we should be focusing more attention on other risks (such as those of global warming). Behnam Taebi and Jan Kloosterman, "To Recycle or Not to Recycle? An Intergenerational Approach to Nuclear Fuel Cycles," *Science & Engineering Ethics* (June 2008), argue that the question of whether to accept reprocessing comes down to choosing between risks for the present generation and risks for future generations.

In November 2005, President Bush directed the Department of Energy to start work toward a reprocessing plant; see Eli Kintisch, "Congress Tells DOE to Take

Fresh Look at Recycling Spent Reactor Fuel," *Science* (December 2, 2005). By April 2008, Senator Pete Domenici of the U.S. Senate Energy and Natural Resources Committee was working on a bill that would set up the nation's first government-backed commercialized nuclear waste reprocessing facilities. Reprocessing spent nuclear fuel will be expensive, but the costs may not be great enough to make nuclear power unacceptable; see "The Economic Future of Nuclear Power" (University of Chicago, August 2004) (www .ne.doe.gov/np2010/reports/NuclIndustryStudy-Summary.pdf). Sarah Widder, "Benefits and Concerns of a Closed Nuclear Fuel Cycle," *Journal of Renewable & Sustainable Energy* (November 2010), notes that the present once-through fuel cycle is not sustainable, but reprocessing remains controversial. Tomaz Zagar et al., "Recycling as an Option of Used Nuclear Fuel Management Strategy," *Nuclear Engineering & Design* (April 2011), recognize the difficulties but argue that a Slovenian reactor would benefit greatly from using reprocessing to close the nuclear fuel cycle.

In February 2006, the United States Department of Energy announced the Global Nuclear Energy Partnership (GNEP) to be operated by the United States, Russia, Great Britain, and France. It would lease nuclear fuel to other nations, reprocess spent fuel without generating material that could be diverted to making nuclear bombs, reduce the amount of waste that must be disposed of, and help meet future energy needs. See Stephanie Cooke, "Just Within Reach?" *Bulletin of the Atomic Scientists* (July/August 2006), and Jeff Johnson, "Reprocessing Key to Nuclear Plan," *Chemical & Engineering News* (June 18, 2007). Critics such as Karen Charman ("Brave Nuclear World, Parts I and II," *World Watch* (May/June and July/August 2006), insist that nuclear power is far too expensive and carries too serious risks of breakdown and exposure to wastes to rely upon, especially when cleaner, cheaper, and less dangerous alternatives exist. Early in 2009, the Department of Energy announced it was closing down the GNEP. Ozzie Zehner, "Nuclear Power's Unsettled Future," *Futurist* (March–April 2012), notes that thanks in part to the 2011 earthquake and tsunami-created nuclear disaster in Fukushima, Japan, public skepticism of the value and safety of nuclear power is high and the industry may develop no further. However, in Japan itself, though there were announcements soon after the Fukushima disaster that all Japanese nuclear power plants would be closed by 2040, the government soon said reactors would be put back to work as they passed new safety tests. So far, most of Japan's reactors are inactive; see "Japan's Nuclear Future: Don't Look Now," *The Economist* (April 20, 2013). The issue is complicated because even though nuclear power has obvious dangers, it also produces large amounts of needed electricity without the side effects of burning fossil fuels (air pollution and carbon emissions), which are themselves dangerous.

The Union of Concerned Scientists, on its "Nuclear Reprocessing: Dangerous Dirty, and Expensive" page (www.ucsusa.org/nuclear_power/nuclear_power_risk/nuclear _proliferation_and_terrorism/nuclear-reprocessing.html), notes that the reprocessing of nuclear spent fuel increases the risks of nuclear proliferation. Both nations and terrorists itch to possess nuclear weapons, whose destructive potential makes present members of the "nuclear club" tremble. Can the risks be controlled? John Deutch, Arnold Kanter, Ernest Moniz, and Daniel Poneman, in "Making the World Safe for Nuclear Energy," *Survival* (Winter 2004/2005), argue that present nuclear nations could supply fuel and reprocess spent fuel for other nations; nations that refuse to participate would be seen as suspect and subject to international action. Nuclear physicist and Princeton professor Frank N. von Hippel, "Rethinking Nuclear Fuel Recycling," *Scientific American* (May 2008), argues that reprocessing nuclear spent fuel is expensive and emits lethal radiation. There is also a worrisome risk that the increased availability of bomb-grade nuclear materials will increase the risk of nuclear war and terrorism. Prudence demands that spent fuel be stored until the benefits of reprocessing exceed the risks (if they ever do). See also Rodney C. Ewing and Frank N. von Hippel, "Nuclear Waste Management in the United States—Starting Over," *Science* (July 10, 2009), and Eli Kintisch, "Waste Panel Expected to Back Interim Storage," *Science* (July 8, 2011).

In the summer of 2009, eleven Harvard graduate students and postdocs went to France to investigate nuclear fuel reprocessing. During the trip, they divided into two groups, one for reprocessing and one against (with two undecided). After the trip, they wrote "Should the United States Resume Reprocessing? A Pro and Con" for the *Bulletin of the Atomic Scientists*. In the YES selection, from that article, Kate J. Dennis, Jason Rugolo, Lee T. Murray, and Justin Parrella argue that nuclear fuel reprocessing extracts more energy from nuclear fuel and reduces the amount of nuclear waste to be disposed of. "If the United States truly wants to proceed with nuclear energy as a viable, low-carbon emitting source of energy, it should pursue reprocessing in combination with the development of fast reactors. Once such a decision is made, the debate should turn to how best to develop cheaper and safer reprocessing options, rather than denying its general benefit." In the NO selection, David M. Romps, Christopher D. Holmes, Kurt Z. House, Benjamin G. Lee, and Mark T. Winkler argue that reprocessing is both dangerous and unnecessary. "It is in the best interests of the United States—from the perspective of waste management, national security, economics, and environmental protection—to maintain its *de facto* moratorium on reprocessing and encourage other countries to follow suit."

YES ↩

Kate J. Dennis et al.

The Case for Reprocessing

The United States should reconsider reprocessing its spent nuclear fuel to obtain the highest efficiency and lowest waste. With the correct and necessary guidelines, closing the nuclear fuel cycle by allowing waste to be turned back into fuel is a viable option and should be a goal for the country.

Critics argue that the high cost of reprocessing spent fuel and fabricating mixed-oxide (MOX) fuel rods—a mixture of uranium and plutonium oxides—far outweighs the benefits. According to MIT's 2003 "Future of Nuclear Power" report, the so-called once-through fuel cycle, where spent fuel is directly deposited in geologic repositories, is four to five times less expensive than the costs associated with reprocessing. We do not argue that reprocessing is cheaper in the short-term, but that it is extremely difficult, if not impossible, to compare the short-term costs associated with reprocessing with the benefits that would occur 50–100 years hence. (A 2006 study by the Boston Consulting Group does find the long-term cost of reprocessing to be almost equivalent to once-through fuel management.) These longer-term benefits include a twofold volumetric reduction in nuclear waste, conservation of uranium resources, and a reduction in the environmental impact of uranium mining. Additionally, if the United States considers building fast reactors in the future, reprocessing becomes a necessary step to remove the plutonium that these reactors generate. Fast reactors consist of a core of fissile plutonium or highly enriched uranium surrounded by a blanket of uranium 238, which captures neutrons escaping from the core and partially transmutes into plutonium 239. The result is a reactor that produces or "breeds" a surplus of fuel. This blanket material, however, must be reprocessed to recover the generated plutonium. The development of such breeder reactors would increase the energy output for a given amount of nuclear fuel 60–100 times, according to a range of estimates, and therefore reduce consumption of natural uranium ore.

Currently, it is unclear if the United States will continue to pursue fast reactors as outlined by former President George W. Bush in the 2006 Global Nuclear Energy Partnership (GNEP), which encouraged the use of nuclear energy abroad and the restart of U.S. reprocessing. But even without GNEP, the Obama administration has continued funding the Advanced Fuel Cycle Initiative, a research project within the Energy Department that is focused on "proliferation-resistant fuel cycles and waste reduction strategies." This seems to indicate research into reprocessing will continue.

Until early 2009, the U.S. nuclear waste plan was to dispose of the country's spent fuel in a permanent geologic site at Yucca Mountain in Nevada. In the 2010 Energy budget, however, the Obama administration effectively shut down Yucca Mountain, stating its funding would be "scaled back to those costs necessary to answer inquiries from the Nuclear Regulatory Commission, while the administration devises a new strategy toward nuclear waste disposal." As a result, we are left with a dispersed, decentralized patchwork of highly radioactive waste sites (interim storage pools and dry-cask storage at nuclear power plants) with the hope that a long-term repository will be opened somewhere, sometime in the future. Otherwise, an alternative waste-disposal strategy must be established, which is no easy task.

To give an idea of what regulatory hurdles any alternate waste-disposal strategy faces, it is helpful to look back at the history of Yucca Mountain. Until recently, the repository was scheduled to open in 2017, but that date itself was delayed numerous times from 1998, the original year it was supposed to open and start accepting U.S. spent fuel. Even if the Obama administration had not ended the project, it was unclear whether Energy's Office of Civilian Radioactive Waste Management would have been able to meet the 2017 deadline, especially with political opposition from key congressional leaders. The Democratic Senate majority leader, Nevada Sen. Harry Reid, for example, is strongly opposed to Yucca Mountain; he has argued instead for on-site, dry-cask storage as a more viable solution for the country's spent nuclear fuel. Although we disagree with the long-term viability of dry-cask storage as a solution (not only does on-site, dry-cask storage create a patchwork of nuclear waste, but it also pushes long-term decision making on nuclear waste strategy to the next generation), the fact remains that even without President Barack Obama's recent cuts, congressional support for the repository site is lacking.

Cost estimates for Yucca Mountain also have risen significantly over time. Estimates to complete the project were $79.3 billion in 2008, much higher than the 1998 estimate of $11.6 billion (in 2000 dollars). Plus, it is increasingly obvious that Yucca Mountain would not

From *Bulletin of the Atomic Scientists*, November/December 2009, pp. 30–36. Copyright © 2009 by Bulletin of the Atomic Scientists. Reprinted by permission of Sage Publications via Rightslink.

have been large enough to accept all of the country's current and future nuclear waste. As of 2008, the United States had generated 58,000 tons of civilian spent fuel, and along with the 13,000 tons of government-sourced spent fuel and high-level waste, the repository's 70,000-ton capacity would have been exceeded on the day it opened. If waste continues to be generated at the rate of approximately 2,500 tons per year, another permanent geologic disposal site, or an increase in the capacity of Yucca Mountain, would have been required. If there are additional nuclear plants built in the future, this problem only will be exacerbated. Although it remains unclear if the Obama administration will support extensive nuclear power investment, Energy Secretary Steven Chu has repeatedly come out in favor of nuclear power. In addition, the 2005 Energy Policy Act provided incentives for the nuclear industry, including tax credits and up to $18.5 billion in loan guarantees for new U.S. reactors. Although we are not aware of any allocations of these funds, it again suggests that there is interest in expanding the civilian nuclear energy industry.

Since 1983, $10.3 billion has been spent to manage and dispose of the country's nuclear waste. That amount has been taken out of the Nuclear Waste Fund, which was established with a $0.001 per kilowatt-hour surcharge on nuclear power generating utilities. The total amount in the fund was $29.6 billion at the end of 2008. With the majority of the $10.3 billion spent developing Yucca Mountain as a repository, it is unclear how much more money will be needed before a nuclear waste storage site is decided upon, built, and opened.

Reprocessing saves valuable repository space. For long-term geologic storage, reductions in waste volume are important. But it is not just the space that the waste would physically take up that is vital, the heat output of the waste also must be taken into consideration, as does the space between waste packages necessary to prevent overheating in the repository. While it is true that high-level waste from reprocessing is hotter than non-reprocessed spent fuel, this does not completely nullify the decrease in waste volume achieved by reprocessing. The heat emitted from post-reprocessing waste decreases by approximately 70 percent during its first 30 years. In other words, such waste initially can be stored either aboveground in well-ventilated storage buildings (as Areva does), or it can be stored in geologic repositories with space between packages left empty and then filled over the years as heat output decreases.

In contrast, spent fuel rods that are directly disposed in repositories cool more slowly and require larger geologic repositories. One estimate, which appears in the book *Megawatts and Megatons* by Richard Garwin and Georges Charpak, suggests that even with the increased heat output of high-level wastes from reprocessing, the amount of space required for a geologic repository can be reduced by *one-half* if the waste is reprocessed. Overall, Garwin and Charpak argue against reprocessing but acknowledge several benefits that we believe outweigh the economic burdens, the most important being that reprocessing can effectively double the capacity of a Yucca Mountain-sized permanent repository.

Reprocessing reduces radioactivity of waste. Reprocessing also reduces the radiotoxicity of high-level waste by one-half to one-tenth when compared with direct burial, and the waste decays to the radioactivity of natural uranium in 10,000 years versus 100,000 years. With the advent of fast reactors, coupled with reprocessing, radiotoxicity of waste would be further reduced with radioactivity reaching the level of natural uranium in only 1,000 years. Given the necessity of any nuclear waste strategy's long-term viability, these reductions are significant advantages for reprocessing.

Our discussion of waste management would not be complete without acknowledging that after reprocessing spent fuel and fabricating MOX fuel rods, the spent MOX fuel rods present a unique problem when dealing with their final disposal. Spent MOX fuel has higher contents of plutonium (plutonium 238 and plutonium 241), americium, and curium than conventional low-enriched uranium (LEU) spent fuel rods, and as a result, the management of spent MOX fuel is more challenging due to cooling and criticality concerns. For interim storage, spent MOX fuel can be dispersed among LEU spent fuel resulting in no change in storage requirements. But in a geologic repository, according to a 2003 International Atomic Energy Agency report on MOX fuel technology, spent MOX fuel would need three times as much space as spent LEU fuel, or require interim storage aboveground for 150 years to reach the same thermal output and then be able to occupy the same amount of space. If we are to assume fast reactors are the long-term goal of the nuclear industry, the optimal and safest use of MOX fuel rods would be to continue recycling them in fast reactors. Yet without that option available, we must acknowledge that some of the gains made by reprocessing are lost in the storage of spent MOX fuel.

Reprocessing does not pose a proliferation threat. A frequent criticism of reprocessing is that separating pure plutonium from spent fuel creates a proliferation and theft risk. Specifically, critics say that spent nuclear fuel without reprocessing is too radioactive to be stolen easily and thus is self protecting. Therefore, some suggest, all spent fuel should remain unreprocessed. Currently, Areva, the only large-scale operator of reprocessing plants in the world, uses the PUREX technique, which separates spent fuel rods into individual streams of uranium, plutonium, and high-level fission products. Although it is true that PUREX results in a pure plutonium stream, the separated plutonium is not considered weapon-grade. Weapon-grade plutonium contains more than 93 percent plutonium 239, while reprocessed fuel contains closer to 50 percent with the remainder being oxygen and other plutonium isotopes, including approximately 15 percent plutonium 241. A nuclear explosive device that used 50 percent plutonium 239 would have an expected yield of about 1 kiloton (approximately 5 percent the power of a weapon-grade plutonium bomb). Even so, such an explosion could wipe out a handful of city blocks (in comparison, the nuclear weapon dropped on Hiroshima was 12–15 kilotons, destroying a radial area of approximately 1.6 kilometers) and is exactly why the implementation of effective security measures is paramount for safe reprocessing. According to an Areva representative, the company, along with French national security personnel, goes to great lengths to secure the transport of reactor-grade plutonium across France. In fact, during conversations with Areva we were told that if a truck were ever hijacked, "it would not get further than 100 meters." The representative was not able to further elaborate, stating that security protocol restricted his ability to clarify.

It is reasonable to worry about the risks presented by transporting nuclear materials and plutonium separated by reprocessing, and this is exactly why we stress that the United States should use and build reprocessing and fuel fabrication facilities in a single location if it chooses to restart its domestic reprocessing program. (Areva's facilities are on opposite ends of France.) Such a combined facility, with all elements in one location, would circumvent the security concerns associated with reprocessing. The United States is currently using this design for its Savannah River Site in Aiken, South Carolina, where it plans to decommission nuclear weapons. This facility will downblend surplus weapon-grade material and use that material in the fabrication of MOX fuel rods. Although the site does not currently reprocess civilian fuel, it seems logical that any civilian program would benefit from a similarly combined facility design.

To further prevent proliferation risks, the United States should develop advanced reprocessing technologies that have been researched under GNEP and will likely continue with the Advanced Fuel Cycle Initiative. These new reprocessing techniques (COEX, UREX+, and NUEX) avoid creating a pure plutonium stream. COEX extracts plutonium and uranium together, while UREX+ and NUEX extract plutonium with some combination of highly radioactive elements that are present in spent fuel. The inclusion of such transuranic elements increases the heat output and radioactive emission rate of the produced waste, necessitating robust radioactive shielding to safely manipulate or handle the material. Such advanced reprocessing techniques coupled with combined reprocessing and fabrication facilities would provide additional layers of security to address the major proliferation concerns associated with reprocessing.

The final word. Reprocessing allows the utilization of more available energy from nuclear fuel than is currently possible in the once-through fuel cycle. Reprocessing represents a path toward decreasing the current nuclear waste burden on geologic disposal sites and on future generations. If the United States truly wants to proceed with nuclear energy as a viable, low-carbon emitting source of energy, it should pursue reprocessing in combination with the development of fast reactors. Once such a decision is made, the debate should turn to how best to develop cheaper and safer reprocessing options, rather than denying its general benefit.

KATE J. DENNIS, JASON RUGOLO, LEE T. MURRAY, AND JUSTIN PARRELLA are PhD candidates at Harvard University's Department of Earth and Planetary Sciences.

David M. Romps et al.

 NO

The Case Against Nuclear Reprocessing

In France, as in the United States and other countries, reprocessing technology was originally pursued to produce plutonium for nuclear weapons. As reprocessing plants churned out their product, the growing stockpiles of plutonium cast a doomsday pall across the globe. Therefore, it is somewhat ironic that reprocessing became part of a utopian dream to provide the world with cheap and nearly limitless energy. The concept was dubbed the "plutonium economy." In this vision, a special kind of nuclear reactor—the fast "breeder" reactor—would burn plutonium to make electricity, but it also would convert non-fissile uranium 238 into plutonium 239, thereby creating or "breeding" more fuel. Before that new fuel could be used, however, it would have to be separated from less useful radioactive material in a nuclear reprocessing plant and then formed into fuel rods in a mixed-oxide (MOX) fuel fabrication plant.

In the 1970s, France embarked on an effort to make the plutonium economy a reality. To do so, it needed a breeder reactor, a reprocessing plant, and a MOX fuel fabrication plant. For the reprocessing plant, France pressed into service its military reprocessing facility at La Hague, blurring the line between its civilian and military nuclear programs. To turn the plutonium into ceramic MOX fuel pellets, it constructed the MELOX plant at Marcoule. All that was left to build was the commercial-scale fast reactor, dubbed Superphénix. Unfortunately the reactor's construction turned out to be much more complicated than anyone imagined, and Superphénix became a notorious flop.

In the aftermath of the failure of fast reactors (not just in France, but throughout the world), the plutonium economy died. This left the French with two very expensive nuclear facilities that had lost their *raison d'être*. Although the French nuclear industry still hopes that commercial fast reactors will someday become viable, they have been forced to argue the merits of reprocessing in the current world of light water reactors fueled by low-enriched uranium.

Today's reprocessing advocates have two main arguments: It reduces hazardous waste and conserves uranium resources. Both of these justifications are seriously flawed. It is also unlikely that fast reactors will be economically competitive in the next several decades, so reprocessing will remain a technology in search of a rationale for years

to come. In the nearer term, the economic costs, environmental harm, and proliferation hazards overwhelm the supposed benefits of reprocessing.

Reprocessing does not save uranium or repository space. Reprocessing proponents claim that the process dramatically reduces the volume of nuclear waste. Indeed, spent nuclear fuel contains roughly 1 percent plutonium, 4 percent high-level radioactive waste, and 95 percent uranium. After reprocessing, the highly radioactive fuel waste and fuel-rod casing material occupy only 20 percent of the original spent-fuel volume. Aboveground, where air currents can cool the high-level waste, the space savings is significant. In fact, all of France's high-level waste sits in a modestly sized building at the La Hague complex. For final geological storage, however, the high temperature of the waste is a more restrictive design consideration than the waste's volume. In order to avoid damaging the geologic repository and risk releasing radioactivity, the high-level waste must be spaced at sufficient intervals to allow for cooling. Even if the reprocessed high-level waste is allowed to cool for 100 years before final disposal, it has been estimated that the repository volume only would be cut by one-half. Whether the United States chooses Yucca Mountain or another site for geologic storage, this modest space savings does not justify the additional costs and hazards of reprocessing because reducing the volume of the repository has an almost negligible effect on the storage costs. According to a recent report from Harvard University's Project on Managing the Atom, even a fourfold decrease in the repository volume would decrease storage costs by less than 15 percent.

The other major claim is that, without reprocessing, there is only enough low-cost uranium for 50 more years of nuclear power at current usage levels. Indeed, from 1965 to 2003, the Organisation for Economic Co-operation and Development's "Red Book" listed known conventional reserves of natural uranium recoverable at less than $130 per kilogram at 3–5 megatons. At the low end of this range, the current global consumption of 0.07 megatons per year would exhaust reserves in 50 years. Reserves, however, will continue to grow in the future, as they have in the past. As prices rise and extraction technology improves, extraction from difficult deposits will become more profitable, increasing the size of known reserves. For example, from 2005 to 2007, the known conventional reserves recoverable at

$130 per kilogram increased by 20 percent to 5.5 megatons. Adding in "Red Book" estimates of so-far undiscovered resources brings the amount of uranium to roughly 17 megatons, which, if mined, would provide a 240-year supply. Reprocessing is a proliferation threat. Reprocessing supporters argue that plutonium from a reprocessing plant is not weapon-grade. The term "weapon-grade" refers to plutonium that is primarily plutonium 239, with less than 6 percent plutonium 240. Plutonium 240 has a high rate of spontaneous fission, which increases the odds of prematurely igniting a fission weapon during the detonation sequence, thereby decreasing its yield. In contrast to weapon-grade plutonium, the plutonium that comes out of France's La Hague facility has about 24 percent plutonium 240. The plutonium bombs detonated during the Trinity test and over Nagasaki used super-weapon-grade plutonium to produce an explosion equivalent to 20 kilotons TNT with a 1.6-kilometer destruction radius. If the same bomb were made with plutonium from the La Hague plant, its expected yield would be around 1 kiloton of TNT equivalent. Although a 1-kiloton explosion might sound less dangerous than a 20-kiloton explosion, its likely blast radius would still be one-half to four-fifths of a kilometer, enough to level most of downtown Boston. Such a device would be 90 times more powerful than the largest conventional weapon in the U.S. arsenal and 500 times more powerful than what destroyed the Alfred P. Murrah Federal Building in Oklahoma City. So while it is true that reactor-grade plutonium is not as powerful an explosive as weapon-grade plutonium, it certainly should be considered "terrorist-grade."

In France, reprocessed plutonium is shipped the entire length of the country from La Hague on the north coast to the MELOX fuel fabrication plant in Marcoule near the south coast. To avoid uncontrolled fission, the plutonium oxide—a fine, yellow powder—is divided into small, thermos-sized canisters. Each canister contains about 2.5 kilograms of material, and only three of these provide enough to make a bomb. Every week, Areva ships dozens of these canisters from La Hague to Marcoule in shielded trucks, presenting an opportunity for theft. During our visit, we observed two such trucks on the highway guarded by four police cars. Given Areva's spin that this is non-weapon-grade plutonium, we wondered whether the police officers were aware that they were guarding enough material to destroy the downtown areas of 30 mid-sized cities.

Unfortunately, this is not the only chance someone has to steal the plutonium. Even with strict monitoring at reprocessing plants, it is not feasible to account for every single gram of plutonium produced. With the large volumes of plutonium being separated and the possibilities for measurement errors or poor bookkeeping, an insider could smuggle out a bomb's worth of plutonium in several months without anyone noticing the missing plutonium.

In addition to theft, there is the danger that reprocessing plants could be used by a host nation to initiate a nuclear weapons program. This has already happened in the case of India. The United States sold reprocessing technology to India with the understanding that it would only be used for civilian nuclear power. The plutonium product, however, was diverted for military purposes. After the detonation of an Indian bomb in 1974, the United States reversed its pro-reprocessing stance and put an end to domestic civilian reprocessing. Although reprocessing is no longer technically outlawed in the United States, a de facto ban has persisted.

In response to proliferation concerns, many reprocessing advocates recommend a new method, called COEX, for future reprocessing facilities. In contrast to the current PUREX process, which produces pure plutonium oxide, COEX would extract plutonium and uranium together to form a mixed plutonium-uranium oxide that can be directly fabricated into MOX fuel. Although this product cannot fuel bombs directly, a malicious organization could later extract the plutonium. Since none of the MOX components are highly radioactive, plutonium separation could be carried out safely in a standard chemical laboratory with existing methods. Even more worrisome, the separation process need not be efficient to obtain a large quantity of material because the COEX product is composed of 50 percent plutonium. In addition, because COEX has never been deployed on an industrial scale, the costs of developing it commercially could be massive. In combination with the unknown operating and environmental costs, COEX is a big gamble for little gain.

Reprocessing is not cost effective. In order for reprocessing to make sense economically, the price of a new MOX fuel rod must be competitive with the price of a new uranium fuel rod, which largely depends on the price of mined uranium. Several studies have concluded that the price of uranium would have to be in the range of $400–$700 per kilogram in order for reprocessed MOX to break even. But for the first half of 2009, the price of uranium oxide has hovered around $100 per kilogram. In fact, uranium prices have reached $300 per kilogram (in 2008 dollars) only twice in history—in the late 1970s during the energy crisis and briefly in the summer of 2007. In other words, uranium has never sustained a price that would make reprocessing profitable. And given the large estimated resources of uranium available at or below $130 per kilogram, it is unlikely that reprocessing will become cost competitive any time in the foreseeable future.

The high expense of reprocessing is rooted in the fact that chemically separating and processing spent nuclear fuel requires large, complex facilities that produce significant quantities of radioactive and chemical wastes. These facilities also must meet modern health and safety standards for dealing with highly toxic plutonium, which adds to the expense. In addition, these facilities produce weapons-usable plutonium and must be operated under military guard, which adds more costs to the process. At the same time, it is difficult to recoup all of these expenses when reprocessing yields so little usable product—only one MOX fuel rod is produced for every seven spent uranium fuel rods reprocessed. That means, for all of the investment and operating costs, reprocessing boosts the usable energy extracted from mined uranium only about 14 percent.

Reprocessing harms the environment. The reprocessing plants at Sellafield in Britain and La Hague in France are two of the largest anthropogenic emitters of radioactivity in the world. Both facilities intentionally discharge significant amounts of radioactive cesium, technetium, and iodine into the surrounding oceans, which show up in seafood harvested in the Irish and North seas. When it comes time to close a reprocessing plant, it requires decades to decontaminate facilities, soil, and groundwater. The United States currently spends billions of dollars each year to rehabilitate the reprocessing facilities at the Hanford site in Washington state and Britain will require similar resources to clean up Sellafield, making these among the most complicated and costly environmental cleanup projects in the world.

The final word. Contrary to the arguments given by its proponents, reprocessing is dangerous and unnecessary.

Although nuclear power will remain a significant source of electricity for the coming decades, spent fuel reprocessing should have no part in this future. It is in the best interests of the United States—from the perspective of waste management, national security, economics, and environmental protection—to maintain its de facto moratorium on reprocessing and encourage other countries to follow suit.

DAVID M. ROMPS is a Research Scientist at Harvard University's Department of Earth and Planetary Sciences.

CHRISTOPHER D. HOLMES is a PhD candidate at Harvard University's Department of Earth and Planetary Sciences.

KURT Z. HOUSE is the President of C12 Energy, based in Cambridge, MA.

BENJAMIN G. LEE is a Postdoctoral Researcher at the National Renewable Energy Lab in Golden, Colorado.

MARK T. WINKLER is a PhD candidate at Harvard University's Department of Physics.

EXPLORING THE ISSUE

Should the United States Reprocess Spent Nuclear Fuel?

Critical Thinking and Reflection

1. Why has fear of nuclear proliferation inhibited nuclear fuel reprocessing?
2. What are the advantages of putting off the decision to reprocess spent nuclear fuel until the necessary technology has been more fully developed?
3. What are the disadvantages of putting off the decision to reprocess spent nuclear fuel until the necessary technology has been more fully developed.
4. Economic incentives have been used to persuade people to accept nearby undesirable facilities such as dumps, power plants, and factories. Would it be ethical to do the same for a nuclear fuel reprocessing plant?

Is There Common Ground?

Both sides agree that nuclear fuel reprocessing can help support the use of nuclear power and reduce the nuclear waste problem. The major difference lies in whether the technology can be implemented immediately or must undergo further research and development. Present and under-development technologies are described on the website of the World Nuclear Association. Visit it at www.world-nuclear.org/info/inf69.html and answer the following questions:

1. What is the purpose of nuclear fuel reprocessing?
2. What is the dominant technology today?
3. What is transmutation and how does it reduce the nuclear waste problem?

Additional Resources

Robert Alvarez, "Improving Spent-Fuel Storage at Nuclear Reactors," *Issues in Science and Technology* (Winter 2012)

Frank N. von Hippel, "Rethinking Nuclear Fuel Recycling," *Scientific American* (May 2008)

Behnam Taebi and Jan Kloosterman, "To Recycle or Not to Recycle? An Intergenerational Approach to Nuclear Fuel Cycles," *Science & Engineering Ethics* (June 2008)

Ozzie Zehner, "Nuclear Power's Unsettled Future," *Futurist* (March–April 2012)

Create Central

www.mhhe.com/createcentral

Internet References . . .

The La Hague Nuclear Reprocessing Plant

www.areva.com/EN/operations-1118/areva-la-hague
-recycling-spent-fuel.html

The Union of Concerned Scientists

www.ucsusa.org/nuclear_power/nuclear_power_risk
/nuclear_proliferation_and_terrorism/nuclear
-reprocessing.html

The World Nuclear Association

www.world-nuclear.org/info/Nuclear-Fuel-Cycle
/Fuel-Recycling/Processing-of-Used-Nuclear-Fuel/#
.UgD9C6xcWF8